PREALGEBRA
College Preparatory Mathematics

PREALGEBRA
College Preparatory Mathematics

J. Louis Nanney
Miami-Dade Community College

John L. Cable
Miami-Dade Community College

Hawkes Publishing

Book Team

Editor *Earl McPeek*
Developmental Editor *Theresa Grutz*
Designer *K. Wayne Harms*
Art Editor *Janice M. Roerig*
Visuals Processor *Andê Meyer*

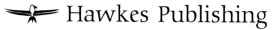
Hawkes Publishing

Cover credit: Robert Phillips

Chapter Openers
One: © Keith Gunman/Photo Researchers; **Two:** © U.S. Trotting Association; **Three:** © NASA; **Four:** © Paul Silverman/Fundamental Photographs; **Five:** © Alexander Lowry/Photo Researchers, Inc.; **Six:** © George Whiteley/Photo Researchers; **Seven:** © Mark Antman/The Image Works; **Eight:** © Paul Silverman/Fundamental Photographs.

Copyright © 1990 by Wm. C. Brown Publishers. All rights reserved.
Copyright © 1996 Hawkes Publishing. All rights reserved.

Library of Congress Catalog Card Number: 89-43038

ISBN 0-697-06428-X

No part of this publication may be reproduced, stored in a retrieval system, or transmitted, in any form or by any means, electronic, mechanical, photocopying, recording, or otherwise, without the prior written permission of the publisher.

Printed in the United States of America by Wm. C. Brown Publishers, 2460 Kerper Boulevard, Dubuque, IA 52001

CONTENTS

Preface ix
Foreword to the Student xiii

Chapter 1 Pretest 2

Whole Numbers 1

1–1 Reading Whole Numbers   4
1–2 Expanded Form   10
1–3 Rounding   14
1–4 Addition of Whole Numbers   17
1–5 Subtraction of Whole Numbers   24
1–6 Multiplication of Whole Numbers   36
1–7 Division of Whole Numbers   45
1–8 Grouping Symbols   60
1–9 Factors and Prime Factorization   67
1–10 Greatest Common Factor and Least Common Multiple   76

Chapter 1 Summary   83
Chapter 1 Review   84
Chapter 1 Test   90

Chapter 2 Pretest   92

Fractions and Mixed Numbers 91

2–1 Simplifying Fractions   94
2–2 Multiplication of Fractions   99
2–3 Division of Fractions   109
2–4 Addition and Subtraction of Fractions   122
2–5 Mixed Numbers   136

Chapter 2 Summary   151
Chapter 2 Review   152
Chapter 2 Test   161
Chapters 1–2 Cumulative Test   163

Chapter 3 Pretest 166

Decimals, Scientific Notation, and Square Roots 165

3–1 Expanded Form and Reading and Writing Decimals 168
3–2 Rounding Decimals 171
3–3 Adding and Subracting Decimal Numbers 173
3–4 Multiplying Decimal Numbers 177
3–5 Dividing Decimal Numbers by Whole Numbers 183
3–6 Dividing by Decimal Numbers 192
3–7 Interchanging Fractions and Decimals 197
3–8 Scientific Notation 202
3–9 Square Roots 208

Chapter 3 Summary 212
Chapter 3 Review 213
Chapter 3 Test 222

Chapter 4 Pretest 224

The Language of Algebra 223

4–1 Algebraic Expressions 226
4–2 Evaluating Literal Expressions 232
4–3 Combining Like Terms 237

Chapter 4 Summary 241
Chapter 4 Review 242
Chapter 4 Test 244
Chapters 1–4 Cumulative Test 245

Chapter 5 Pretest 248

Solving Equations 247

5–1 Conditional and Equivalent Equations 250
5–2 The Division Rule 254
5–3 The Subtraction Rule 260
5–4 The Addition Rule 265
5–5 The Multiplication Rule 269
5–6 Combining Rules for Solving Equations 274

Chapter 5 Summary 277
Chapter 5 Review 278
Chapter 5 Test 282

Chapter 6 Pretest 284

CHAPTER 6 Ratio, Proportion, and Percent 283

6–1 Ratios 286
6–2 Proportions 289
6–3 Problems Involving Proportions 292
6–4 Metric Measurement 296
6–5 Percent 299
6–6 Problems Involving Percents 302

Chapter 6 Summary 306
Chapter 6 Review 307
Chapter 6 Test 312
Chapters 1–6 Cumulative Test 313

Chapter 7 Pretest 316

CHAPTER 7 Signed Numbers 315

7–1 The Meaning of Signed Numbers 318
7–2 Combining Signed Numbers 323
7–3 Rules for Combining Signed Numbers 329
7–4 Combining Like Terms 336
7–5 A Word About Subtraction 338
7–6 Multiplication and Division of Signed Numbers 341
7–7 Exercises Using Signed Numbers 349

Chapter 7 Summary 351
Chapter 7 Review 352
Chapter 7 Test 355

Chapter 8 Pretest 358

CHAPTER 8 Equations and Inequalities Involving Signed Numbers 357

8–1 Solving Equations Involving Signed Numbers 360
8–2 Literal Equations 364
8–3 Graphing Inequalities 369
8–4 Solving Inequalities 375

Chapter 8 Summary 382
Chapter 8 Review 383
Chapter 8 Test 386
Chapters 1–8 Cumulative Test 387

Appendix A Useful Facts from Plane Geometry   *389*
   A–1   Definitions   *389*
   A–2   Facts and Formulas   *395*
   A–3   Examples and Applications   *397*

Appendix B Using a Calculator   *401*
   B–1   Arithmetic Operations   *401*

Appendix C Answers   *405*

Index   *429*

PREFACE

Many students who enter college do not have the basic skills necessary to be successful in a regular course in algebra. *Prealgebra: College Preparatory Mathematics* was written for these students. For those who have been away from school for awhile this text will provide a needed review of forgotten concepts. For others, who did not take the proper courses in high school or who did not sufficiently comprehend the content of the courses taken, this text represents a necessary learning experience. In any case, mastery of the material in *Prealgebra: College Preparatory Mathematics* will give the student an excellent foundation for a course in elementary algebra.

After a thorough presentation of the basic skills of arithmetic, the elementary concepts of algebra are introduced. Our goal, of an orderly transition from arithmetic to algebra, is accomplished in two ways. First, whenever possible, we utilize formats usually reserved for algebra in the development of the skills of arithmetic. The solution of proportions is a good example of this. Also, positive whole number exponents are used early in chapter 1 for expanded numerals and negative exponents as well as scientific notation are introduced in the chapter on decimals. Second, and perhaps the most unique feature of this text, is our introduction to algebra. The basic concepts of algebra, including the skills necessary for solving first-degree equations, are presented *before* signed numbers. Then these concepts are reinforced using signed numbers. Thus the student avoids the trauma of too many new ideas at the same time. We have taught using this order of presentation and found it highly successful.

Pedagogical Features

Format

Each section of the text begins with clear, concise, and mathematically correct explanations and/or definitions. These are followed by many detailed examples and an abundance of exercises. We have endeavored to present only one new idea at a time. Our goal is a text that is easy to read and understand and at the same time provides the student with the necessary foundation for further studies in mathematics. We feel this format has accomplished this goal.

Objectives

Each section of the text contains a boxed-in statement of the objectives. Thus, a clearly stated purpose for that section is available to both student and instructor.

Geometry

Examples and exercises using facts and formulas from plane geometry occur throughout the text. In each example or exercise the definition or formula to be used is given. Thus, an abundance of plane geometry facts are presented and used. An appendix is provided that contains definitions, formulas, and examples from plane geometry appropriate for a text of this level.

Applications

Applications are presented throughout the text wherever possible. Those in the early sections involve only one operation, while later applications require multi-calculations. Thus, the student is informally introduced to word problems and gains practice in solving them, although a formal presentation is left for a subsequent algebra course.

Pretest/Chapter Test

Each chapter begins with a pretest and ends with a chapter test. These tests are similarly constructed so that the student may compare results and evaluate progress. Answers to these tests are keyed to sections within the chapter as an aid to any necessary review.

Chapter Summary

Each chapter is followed by a summary that lists key words, ideas, rules, and definitions covered in the chapter.

Chapter Review

After the chapter summary, an abundance of exercises is provided as a chapter review. Answers to these exercises are keyed to sections within the chapter so that topics needing further study can easily be identified.

Cumulative Tests

Cumulative tests are provided after even-numbered chapters. Answers to these tests are keyed to the chapter and section for any needed review.

Warnings

We have drawn on many years of teaching experience to identify common errors students are likely to make. These are noted as WARNINGS! and should be a special aid to the student.

Answers

Answers to *all* exercises are provided in appendix C at the end of the book.

Acknowledgments

A text of this type cannot be produced without the aid of many people. We wish to express our thanks to those at Wm. C. Brown Publishers who have been involved in this endeavor. We are especially grateful to the following reviewers whose many suggestions have been of great value to us.

Reviewers

Jane C. Beattie
University of South Carolina at Aiken

Julia A. Brown
Atlantic Community College

John W. Carman
San Diego Mesa College

Deann J. Christianson
University of the Pacific

Sally Copeland
Johnson County Community College

Linda Crabtree
Longview Community College

Michele Satty Gage
New York City Technical College

Elaine Klett
Brookdale Community College

Lloyd Roeling
University of Southwestern Louisiana

Mary A. Scholes
Boise State University

Deborah S. Temperly
Northwood Institute

Jeannine G. Vigerust
New Mexico State University

FOREWORD TO THE STUDENT

Why Study Mathematics?

We know that you have probably asked yourself, "Why should I take this course?" or "What do I need with this?" We can easily answer by saying, "You will need this material later on." True as this may be, it is not really a satisfying answer. We believe there are good reasons why you should try to learn all you can from this text. Following are some of them.

First this material may be a prerequisite for courses you must take in your chosen field of study. If this is the case, you already know why you are taking this course.

Even if you don't plan to take other courses for which this is a prerequisite, there is still a reason for learning this material. Not everyone enters college with definite plans, and a large number of those who do will change those plans somewhere along the line. So the second reason is *to keep your options open.* Lack of a background in mathematics can severely narrow your choices.

Tests pertaining to mathematical skills are often used for hiring, promotions, admission to special schools, and the like. You may have to pass these tests even though they may have nothing to do with the desired position. We do not like this misuse of mathematics, but the fact remains that such tests are given and probably will continue to be given. So a third reason for learning this material is to help you survive this barrage of tests to reach your goal even in unrelated areas.

How to Use This Text

This text has many features that are specifically designed to make your learning experience easier. We present only one new idea at a time and have made the explanations complete and easy to read.

Objectives

Each section begins with a boxed-in statement of objectives. This tells you what you should learn in that section. After completing the section you should again read the objectives to make sure you have accomplished them.

Examples

Within each section you will find several worked-out examples. Reviewing them step-by-step will help clarify the explanations and give you models by which to work the exercises.

Exercises

You will find an abundance of exercises. These are designed to reinforce the explanations so that you may know you have mastered the material within the section.

Answers

Answers to all problems are given in appendix C at the end of the text. This allows you to check your answers after you have worked a problem set. In the event you have an incorrect answer you will many times be able to identify your mistake and avoid making a similar error in the future. You are urged to make proper use of the answer section.

Warnings

From our years of experience in the classroom we have learned where students are most likely to make errors. We have used this knowledge to place warnings throughout the text. These warnings are there to help prevent you from making the same mistakes as others have. We urge you to pay special attention to them.

Chapter Reviews

An extensive chapter review follows each chapter. The answers to these exercises are keyed to the sections within the chapter that contain that type of problem. Thus, if you should have trouble working a particular problem, or if you do not get the correct answer, you can easily go back to the appropriate section and review.

Pretests and Chapter Tests

You will find a pretest at the beginning of each chapter and a chapter test at the end. When you compare results from each of these tests, you will recognize how much you have learned from the chapter. You may be able to do very little on the pretest, but you should be able to make an excellent score on the chapter test. The answers to the pretests and chapter tests are also keyed in the same manner as those to the chapter reviews.

Cumulative Tests

Cumulative tests occur throughout the text. These tests are excellent for review since they contain problems that reinforce concepts from previous chapters. This gives you the opportunity to tie together ideas and recognize the continuity of the subject matter.

How to Study Mathematics

Reading a mathematics text is far different from reading other material, such as a newspaper or a novel. Each sentence must be read carefully as its meaning may be necessary to the understanding of the next sentence. So our first suggestion is that you read carefully and then re-read if necessary until you comprehend the meaning of what you are reading.

Someone has said that the mathematics you learn must work its way up your arm before it can get to your brain. In other words, you only learn mathematics by *doing* mathematics. Once you comprehend a concept, you must work out many, many exercises until that concept is so fixed in your mind that you will not forget it. This is true for the most basic facts of arithmetic as well as the more complicated ideas that come later. So our advice is that you spend the time to practice by working all exercises in this text as well as reworking the examples.

Whenever you find a problem that you cannot seem to solve, go back and try to find a worked-out example that will help you understand how to proceed. Always check your computations for errors. You may have approached the problem correctly but made a simple error in some basic fact. If you have worked an exercise and arrived at a wrong answer, it may be helpful to compare your answer to the correct answer. Many times this will help you find your error.

When you take an examination, we suggest that you first look over the entire set of problems. Work first those problems that you see are going to be easy for you and leave the more difficult ones for last. This accomplishes two things. First you won't spend all your time on a few problems when they all may carry the same point value. Second, by working those problems that come easy to you, the anxious feeling that often comes with an exam may be lessened.

Our goal is to provide a text that will give you the proper foundation for success in future studies in mathematics. Your goal is, of course, to obtain that foundation. We are certain that if you use this text as we have suggested, your goal will be realized. Good Luck!

CHAPTER 1 **PRETEST**

NAME: _____

CLASS / SECTION: _____   DATE: _____

A N S W E R S

Answer as many of the following problems as you can before starting this chapter. When you finish the chapter, take the test at the end and compare your scores to see how much you have learned.

1. _____

2. _____

3. _____

4. _____

5. _____

6. _____

7. _____

8. _____

9. _____

10. _____

11. _____

12. _____

13. _____

14. _____

15. _____

16. _____

17. _____

18. _____

19. _____

20. _____

SCORE: _____

1. What is the place value of the digit 2 in the number 20,158?

2. Write the number 38,026 in words.

3. Write the number 7,253 in expanded form.

4. Round 624 to the nearest ten.

5. Round 14,602 to the nearest thousand.

6. Add: 315
        687

7. Add: 241
         356
         519

8. Find the sum of 563; 4,207; 73; and 2,064.

9. Subtract: 693
             −495

10. Subtract: 702
              −594

11. Find the difference of 12,350 and 7,341.

12. Multiply: 574
                38

13. Multiply: 2,369
                 305

14. Divide: $7\overline{)406}$

15. Divide: $31\overline{)1{,}413}$

16. Divide: $14\overline{)28{,}546}$

17. Evaluate: $5[2(7 - 4) - 4]$

18. Find the prime factorization of 990.

19. Find the greatest common factor of 189 and 210.

20. Find the least common multiple of 14 and 21.

2

CHAPTER 1

Whole Numbers

The height of Mt. Everest is 29,028 feet and that of Mt. Whitney is 14,494 feet. How much taller is Mt. Everest?

When the concept of *number* first entered the mind of primitive man, it was most probably for the purpose of counting. These *counting numbers* together with the later introduction of zero became the foundation on which all mathematics is built.

Numeral and *number* are generally used as if they have the same meaning. Actually **number** is an abstract idea and **numeral** is a symbol used to represent the abstract idea. In this text we will not attempt to distinguish between these meanings and will most often use the word *number*.

The set of **whole numbers** includes the number zero and all numbers used for counting. We can write the set as {0,1,2,3,4, . . .}. (The three dots indicate that the set continues in the pattern shown.)

All other sets of numbers used in mathematics are defined in terms of the set of whole numbers. This statement may mean very little at this time but it should indicate the importance of the set of whole numbers. They are so important, in fact, that if you are not proficient in the use of whole numbers, it is impossible to be proficient in the use of the other sets of numbers.

In this chapter we will learn to read and write whole numbers and to perform the four basic operations on them.

 1-1 READING WHOLE NUMBERS

OBJECTIVES

In this section you will learn to:
1. Determine the place value of a digit.
2. Write a number in words.

The digits 0, 1, 2, 3, 4, 5, 6, 7, 8, 9 are used to write all whole numbers. The reason only ten digits are necessary is that each digit has two values, a digital value and a place value.

For example, observe the following three numbers.

    7
   75
 756

In each of these numbers the 7 has the same digital value but a different place value.

In the first number the 7 represents seven ones. In the second number it represents seven tens. In the third it represents seven hundreds. This **place value** is the basis of our system of writing numbers and allows us to use only ten digits to write all whole numbers.

Reading whole numbers is simplified if we note that each group of three numbers is given the place value names as shown in this table.

Hds Tens Ones	Hds Tens Ones	Hds Tens Ones	Hds Tens Ones	Hds Tens Ones
trillions	billions	millions	thousands	units

Groups above trillions are rarely used.

1-1 READING WHOLE NUMBERS

Example 1 Consider the number 5,814,263.

```
5,    8,              1,             4,         2,        6,    3
millions  hundred thousands  ten thousands  thousands  hundreds  tens  ones
```

    a. The place value of the digit 2 is hundreds.

    b. The place value of the digit 1 is ten thousands.

    c. The place value of the digit 8 is hundred thousands.

    d. The place value of the digit 5 is millions.

If we have only one, two, or three digits, we do not usually use the name of the group. For instance, 325 would usually be read as "three hundred twenty-five" rather than "three hundred twenty-five units." However, 325,000 is read as "three hundred twenty-five thousand." All group names other than the "units" group must be read.

Notice that the word "and" is not used in reading whole numbers.

Example 2 216,358,234 is read as "two hundred sixteen million, three hundred fifty-eight thousand, two hundred thirty-four."

Example 3 561,000 is read as "five hundred sixty-one thousand."

Example 4 561,000,000 is read as "five hundred sixty-one million."

Example 5 561,561,561 is read as "five hundred sixty-one million, five hundred sixty-one thousand, five hundred sixty-one."

Example 6 "Three hundred eighty-two thousand, twenty" written as a number is 382,020.

Example 7 "Two hundred million, seventy-five thousand, six" written as a number is 200,075,006.

▼ EXERCISE SET 1-1-1

For problems 1–8 refer to the number 714,385,602.

1. What is the place value of the digit 8?

2. What is the place value of the digit 5?

3. What is the place value of the digit 6?

4. What is the place value of the digit 3?

ANSWERS

1. _____
2. _____
3. _____
4. _____

ANSWERS

5. What is the place value of the digit 4?

6. What is the place value of the digit 2?

5. _____

6. _____

7. What is the place value of the digit 1?

8. What is the place value of the digit 0?

7. _____

8. _____

In problems 9–28 write the number in words.

9. _____

9. 321

10. 438

10. _____

11. _____

11. 2,129

12. 5,280

12. _____

13. _____

13. 10,124

14. 11,432

14. _____

15. _____

15. 100,241

16. 103,375

16. _____

17. _____

17. 320,004

18. 619,200

18. _____

19. _____

19. 400,001

20. 562,040

20. _____

21. 2,503,100

22. 6,020,327

23. 15,200,621

24. 352,352,352

25. 763,763,763

26. 349,216,123

27. 22,519,054,111

28. 83,000,500,009

Write the following words in numbers.

29. Three hundred twenty-six

30. Four hundred eight

31. Nine hundred five

32. One thousand forty

33. One thousand, eight hundred one

34. Thirteen thousand, four hundred nineteen

ANSWERS

21. _____

22. _____

23. _____

24. _____

25. _____

26. _____

27. _____

28. _____

29. _____

30. _____

31. _____

32. _____

33. _____

34. _____

ANSWERS

35. _____

36. _____

37. _____

38. _____

39. _____

40. _____

41. _____

42. _____

43. _____

44. _____

45. _____

46. _____

47. _____

48. _____

35. Twenty-six thousand, forty-two

36. One hundred twenty-five thousand, four

37. Two hundred six thousand, two hundred one

38. Five hundred thousand, three

39. Six hundred ten thousand, fifty

40. Three million, forty thousand, six hundred nineteen

41. Five million, three thousand, two hundred seven

42. Thirty-one million, four hundred eleven thousand, nine hundred sixteen

43. Sixty-eight million, forty-five thousand, three hundred eight

44. One hundred twenty-two million, six hundred ten thousand, five

45. Five hundred eleven million, four hundred twelve thousand, one hundred

46. Fifty-eight billion, twenty million, fourteen thousand, three hundred twelve

In problems 47–52 write the number in the sentence in words.

47. The price of a new Cadillac was $26,798.

48. The height of Mt. Everest is 29,028 feet.

1-1 READING WHOLE NUMBERS

49. The reading on the odometer of a car was 31,045 miles.

50. In one season, quarterback Dan Marino of the Miami Dolphins football team passed for a total of 5,084 yards.

51. The play *A Chorus Line* was performed 4,552 times on Broadway.

52. The film *E. T. The Extra-Terrestrial* grossed $227,960,804.

In problems 53–58 write the number in the sentence using digits.

53. The distance from Boston to Los Angeles is two thousand, five hundred ninety-six miles.

54. The depth of Great Slave Lake in Canada is two thousand, fifteen feet.

55. A man wrote a check for fifty-two thousand, nine hundred four dollars.

56. The football stadium at Ohio State University has a capacity of eighty-three thousand, one hundred people.

57. The State Russian Museum in Leningrad has two hundred fifty thousand objects of art on display.

58. In 1980 the population of metropolitan Dallas–Fort Worth was estimated at two million, nine hundred seventy-four thousand, eight hundred seventy-eight.

ANSWERS

49. _____

50. _____

51. _____

52. _____

53. _____

54. _____

55. _____

56. _____

57. _____

58. _____

 1-2 EXPANDED FORM

OBJECTIVES

In this section you will learn to write any whole number in expanded form.

In section 1-1 we noted that each digit in a number has a place value and a digital value. The knowledge of place value will be necessary as we begin to work with the basic operations.

The following table shows an important relationship.

Ten Thousands	Thousands	Hundreds	Tens	Ones
10 × 10 × 10 × 10	10 × 10 × 10	10 × 10	10 × 1	1

Notice that each place value is ten times the place value to its right. For this reason we say we have a **base ten** number system.

At this point we wish to introduce a new notation that will be used in writing numbers in expanded form.

Rather than writing an expression such as 10 × 10 × 10, for 1,000, mathematicians use a "shorthand" notation called an **exponent.** Using this notation, 10 × 10 × 10 is written as 10^3. The small superscript indicates that 10 is multiplied by itself three times.

Example 1 $10^4 = 10 \times 10 \times 10 \times 10$

WARNING

Make sure you understand that 10^4 does *not* mean 10 × 4.

Example 2 $3^5 = 3 \times 3 \times 3 \times 3 \times 3$

Example 3 $2^3 = 2 \times 2 \times 2$

The exponent is written smaller and a half-space above the number to be multiplied. 10^4 is read as "ten to the fourth power" or "ten exponent four." The second power and the third power have special names that are sometimes used. 5^2 can be read as "five squared." In other words, to "square" a number, we multiply it by itself. 5^3 can be read as "five cubed." Any power above the third has no special name. 5^4 is read as "five to the fourth power" or "five exponent four."

Example 4   Evaluate "three cubed."

$$3^3 = 3 \times 3 \times 3 = 27$$

Example 5   Evaluate 4^2.

$$4^2 = 4 \times 4 = 16$$

Example 6 Evaluate "nine squared."

$$9^2 = 9 \times 9 = 81$$

When no exponent is written, it is understood to be the number 1.

Example 7 $5^1 = 5$

Example 8 $10^1 = 10$

▼ EXERCISE SET 1-2-1

Evaluate the following.

1. 3^2
2. 5^2
3. 2^4
4. 3^4
5. 1^5
6. 2^5
7. 8^1
8. 10^1
9. 10^2
10. 10^3
11. 10^4
12. 10^5

ANSWERS

1. _____
2. _____
3. _____
4. _____
5. _____
6. _____
7. _____
8. _____
9. _____
10. _____
11. _____
12. _____

Before we can use exponents to write whole numbers in expanded form, a special definition is necessary.

Definition Any number (except zero) raised to the zero power is equal to 1.

Example 9 $3^0 = 1$

Example 10 $5^0 = 1$

Example 11 $10^0 = 1$

Example 12 $1^0 = 1$

The following table will be helpful as we write numbers in expanded form.

Millions	One Hundred Thousands	Ten Thousands	Thousands	Hundreds	Tens	Ones
10^6	10^5	10^4	10^3	10^2	10^1	10^0

Example 13 Write 536 in expanded form.

$$536 = 5 \text{ hundreds} + 3 \text{ tens} + 6 \text{ ones}$$
$$= 5 \times 100 + 3 \times 10 + 6 \times 1$$
$$= 5 \times 10^2 + 3 \times 10^1 + 6 \times 10^0$$

Example 14 Write 1,784 in expanded form.

$$1,784 = 1 \text{ thousand} + 7 \text{ hundreds} + 8 \text{ tens} + 4 \text{ ones}$$
$$= 1 \times 1,000 + 7 \times 100 + 8 \times 10 + 4 \times 1$$
$$= 1 \times 10^3 + 7 \times 10^2 + 8 \times 10^1 + 4 \times 10^0$$

Example 15 Write 5,964 in expanded form.

$$5,964 = 5 \times 10^3 + 9 \times 10^2 + 6 \times 10^1 + 4 \times 10^0$$

Example 16 Write 1,780,263 in expanded form.

$$1,780,263 = 1 \times 10^6 + 7 \times 10^5 + 8 \times 10^4 + 0 \times 10^3 + 2 \times 10^2 + 6 \times 10^1 + 3 \times 10^0$$

EXERCISE SET 1-2-2

Write the following numbers in expanded form using exponents.

1. 251

2. 341

3. 619

4. 514

5. 189

6. 135

1-2 EXPANDED FORM

7. 2,458
8. 4,362
9. 7,625
10. 2,957
11. 30
12. 80
13. 210
14. 350
15. 704
16. 601
17. 500
18. 900
19. 3,041
20. 4,029
21. 5,398
22. 3,782
23. 6,000
24. 9,000
25. 23,426
26. 39,582
27. 4,502
28. 7,803
29. 81,294
30. 63,514

ANSWERS

7. _____
8. _____
9. _____
10. _____
11. _____
12. _____
13. _____
14. _____
15. _____
16. _____
17. _____
18. _____
19. _____
20. _____
21. _____
22. _____
23. _____
24. _____
25. _____
26. _____
27. _____
28. _____
29. _____
30. _____

14 CHAPTER 1 WHOLE NUMBERS

A N S W E R S

31. _____

32. _____

33. _____

34. _____

35. _____

36. _____

37. _____

38. _____

39. _____

40. _____

31. 15,804

32. 19,603

33. 23,038

34. 44,016

35. 172,254

36. 132,693

37. 350,400

38. 520,500

39. 5,461,006

40. 8,319,001

1-3 ROUNDING

OBJECTIVES

In this section you will learn to round any whole number to any place value.

In many situations it may be desirable to use a close estimate of a number rather than the exact number. However, there must be exact rules for obtaining this estimate. The process is known as *rounding*.

For instance, suppose an automobile is priced at $12,859. Would you say it cost "about $13,000"? Why? This illustrates the basic idea of **rounding.** We want to go to the *nearest* number. Sometimes we want the *nearest ten, nearest hundred,* or *nearest thousand,* and so on. So we need a rule that will work for all cases.

> **Rule** To round a number to any place value if the number in the next place to the right is 5 or more, the number in the desired place is increased by one. If it is less than 5, the number in the desired place remains the same. All numbers to the right of the desired place are changed to zero.

Example 1 Round 368 to the nearest ten.

Since 8 is in the next place to the right of the tens place, 368 rounds as 370 to the nearest ten.

Example 2 Round 6,439 to the nearest hundred.

Since 3 is in the next place to the right of the hundreds place, 6,439 rounds as 6,400 to the nearest hundred.

Example 3 Round 25,738 to the nearest thousand.

Since 7 is in the next place to the right of the thousands place, 25,738 rounds as 26,000 to the nearest thousand.

Example 4 Round 25,738 to the nearest ten thousand.

Since 5 is in the next place to the right of the ten thousands place, 25,738 rounds as 30,000 to the nearest ten thousand.

Example 5 Round 5,538,278 to the nearest million.

Since 5 is in the next place to the right of the millions place, 5,538,278 rounds as 6,000,000 to the nearest million.

Example 6 Round 999 to the nearest hundred.

Since 9 is in the next place to the right of the hundreds place, we add one to the hundreds place giving ten hundreds. We therefore place a one in the thousands place. Thus 999 rounds as 1,000 to the nearest hundred.

▼ EXERCISE SET 1-3-1

Round the numbers in problems 1–10 to the nearest ten.

1. 63	**2.** 86	**3.** 28	**4.** 44	**5.** 517
6. 388	**7.** 293	**8.** 794	**9.** 4,206	**10.** 1,387

Round the numbers in problems 11–20 to the nearest hundred.

11. 384	**12.** 596	**13.** 249	**14.** 156	**15.** 2,352
16. 5,233	**17.** 1,063	**18.** 3,049	**19.** 6,954	**20.** 4,980

Round the numbers in problems 21–30 to the nearest thousand.

21. 58,473	**22.** 23,258	**23.** 25,581	**24.** 88,741	**25.** 34,699

ANSWERS

1. _____
2. _____
3. _____
4. _____
5. _____
6. _____
7. _____
8. _____
9. _____
10. _____
11. _____
12. _____
13. _____
14. _____
15. _____
16. _____
17. _____
18. _____
19. _____
20. _____
21. _____
22. _____
23. _____
24. _____
25. _____

ANSWERS

26. 36,282   27. 40,455   28. 70,501   29. 19,458   30. 59,628

Round the numbers in problems 31–40 to the nearest ten thousand.

31. 135,000   32. 315,000   33. 455,133   34. 629,514   35. 263,594

36. 583,685   37. 337,999   38. 142,815   39. 595,416   40. 294,956

In problems 41–45 round 158,463 to the indicated place value.

41. nearest ten   42. nearest thousand

43. nearest hundred   44. nearest ten thousand

45. nearest hundred thousand

In problems 46–50 round 295,997 to the indicated place value.

46. nearest ten   47. nearest ten thousand

48. nearest hundred   49. nearest thousand

50. nearest hundred thousand

In problems 51–56 round 5,555,555 to the indicated place value.

51. nearest ten   52. nearest hundred thousand

53. nearest hundred   54. nearest thousand

55. nearest ten thousand 56. nearest million

In problems 57–60 round 9,999 to the indicated place value.

57. nearest ten   58. nearest hundred

59. nearest thousand   60. nearest ten thousand

1-4 ADDITION OF WHOLE NUMBERS

The first operation on whole numbers that we will examine is addition. The facts you have learned so far about whole numbers will be of great value when performing operations on them.

One basic fact holds true throughout all of mathematics. *Only like quantities can be added.* When applying this principle to the addition of whole numbers, we will keep in mind that ones can only be added to ones, tens to tens, hundreds to hundreds, and so on.

OBJECTIVES

In this section you will learn to add any group of whole numbers using the addition algorithm.

To prepare to add any group of numbers rapidly and correctly, we must first make sure we know all the basic addition facts. We know that all numbers are written with the digits 0, 1, 2, 3, 4, 5, 6, 7, 8, 9. If we add any of these digits two at a time in all possible ways, we have the basic addition facts. It is extremely important that you know all of these basic facts. The authors have found that students not proficient in the basic facts of arithmetic have difficulty in all aspects of future studies in mathematics.

We will now look at an example of adding whole numbers. Let's examine closely what happens as we add 574 and 299. Recalling our basic principle that only like quantities can be combined, we will first write the numbers in expanded form.

$$574 = 5 \text{ hundreds} + 7 \text{ tens} + 4 \text{ ones}$$
$$299 = 2 \text{ hundreds} + 9 \text{ tens} + 9 \text{ ones}$$

Using our addition facts and adding hundreds to hundreds, tens to tens, and ones to ones, we obtain

$$7 \text{ hundreds} + 16 \text{ tens} + 13 \text{ ones}.$$

We note that

$$16 \text{ tens} = 1 \text{ hundred} + 6 \text{ tens}.$$

Also,

$$13 \text{ ones} = 1 \text{ ten} + 3 \text{ ones}.$$

So 7 hundreds + 16 tens + 13 ones = 7 hundreds + 1 hundred + 6 tens
 + 1 ten + 3 ones
 = 8 hundreds + 7 tens + 3 ones,

which is written as 873. Thus the sum of 574 and 299 is 873.

This method may explain how addition actually takes place, but you would agree that we are fortunate to have shorter methods. Let's look at the more common (and shorter) way of adding 574 and 299. This short method is called the **addition algorithm**. An **algorithm** is a step-by-step procedure for obtaining an answer. An algorithm should reduce computation time.

Example 1   Add 574 and 299 using the addition algorithm.

First the numbers are written in a column, making sure the ones place is under the ones, tens under tens, and so on.

$$\begin{array}{r} 574 \\ \underline{299} \end{array}$$

Now we add the 9 + 4 obtaining 13. We recognize this as 1 ten and 3 ones, so we place the 3 in the ones column and the 1 in the tens column.

$$\begin{array}{r} 1 \longleftarrow \text{1 in the tens column} \\ 574 \\ \underline{299} \\ 3 \longleftarrow \text{3 in the ones column} \end{array}$$

This process is referred to as *carrying,* and we say "write down the three and carry the one." Of course you could "carry" mentally, but when checking, it is helpful to have written down the number carried.

To proceed, we add the 9 + 7 + 1 of the tens column obtaining 17 tens. We record the 7 in the tens column and carry the 1 to the hundreds column. We finally add the 2 + 5 + 1 in the hundreds column to obtain 8 hundreds and record this in the hundreds column.

$$\begin{array}{r} \text{1 in the hundreds column} \longrightarrow 11 \\ 574 \\ \underline{299} \\ \text{8 in the hundreds column} \longrightarrow 873 \\ \text{7 in the tens column} \end{array}$$

Example 2   Find the sum of 17; 325; and 983.

Again, be careful to write ones, tens, and so on, in their proper places so that we will be adding like quantities.

$$\begin{array}{r} 11 \\ 17 \\ 325 \\ \underline{983} \\ 1{,}325 \end{array}$$

We now consider two important properties of addition.

Commutative Property of Addition   If a and b represent any two numbers, then $a + b = b + a$.

Example 3 $5 + 4 = 4 + 5$

In other words, we will obtain the same result (sum) if we add any two numbers in any order.

Of course, you probably have realized by now that addition only applies to two numbers at a time. The proper mathematical term is that addition is a **binary** (that is, two number) operation. So when we desire the sum of three or more numbers, we must combine them two at a time. This brings us to the second important property.

> **Associative Property of Addition** If a, b, and c represent any three numbers, then $(a + b) + c = a + (b + c)$.

Example 4 $(5 + 4) + 3 = 9 + 3 = 12$
$5 + (4 + 3) = 5 + 7 = 12$

In other words, it does not matter which two numbers are added first.

This is sometimes referred to as the **grouping principle.** We can use this property and the commutative property to increase our speed in adding numbers.

Example 5 Add $5 + 3 + 7 + 6 + 5 + 4 + 9$.

We first place the numbers in column form. As we add from bottom to top, we might "pick out" combinations that add to ten. In this case the 6 and 4, the two 5s, and the 7 and 3.

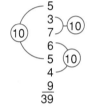

Example 6 Add $8 + 4 + 9 + 2 + 1 + 7 + 6$.

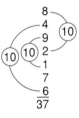

> **WARNING**
>
> If you skip around, don't overlook a number. This process must be practiced to be effective.

Example 7 Alex purchased 5 pounds of peanuts, 3 pounds of cashews, 2 pounds of almonds, and 4 pounds of pecans. What is the total number of pounds of nuts that he purchased?

When you encounter a problem that is stated in words, be certain that you read very carefully and answer the question that is asked. In this case we are to find a *total* that tells us we must use addition.

To find the total we need to add the number of pounds of each type of nut purchased.

$$\begin{array}{r} 5 \\ 3 \\ 2 \\ \underline{4} \\ 14 \end{array}$$

Thus he purchased a total of 14 pounds.

EXERCISE SET 1-4-1

ANSWERS

1. _____
2. _____
3. _____
4. _____
5. _____
6. _____
7. _____
8. _____
9. _____
10. _____
11. _____
12. _____
13. _____
14. _____
15. _____
16. _____
17. _____
18. _____
19. _____
20. _____
21. _____
22. _____
23. _____
24. _____
25. _____
26. _____
27. _____
28. _____
29. _____
30. _____
31. _____
32. _____
33. _____
34. _____
35. _____

Add the following numbers using the principles learned in this section.

1. 2 8 5 <u>5</u>	**2.** 3 7 4 <u>6</u>	**3.** 9 1 3 <u>7</u>	**4.** 8 2 4 <u>6</u>	**5.** 7 5 3 <u>5</u>
6. 5 2 3 8 <u>7</u>	**7.** 1 8 3 9 <u>2</u>	**8.** 7 9 5 5 <u>3</u>	**9.** 2 1 9 7 <u>8</u>	**10.** 5 4 8 6 <u>5</u>
11. 7 5 3 5 6 <u>4</u>	**12.** 3 4 5 7 4 <u>1</u>	**13.** 2 3 8 3 1 <u>4</u>	**14.** 5 6 5 7 2 <u>3</u>	**15.** 3 9 4 6 1 <u>5</u>
16. 7 8 1 3 1 <u>9</u>	**17.** 6 5 8 1 4 <u>7</u>	**18.** 5 4 5 2 8 <u>1</u>	**19.** 4 3 5 7 6 <u>2</u>	**20.** 7 4 3 1 6 <u>8</u>
21. 54 <u>32</u>	**22.** 75 <u>24</u>	**23.** 13 <u>56</u>	**24.** 33 <u>26</u>	**25.** 38 <u>61</u>
26. 28 <u>32</u>	**27.** 49 <u>21</u>	**28.** 32 <u>48</u>	**29.** 35 <u>55</u>	**30.** 23 <u>57</u>
31. 43 <u>72</u>	**32.** 86 <u>31</u>	**33.** 52 <u>67</u>	**34.** 35 <u>94</u>	**35.** 73 <u>56</u>

1-4 ADDITION OF WHOLE NUMBERS

					ANSWERS
36. 58 46	**37.** 83 29	**38.** 65 35	**39.** 29 71	**40.** 38 69	36. _____ 37. _____ 38. _____ 39. _____ 40. _____
41. 23 45 53	**42.** 42 38 44	**43.** 35 13 67	**44.** 24 83 16	**45.** 39 25 71	41. _____ 42. _____ 43. _____ 44. _____ 45. _____
46. 19 28 62	**47.** 55 48 25	**48.** 57 39 86	**49.** 34 48 97	**50.** 26 73 58	46. _____ 47. _____ 48. _____ 49. _____ 50. _____
51. 536 243	**52.** 614 325	**53.** 867 112	**54.** 745 254	**55.** 368 631	51. _____ 52. _____ 53. _____ 54. _____ 55. _____
56. 325 436	**57.** 842 139	**58.** 546 317	**59.** 649 238	**60.** 767 218	56. _____ 57. _____ 58. _____ 59. _____ 60. _____
61. 285 646	**62.** 327 585	**63.** 396 517	**64.** 639 267	**65.** 548 357	61. _____ 62. _____ 63. _____ 64. _____ 65. _____
66. 568 496	**67.** 783 269	**68.** 646 798	**69.** 747 685	**70.** 489 678	66. _____ 67. _____ 68. _____ 69. _____ 70. _____
71. 63 456 547	**72.** 38 419 762	**73.** 84 287 926	**74.** 96 358 834	**75.** 27 518 453	71. _____ 72. _____ 73. _____ 74. _____ 75. _____
76. 16 385 958	**77.** 75 796 489	**78.** 37 677 586	**79.** 49 557 316	**80.** 56 768 493	76. _____ 77. _____ 78. _____ 79. _____ 80. _____

CHAPTER 1 WHOLE NUMBERS

ANSWERS

81. _____
82. _____
83. _____
84. _____
85. _____
86. _____
87. _____
88. _____
89. _____
90. _____
91. _____
92. _____
93. _____
94. _____
95. _____
96. _____
97. _____
98. _____
99. _____
100. _____
101. _____
102. _____
103. _____
104. _____
105. _____
106. _____

81. 219
564
<u>835</u>

82. 614
328
<u>597</u>

83. 325
486
<u>578</u>

84. 753
249
<u>368</u>

85. 527
786
<u>975</u>

86. 409
348
<u>794</u>

87. 829
507
<u>695</u>

88. 185
647
<u>576</u>

89. 436
287
<u>548</u>

90. 344
698
<u>753</u>

91. 2,511
5,699
<u>3,068</u>

92. 1,554
2,795
<u>4,106</u>

93. 5,468
2,456
<u>1,849</u>

94. 6,808
5,099
<u>2,563</u>

95. 3,564
7,482
<u>6,928</u>

96. 7,165
14
2,829
<u>568</u>

97. 2,436
986
4,205
<u>87</u>

98. 24,682
19,428
564
<u>83,046</u>

99. 98
12,564
759
<u>5,685</u>

100. 32,048
673
8,567
<u>14,412</u>

Write the following in column form and add.

101. Find the sum of 46; 198; 2,305; and 1,795.

102. Find the sum of 463; 3,488; 76; and 5,654.

103. Find the sum of 39; 7,255; 664; and 2,325.

104. Find the sum of 719; 1,682; 908; and 4,265.

105. Find the sum of 3,069; 624; 1,253; and 376.

106. Find the sum of 58; 472; 3,804; and 2,196.

107. Find the sum of 515; 724; 36; 1,485; and 99,611.

108. Find the sum of 36; 843; 6,207; 3,124; and 618.

109. Find the sum of 54,213; 6,145; 40,517; 25; and 473.

110. Find the sum of 364,298; 5,622; 541,376; 474; and 100,716.

111. Helen bought a book for $15, a pen for $4, and a calculator for $21. What was the total cost of the three items?

112. Jim earned $164 the first week, $159 the second week, and $208 the third week. What were his total earnings for the three weeks?

113. A man mixed 35 pounds of one type of grass seed with 48 pounds of another type of grass seed. How many pounds of mixture does he have?

114. A man weighing 195 pounds and a woman weighing 120 pounds ride in a car that weighs 3,248 pounds without any passengers. What is the total weight of the car and passengers?

115. A merchant pays $65 for an item and wishes to make a profit of $23 when selling it. How much should the item sell for?

116. A man borrows $2,690 for one year and agrees to pay $325 interest. How much does he need to pay at the end of the year?

ANSWERS

107. _____

108. _____

109. _____

110. _____

111. _____

112. _____

113. _____

114. _____

115. _____

116. _____

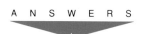

117. _____

118. _____

119. _____

120. _____

117. A woman drove from Mobile to Atlanta, a distance of 353 miles. She then drove 676 miles to Baltimore. What was the total distance that she drove?

118. During one month a person's electric bill was $176, the telephone bill was $38, and the water bill was $17. What was the total of the three bills?

119. In the presidential election of 1960 John Kennedy received 303 electoral votes, Richard Nixon received 219, and Harry Byrd received 15. What was the total number of electoral votes cast in the election?

120. In California the estimated 1986 population of Los Angeles was 3,215,506. That of San Diego was 1,013,400 and San Jose was 713,385. Find the total population of the three cities.

1-5 SUBTRACTION OF WHOLE NUMBERS

In this section you will learn to:
1. Subtract one whole number from another.
2. Check by writing the subtraction problem as an addition problem.

In section 1-4 the basic facts of addition were emphasized. If you know those facts well, you already know the basic facts necessary for subtraction. We use the symbol "$-$" to indicate subtraction of whole numbers.

Example 1 $9 - 6 = 3$

This number sentence states that "nine minus six is three" or "the difference of nine and six is three."

But how do we know that nine minus six is three? We are actually using the *addition* fact that $6 + 3 = 9$. In other words, the question "What is the difference of nine and six?" is the same as "What number added to six will give a result of nine?" A formal statement of this fact is given in the following rule.

Rule   If a, b, and c represent any whole numbers and if $a - b = c$, then it is also true that $b + c = a$.

In other words, we only need the facts from addition to perform the operation of subtraction.

1-5 SUBTRACTION OF WHOLE NUMBERS

Example 2 Find $784 - 432$.

As in addition, we will place the numbers in columns.

$$\begin{array}{r} 784 \\ -432 \end{array}$$

Remember that only like quantities can be added. This is also true for subtraction. We subtract ones from ones, tens from tens, and so on. Thus we obtain

$$\begin{array}{r} 784 \\ -432 \\ \hline 352. \end{array}$$

We can check our answer by adding 432 and 352 to obtain 784.

$$\textit{Check:} \quad \begin{array}{r} 432 \\ +352 \\ \hline 784 \end{array}$$

It is a good practice to always check your answer.

Example 3 Find the difference of 425 and 213.

The number being subtracted follows the word "and," so the subtraction should be written as $425 - 213$.

$$\begin{array}{r} 425 \\ -213 \\ \hline 212 \end{array}$$

$$\textit{Check:} \quad \begin{array}{r} 213 \\ +212 \\ \hline 425 \end{array}$$

▸ EXERCISE SET 1-5-1

Subtract and check.

1. $8 - 5$
2. $9 - 4$
3. $5 - 2$
4. $3 - 1$

5. $7 - 4$
6. $6 - 5$
7. $5 - 5$
8. $7 - 7$

9. $18 - 7$
10. $19 - 6$
11. $14 - 3$
12. $19 - 7$

13. $15 - 5$
14. $27 - 7$
15. $45 - 23$
16. $63 - 31$

ANSWERS

1. _____
2. _____
3. _____
4. _____
5. _____
6. _____
7. _____
8. _____
9. _____
10. _____
11. _____
12. _____
13. _____
14. _____
15. _____
16. _____

ANSWERS

17. _____
18. _____
19. _____
20. _____
21. _____
22. _____
23. _____
24. _____
25. _____
26. _____
27. _____
28. _____
29. _____
30. _____
31. _____
32. _____
33. _____
34. _____
35. _____
36. _____
37. _____
38. _____
39. _____
40. _____
41. _____
42. _____
43. _____
44. _____
45. _____
46. _____
47. _____
48. _____

17. 57 − 35

18. 49 − 25

19. 86 − 54

20. 58 − 46

21. 74
   −53

22. 46
   −33

23. 87
   −36

24. 76
   −44

25. 38
   −27

26. 59
   −48

27. 57
   −37

28. 68
   −48

29. 584
   −263

30. 453
   −241

31. 629
   −518

32. 835
   −421

33. 752
   −431

34. 694
   −523

35. 593
   −361

36. 879
   −565

37. 936
   −416

38. 745
   −525

39. 647
   −245

40. 753
   −351

41. Find the difference of 9 and 2.

42. Find the difference of 25 and 13.

43. Find the difference of 86 and 54.

44. Find the difference of 263 and 142.

45. Find the difference of 574 and 473.

46. Find the difference of 958 and 557.

47. Find the difference of 396 and 153.

48. Find the difference of 477 and 256.

1–5 SUBTRACTION OF WHOLE NUMBERS   **27**

49. Find the difference of 958 and 658.

50. Find the difference of 894 and 594.

49. _____

50. _____

In the previous exercise set all digits in the top number (**minuend**) were larger than those in the bottom number (**subtrahend**). When this is not true, we must use a technique known as **borrowing.** For example, suppose we wish to find the difference of 54 and 9. We first place the two numbers in columns.

$$\begin{array}{r} 54 \\ -9 \\ \hline \end{array}$$

We now have a problem. We know that 54 is larger than 9 but we cannot subtract in the ones column. That is, we cannot subtract 9 from 4. Solving this problem requires *borrowing* and we proceed as follows.

From previous work we know that 54 = 5 tens + 4 ones, and that this number would be the same as 4 tens + 14 ones. Then

$$\begin{array}{r} 54 = 4\ \text{tens} + 14\ \text{ones} \\ -9 = \phantom{4\ \text{tens} + 1}-\ 9\ \text{ones} \\ \hline 4\ \text{tens} + 5\ \text{ones.} \end{array}$$

The result is
So 54 − 9 = 45.

This entire process is simplified by writing the problem in this manner.

$$\begin{array}{r} {\scriptstyle 4\ 14} \\ \cancel{5}4 \\ -\ 9 \\ \hline 45 \end{array}$$

We borrow 1 from the 5 in the tens column (we cross out the 5 to indicate we have borrowed 1). This becomes 10 in the ones column. Together with the 4 already in the ones column, we then have a total of 14 ones. We then use our addition facts to subtract.

$$\begin{array}{rr} \textit{Check:} & 9 \\ & +45 \\ \hline & 54 \end{array}$$

Example 4   Subtract:
$$\begin{array}{r} 238 \\ -\ 59 \\ \hline \end{array}$$

First, since we cannot subtract 9 from 8, we *borrow* one 10 from the tens column. We thus have 18 ones. Now, subtracting 9 from 18 gives us 9.

$$\begin{array}{r} {\scriptstyle 2\ 18} \\ 2\cancel{3}8 \\ -\ 59 \\ \hline 9 \end{array}$$

Working with the tens digits, we cannot subtract 5 tens from 2 tens, so we must borrow one 100 from the hundreds column to obtain 12 tens. We then complete the subtraction to obtain

$$\begin{array}{r} \overset{12}{}\overset{12}{}\overset{18}{} \\ 2\cancel{3}8 \\ -59 \\ \hline 179 \end{array}.$$

Check:
$$\begin{array}{r} 59 \\ +179 \\ \hline 238 \end{array}$$

It is more efficient if we can do the borrowing mentally. However, you should write whatever is necessary to assist you in obtaining the correct answer.

Example 5 Subtract:
$$\begin{array}{r} 834 \\ -265 \end{array}$$

$$\begin{array}{r} \overset{7}{}\overset{2}{} \\ \cancel{8}\cancel{3}4 \\ -265 \\ \hline 569 \end{array}$$

We borrowed from the 3 tens to obtain 14 ones but did not write the 14. We also borrowed from the 8 hundreds to obtain 12 tens but did not write the 12. However, it is always a good idea to strike through as we did the 3 and 8 so that we will not forget that we have borrowed.

Check:
$$\begin{array}{r} 265 \\ +569 \\ \hline 834 \end{array}$$

EXERCISE SET 1-5-2

ANSWERS

Subtract and check.

1. $23 - 8$ 2. $13 - 6$ 3. $14 - 9$ 4. $36 - 9$

5. $53 - 7$ 6. $61 - 5$ 7. $42 - 5$ 8. $54 - 8$

9. $34 - 16$ 10. $44 - 27$ 11. $\begin{array}{r} 45 \\ -18 \end{array}$ 12. $\begin{array}{r} 36 \\ -19 \end{array}$

1. _____
2. _____
3. _____
4. _____
5. _____
6. _____
7. _____
8. _____
9. _____
10. _____
11. _____
12. _____

1-5 SUBTRACTION OF WHOLE NUMBERS

13. 31 −13	**14.** 53 −35	**15.** 48 −29	**16.** 37 −18	
17. 52 −35	**18.** 61 −38	**19.** 236 − 48	**20.** 325 − 39	
21. 524 − 56	**22.** 322 − 35	**23.** 425 − 96	**24.** 643 − 76	
25. 283 − 95	**26.** 234 − 55	**27.** 542 −163	**28.** 436 −158	
29. 333 −157	**30.** 555 −268	**31.** 341 −199	**32.** 465 −299	
33. 812 −648	**34.** 713 −569	**35.** 473 −395	**36.** 531 −463	
37. 352 −264	**38.** 258 −179	**39.** 316 −198	**40.** 413 −284	

41. Find the difference of 35 and 7.

42. Find the difference of 43 and 8.

43. Find the difference of 56 and 28.

44. Find the difference of 43 and 26.

45. Find the difference of 234 and 68.

46. Find the difference of 432 and 79.

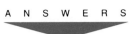

ANSWERS

13. _____
14. _____
15. _____
16. _____
17. _____
18. _____
19. _____
20. _____
21. _____
22. _____
23. _____
24. _____
25. _____
26. _____
27. _____
28. _____
29. _____
30. _____
31. _____
32. _____
33. _____
34. _____
35. _____
36. _____
37. _____
38. _____
39. _____
40. _____
41. _____
42. _____
43. _____
44. _____
45. _____
46. _____

ANSWERS

47. _____

48. _____

49. _____

50. _____

47. Find the difference of 324 and 165.

48. Find the difference of 535 and 367.

49. Find the difference of 611 and 573.

50. Find the difference of 711 and 695.

When zeros occur in the minuend, the problem is a little more complex.

Example 6 Subtract: $\begin{array}{r}703\\-264\end{array}$

Since we cannot subtract 4 ones from 3 ones, we must borrow a 10 from the tens column. But there aren't any tens in that column. What we must do is borrow 1 hundred from the hundreds column and enter it into the tens column as 10 tens.

$$\begin{array}{r}\overset{6\;10}{\cancel{7}03}\\-264\end{array}$$

We can now borrow 1 ten from the tens column and obtain 13 ones in the ones column. This enables us to subtract.

$$\begin{array}{r}9\\6\;\cancel{10}\;13\\703\\-264\\\hline 439\end{array}$$

Check: $\begin{array}{r}264\\+439\\\hline 703\end{array}$

Example 7 Subtract: $\begin{array}{r}2{,}003\\-514\end{array}$

This time we must first borrow one from the thousands place to obtain 10 hundreds, then one from the hundreds place to get 10 tens, and then one from the tens place to have enough ones to subtract.

$$\begin{array}{r}99\\1\;\cancel{10}\;\cancel{10}\;13\\2{,}003\\-514\\\hline 1{,}489\end{array}$$

Check: $\begin{array}{r}514\\+1{,}489\\\hline 2{,}003\end{array}$

1-5 SUBTRACTION OF WHOLE NUMBERS

Example 8 Subtract: 729
             −254

$$\begin{array}{r} {\scriptstyle 6\ 12} \\ \cancel{7}29 \\ -254 \\ \hline 475 \end{array}$$

Notice that in this example we did not need to borrow in order to subtract in the ones column, but had to borrow a hundred so we could subtract in the tens column.

It is only necessary to borrow when the digit in any column of the minuend is smaller than the digit in the corresponding column in the subtrahend.

Example 9 A television set, regularly priced at $565, is reduced by $78. What is the new price?

In this problem the key words "reduced by" tell us that we must subtract.

To find the new price we must subtract 78 from 565.

$$\begin{array}{r} 565 \\ -\ 78 \\ \hline 487 \end{array}$$

Thus the new price is $487.

Example 10 Bill's grandfather sent him $100 as a birthday gift. His grandfather's instructions were "Buy yourself a pair of gym shoes and shorts and then place the remainder in your savings account." Bill bought shoes for $32 and shorts for $16. How much did he have left to place in his savings account?

This word problem is more involved than those you have solved in previous sections. As in all word problems, read and read again if necessary until you clearly understand the principles involved.

First we see that we must find how much money Bill spent. To do this we add the amounts spent on shoes and shorts.

amount for shoes	$32
amount for shorts	$16
total amount spent	$48

If the problem had asked "How much did Bill spend?" we would have our answer. But the question is "How much is left of the $100 for savings?" To find this amount we must subtract the amount spent from the original amount.

original amount	$100
amount spent	−$ 48
amount left for savings	$ 52

WARNING

When solving a word problem—*always* look back to see that you have answered the question asked.

EXERCISE SET 1-5-3

Subtract and check.

1. 30 − 8
2. 40 − 7
3. 50 − 6
4. 20 − 4

5. 203 − 47
6. 504 − 39
7. 605 − 48
8. 702 − 46

9. 306 − 57
10. 305 − 28
11. 408 − 69
12. 204 − 75

13. 506 − 19
14. 708 − 59
15. 307 − 118
16. 405 − 217

17. 801 − 423
18. 501 − 232
19. 701 − 624
20. 301 − 285

21. 2,005 − 386
22. 2,004 − 226
23. 5,004 − 316
24. 3,005 − 218

25. 1,003 − 724
26. 4,002 − 353
27. 2,021 − 563
28. 1,034 − 468

29. 4,203 − 147
30. 3,405 − 386
31. 461 − 153
32. 482 − 273

				ANSWERS
33. 468 − 93	**34.** 524 − 62	**35.** 718 −307	**36.** 324 −116	33. _____
				34. _____
37. 345 −126	**38.** 857 −318	**39.** 527 −386	**40.** 934 −671	35. _____
				36. _____
				37. _____
				38. _____
				39. _____
41. 768 −454	**42.** 375 −162	**43.** 4,007 − 998	**44.** 6,002 − 913	40. _____
				41. _____
				42. _____
				43. _____
45. 408 −299	**46.** 707 −399	**47.** 3,054 − 926	**48.** 2,045 − 917	44. _____
				45. _____
				46. _____
				47. _____
				48. _____
49. 4,060 − 792	**50.** 5,040 − 839	**51.** 573 −263	**52.** 681 −480	49. _____
				50. _____
				51. _____
				52. _____
53. 316 −118	**54.** 417 −219	**55.** 537 −428	**56.** 345 −236	53. _____
				54. _____
				55. _____
				56. _____
57. 54,289 −15,136	**58.** 35,001 − 4,985	**59.** 5,017 −4,259	**60.** 6,018 −5,739	57. _____
				58. _____
				59. _____
				60. _____
				61. _____
61. 54,653 −43,421	**62.** 28,659 −17,514	**63.** 5,998 −2,999	**64.** 3,997 −1,998	62. _____
				63. _____
				64. _____
				65. _____
65. 378 −279	**66.** 496 −399	**67.** 10,000 − 9,768	**68.** 10,000 − 9,492	66. _____
				67. _____
				68. _____
				69. _____
				70. _____
69. 783 −274	**70.** 564 −256	**71.** 3,054 −1,983	**72.** 4,028 −2,935	71. _____
				72. _____

ANSWERS	
73.	
74.	
75.	
76.	
77.	
78.	
79.	
80.	
81.	
82.	
83.	
84.	
85.	
86.	
87.	
88.	
89.	
90.	
91.	
92.	
93.	
94.	
95.	
96.	
97.	
98.	
99.	
100.	

73. 4,307 − 298

74. 2,804 − 796

75. 2,643 − 1,834

76. 3,465 − 2,628

77. 215 − 107

78. 618 − 509

79. 4,113 − 345

80. 3,116 − 458

81. 55,007 − 43,198

82. 68,003 − 16,917

83. 115 − 96

84. 117 − 38

85. 1,004 − 638

86. 1,008 − 749

87. 3,015 − 1,807

88. 6,014 − 4,608

89. 1,483 − 849

90. 1,635 − 716

91. Find the difference of 306 and 58.

92. Find the difference of 701 and 43.

93. Find the difference of 754 and 523.

94. Find the difference of 935 and 724.

95. Find the difference of 1,036 and 938.

96. Find the difference of 2,042 and 744.

97. Find the difference of 1,000 and 564.

98. Find the difference of 1,000 and 405.

99. Find the difference of 6,040 and 5,234.

100. Find the difference of 8,050 and 7,953.

1-5 SUBTRACTION OF WHOLE NUMBERS

101. There are 31 days in March. After March 8, how many days remain in the month?

102. A basketball team scored 110 points in a game. If they scored 48 points during the first half, how many points did they score in the second half?

103. A student read 128 pages of a book containing 415 pages. How many pages remain to be read?

104. It was announced five days before a football game that there were 4,168 unsold tickets for a stadium that has a seating capacity of 83,104. How many tickets had been sold?

105. The temperature rose from 38 degrees in the morning to 64 degrees in the afternoon. How many degrees did the temperature rise?

106. The regular price of an item is $123. A store is having a sale and offers a discount of $35. What is the sale price of the item?

107. The Holland Tunnel is 8,557 feet long. The Hampton Roads Tunnel, near Norfolk, has a length of 7,479 feet. How much longer is the Holland Tunnel?

108. A car with a regular price of $15,098 is offered on sale for $13,999. How much is the discount?

109. The Empire State Building in New York is 1,250 feet tall. The Transamerica Pyramid in San Francisco is 853 feet tall. How much taller is the Empire State Building?

110. The height of Mt. Everest is 29,028 feet and that of Mt. Whitney is 14,494 feet. How much taller is Mt. Everest?

111. A student purchased a book for $23 and a calculator for $19. How much change does the student receive from $50?

112. A trucker drives 427 miles the first day and 515 miles the second day. If the total mileage of the trip is 1,297 miles, how many miles must the trucker drive the third day to complete the trip?

ANSWERS

101. _____
102. _____
103. _____
104. _____
105. _____
106. _____
107. _____
108. _____
109. _____
110. _____
111. _____
112. _____

1-6 MULTIPLICATION OF WHOLE NUMBERS

OBJECTIVES

In this section you will learn to multiply any two whole numbers using the long multiplication algorithm.

The operation of multiplication can be thought of as repeated addition. For instance, if we wish to multiply 7 × 5 we can think of this as seven 5s, or 5 + 5 + 5 + 5 + 5 + 5 + 5. This sum is 35, so we say 7 × 5 = 35. We could also interpret 7 × 5 as five 7s, or 7 + 7 + 7 + 7 + 7, which also has a sum of 35.

The fact that seven 5s (7 × 5) gives the same result as five 7s (5 × 7) illustrates an important property of multiplication called the commutative property. A formal statement of this property follows.

> **Commutative Property of Multiplication** If a and b represent any two numbers, then $a \times b = b \times a$.

When preparing for the operation of addition, we pointed out that there are a limited number of basic facts necessary for being competent in addition. The same is true for multiplication. Since all whole numbers are written by the use of ten digits, we need only to know the results of multiplying any two of these digits. Once again we must emphasize the importance of knowing the basic facts. Satisfactory progress cannot be made in the operation of multiplying whole numbers unless you are proficient in these facts.

A process (or algorithm) of multiplying any two whole numbers is called **long multiplication.** Suppose we want to find the product of 7 × 32. If we look at this as repeated addition, we could say we want the sum of thirty-two 7s. Now, if we remember that 32 = 30 + 2, we could say that 7 × 32 = 30 sevens + 2 sevens. In other words, 7 × 32 = 7 × (30 + 2), which is also (7 × 30) + (7 × 2). A very important property, called the distributive property of multiplication over addition, is illustrated by this fact.

> **Distributive Property of Multiplication over Addition** If a, b, and c are any three numbers, then $a \times (b + c) = (a \times b) + (a \times c)$.

Sometimes this property or law is referred to simply as the distributive property. Now let's apply this property in the following example.

Example 1 Multiply 7 × 32.

7 × 32 = 7 × (30 + 2) = (7 × 30) + (7 × 2)
Now 7 × 30 = 7 × 3 tens = 21 tens = 210.
Also, 7 × 2 = 14 from our basic multiplication facts. So we have

7 × 32 = (7 × 30) + (7 × 2) = 210 + 14 = 224.

1-6 MULTIPLICATION OF WHOLE NUMBERS

Now let's examine the same problem using the multiplication algorithm. We first write the two numbers to be multiplied in columns.

$$\begin{array}{r} 32 \\ \times\ 7 \\ \hline \end{array}$$

We now multiply 7 × 2 and obtain 14.

$$\begin{array}{r} 32 \\ \times\ 7 \\ \hline 14 \end{array}$$

We also multiply 7 × 3 and get 21. But we must note that the 3 in 32 is in the tens column, so the 21 is 21 tens, making it necessary to place the 1 of 21 in the tens column. We then add the 14 ones and 21 tens to obtain the answer of 224.

$$\begin{array}{r} 32 \\ \times\ 7 \\ \hline 14 \\ 21 \\ \hline 224 \end{array}$$

We could further refine the work involved by a process called **carrying.** We first say that 7 × 2 = 14. We then place the 4 in the ones column and carry the 1 to the tens column.

$$\begin{array}{r} 1 \\ 32 \\ \times\ 7 \\ \hline 4 \end{array}$$

We now multiply 7 × 3 to get 21 and add the 1 carried to get 22.

$$\begin{array}{r} 1 \\ 32 \\ \times\ 7 \\ \hline 224 \end{array}$$

It will be helpful if you learn to carry mentally as soon as possible.

Example 2   Find the product of 27 × 32.

We first place the numbers in columns.

$$\begin{array}{r} 32 \\ \times 27 \\ \hline \end{array}$$

We now multiply 7 × 32 to obtain 224 as in the previous example.

$$\begin{array}{r} 32 \\ \times 27 \\ \hline 224 \end{array}$$

We next multiply by the 2 of 27. Since 2 is in the tens column, we are actually multiplying by 2 tens or 20. This requires that our first number $2 \times 2 = 4$ be placed in the tens column.

$$\begin{array}{r} 32 \\ \times 27 \\ \hline 224 \\ 4 \end{array}$$

We must now multiply the 3 tens by the 2 tens or $2 \times 10 \times 3 \times 10$. This gives us 6×100 or 600. So we say that $2 \times 3 = 6$, but we place the 6 in the hundreds column. Adding now gives 864 as the product of 27 and 32.

$$\begin{array}{r} 32 \\ \times 27 \\ \hline 224 \\ 64 \\ \hline 864 \end{array}$$

Example 3 Find the product of 283 and 42.

$$\begin{array}{r} 283 \\ \times\ 42 \\ \hline 566 \end{array}$$
- 2×3
- $2 \times 8 = 16$; write the 6 and carry 1.
- $2 \times 2 = 4$, plus the 1 carried to give 5.

$$\begin{array}{r} 283 \\ \times\ 42 \\ \hline 566 \\ 1132 \end{array}$$
- $4 \times 3 = 12$; write the 2 and carry 1.
- $4 \times 8 = 32$, plus 1 carried equals 33. Write the 3 and carry 3.
- $4 \times 2 = 8$, plus 3 carried equals 11.

Adding now gives 11,886 as the product of 283 and 42.

$$\begin{array}{r} 283 \\ \times\ 42 \\ \hline 566 \\ 11\ 32 \\ \hline 11{,}886 \end{array}$$

> ▼ **WARNING**
>
> In multiplication you will find that neatness is a great help. In fact many errors are made by not keeping columns straight or by placing numbers in the wrong column. The best rule to remember is: place the rightmost digit of the product in the same column as the digit by which you are multiplying.

Example 4 Find the product of 347 and 432.

```
      347
    ×432
      694  ←—— 4 is placed in the same column as 2.
    10 41  ←—— 1 is placed in the same column as 3.
   138 8   ←—— 8 is placed in the same column as 4.
   149,904
```

Example 5 Find the product of 385 and 203.

Note that we need not multiply by the 0 of 203 since only zeros will result. Notice again that $2 \times 5 = 10$, so the 0 (rightmost digit of the product) is placed in the same column as the 2.

```
      385
    ×203
    1 155
    77 0
    78,155
```

Example 6 Find the product of 2,165 and 3,002.

```
     2,165
    ×3,002
     4 330
    6 495
    6,499,330
```

Example 7 Find the product of 241 and 382. Round the answer to the nearest ten.

```
      241
    ×382
      482
    19 28
    72 3
    92,062
```

Rounding the product to the nearest ten gives 92,060.

Example 8 Find the sum of 252×43 and $1,405 \times 302$.

Read carefully! When you see the word *sum*, you know you must add. But don't make the mistake of reading "sum" and then adding all numbers you see in the problem. This problem involves the operation of addition as well as of multiplication. We are asked for the sum of two products.

We first find the indicated products.

```
      252                    1,405
    ×  43                  ×  302
      756                    2 810
    10 08                    421 5
    10,836                   424,310
```

The final step is to find the sum of these two products.

$$\begin{array}{r} 10{,}836 \\ +424{,}310 \\ \hline 435{,}146 \end{array}$$

Example 9 If a car travels at a constant rate of 55 miles per hour for 4 hours, how far has it traveled?

To determine the answer we must find the product of 55 × 4.

$$\begin{array}{r} 55 \\ \times\ 4 \\ \hline 220 \end{array}$$

The car has traveled 220 miles.

EXERCISE SET 1-6-1

ANSWERS

Find the following products.

1. 35 × 5	2. 26 × 3	3. 52 × 7	4. 39 × 8
5. 73 × 4	6. 36 × 5	7. 24 × 5	8. 18 × 9
9. 17 × 9	10. 25 × 8	11. 35 × 6	12. 59 × 4
13. 21 ×34	14. 16 ×32	15. 43 ×21	16. 45 ×31
17. 28 ×19	18. 37 ×18	19. 45 ×24	20. 35 ×26
21. 72 ×24	22. 63 ×24	23. 83 ×37	24. 59 ×83

1-6 MULTIPLICATION OF WHOLE NUMBERS

				ANSWERS
25. 234 × 16	26. 521 × 24	27. 314 × 35	28. 218 × 25	25. ___ 26. ___ 27. ___ 28. ___
29. 374 × 36	30. 267 × 37	31. 498 × 87	32. 489 × 98	29. ___ 30. ___ 31. ___ 32. ___
33. 478 × 69	34. 687 × 76	35. 914 × 35	36. 816 × 75	33. ___ 34. ___ 35. ___ 36. ___
37. 39 × 10	38. 41 × 10	39. 38 × 10	40. 23 × 10	37. ___ 38. ___ 39. ___ 40. ___
41. 154 × 10	42. 523 × 10	43. 430 × 10	44. 610 × 10	41. ___ 42. ___ 43. ___ 44. ___
45. 124 × 232	46. 314 × 163	47. 235 × 124	48. 123 × 412	45. ___ 46. ___ 47. ___ 48. ___
49. 328 × 439	50. 765 × 278	51. 397 × 256	52. 847 × 965	49. ___ 50. ___ 51. ___ 52. ___
53. 325 × 104	54. 634 × 205	55. 426 × 405	56. 813 × 206	53. ___ 54. ___ 55. ___ 56. ___
57. 416 × 100	58. 328 × 100	59. 542 × 100	60. 264 × 100	57. ___ 58. ___ 59. ___ 60. ___
61. 724 × 100	62. 813 × 100	63. 640 × 100	64. 320 × 100	61. ___ 62. ___ 63. ___ 64. ___

ANSWERS

65. _____
66. _____
67. _____
68. _____
69. _____
70. _____
71. _____
72. _____
73. _____
74. _____
75. _____
76. _____
77. _____
78. _____
79. _____
80. _____
81. _____
82. _____
83. _____
84. _____
85. _____
86. _____
87. _____
88. _____
89. _____
90. _____
91. _____
92. _____
93. _____
94. _____
95. _____
96. _____
97. _____
98. _____
99. _____
100. _____
101. _____
102. _____
103. _____
104. _____

65. 2,325 × 164
66. 1,419 × 236
67. 5,281 × 164
68. 3,512 × 283

69. 3,014 × 213
70. 7,304 × 219
71. 3,050 × 264
72. 1,040 × 386

73. 4,625 × 304
74. 2,419 × 507
75. 4,124 × 604
76. 3,114 × 806

77. 2,041 × 703
78. 5,052 × 209
79. 5,206 × 314
80. 6,405 × 519

81. 3,001 × 425
82. 2,008 × 196
83. 4,003 × 509
84. 3,006 × 704

85. 2,153 × 1,409
86. 6,244 × 1,306
87. 4,116 × 1,214
88. 6,224 × 1,049

89. 2,512 × 1,000
90. 3,416 × 1,000
91. 5,428 × 1,000
92. 7,038 × 1,000

93. 3,205 × 10
94. 5,214 × 10
95. 3,205 × 100
96. 5,214 × 100

97. 3,205 × 1,000
98. 5,214 × 1,000
99. 5,900 × 1,000
100. 6,300 × 1,000

In problems 101–4 round the product to the nearest ten.

101. 38 × 46
102. 207 × 56
103. 326 × 34
104. 608 × 121

In problems 105–8 round the product to the nearest hundred.

105. 27
     ×63

106. 512
     × 38

107. 714
     × 25

108. 273
     ×104

In problems 109–12 round the product to the nearest thousand.

109. 78
     ×34

110. 216
     × 95

111. 326
     ×104

112. 123
     ×216

In problems 113–16 round the product to the nearest ten thousand.

113. 413
     × 67

114. 512
     × 86

115. 1,204
     × 516

116. 3,029
     × 468

117. Find the sum of 324 × 116 and 1,204 × 639.

118. Find the sum of the following products.
     63   560   384   401
     ×81   × 34   × 76   ×306

119. Find the difference of 165 × 73 and 403 × 25.

120. Find the difference of 2,419 × 201 and 1,264 × 306.

121. A man buys six bags of grass seed at $15 per bag. What is the total cost?

122. If a person earns $275 per week, how much is earned in 26 weeks?

123. If you drove a constant rate of 45 miles per hour, how far would you travel in 14 hours?

124. If a certain car gets 32 miles per gallon, how far can it go on 15 gallons?

125. If fencing costs $4 per foot, what is the cost of 193 feet?

126. Each home in a homeowners association is assessed $192 per year. If there are 47 homes in the association, how much money should be collected per year?

ANSWERS

105. _____
106. _____
107. _____
108. _____
109. _____
110. _____
111. _____
112. _____
113. _____
114. _____
115. _____
116. _____
117. _____
118. _____
119. _____
120. _____
121. _____
122. _____
123. _____
124. _____
125. _____
126. _____

ANSWERS

127. _____

128. _____

129. _____

130. _____

127. The area of a rectangle is found by multiplying its length by its width. Find the area of a rectangle if its length is 28 feet and its width is 17 feet. (Give the answer in square feet.)

128. A rectangular room measures 16 feet by 21 feet. How many square feet of tile are needed to cover the floor?

129. A car is priced at $8,995 cash or 36 payments of $328 each. What is the total of the payments? How much is saved by paying cash?

130. A woman owes $3,200. She pays $112 per month for 24 months. How much does she still owe?

In the preceding exercise set, some of the problems involved multiplying by a power of ten (that is, 10, 100, 1,000, and so on). A short method of multiplying by a power of ten is given below.

> **Rule** The product of a whole number and a power of ten is that number followed by the number of zeros in the power of ten.

Instead of placing the numbers to be multiplied in columns, we write them horizontally as shown in the following examples.

Example 10 Find the product of 24 and 10.

The product is 24 followed by one zero.

$$24 \times 10 = 240$$

When writing a product horizontally, parentheses are often used instead of the symbol $\times$ to indicate multiplication. So instead of 24×10, we can write 24(10).

Example 11 Find the product of 65 and 100.

The product is 65 followed by two zeros.

$$65(100) = 6,500$$

Example 12 Find the product of 214(10,000).

The product is 214 followed by four zeros.

$$214(10,000) = 2,140,000$$

EXERCISE SET 1-6-2

Find the following products.

1. 12(10)
2. 81(10)
3. 211(10)
4. 601(10)
5. 32(100)
6. 316(100)
7. 205(100)
8. 1,216(100)
9. 132(1,000)
10. 204(1,000)
11. 16(10,000)
12. 345(10,000)
13. 503(100,000)
14. 62(100,000)
15. 2,091(1,000)
16. 5,117(1,000)
17. 20(100)
18. 80(1,000)
19. 610(10,000)
20. 100(10,000)

ANSWERS

1. _____
2. _____
3. _____
4. _____
5. _____
6. _____
7. _____
8. _____
9. _____
10. _____
11. _____
12. _____
13. _____
14. _____
15. _____
16. _____
17. _____
18. _____
19. _____
20. _____

1-7 DIVISION OF WHOLE NUMBERS

The last of the four basic operations on whole numbers that we will study is division. Just as we found subtraction defined in terms of addition, we will find that division can be defined in terms of multiplication.

Example 1 Suppose that a man sent $15 to be divided equally among his five grandchildren. How much would each grandchild receive?

The problem is simply, "How much is 15 divided by 5, or $15 \div 5$?" The answer is 3.

But the real question in this simple problem is "How do we know that $15 \div 5 = 3$?" The answer to this question actually gives us the definition of division, for we know that $15 \div 5 = 3$ because $5 \times 3 = 15$. In other words, we use our multiplication facts to divide.

OBJECTIVES

In this section you will learn to:
1. Rewrite a division problem as a multiplication problem.
2. Perform division using the long-division algorithm.

> **Definition** If a, b, and c represent whole numbers and if $a \div b = c$, then it must be true that $a = b \times c$.

Since you already know the basic facts of multiplication, it is not necessary to learn another set of basic facts for division.

Example 2 Solve $42 \div 7 = ?$

We know that $42 \div 7 = 6$ because $7 \times 6 = 42$.

Example 3 Solve $72 \div 9 = ?$

$72 \div 9 = 8$ because $9 \times 8 = 72$.

Other ways of indicating $72 \div 9$ are $\dfrac{72}{9}$ and $9\overline{)72}$.

In the example $72 \div 9 = 8$, 72 is called the *dividend,* 9 is called the *divisor,* and 8 is called the *quotient.*

Dividend is the number being divided.

Divisor is the number doing the dividing.

Quotient is the answer.

Of course, not all division problems will come out evenly. For example, if we wish to find $45 \div 7$, we try to find a number that will multiply by 7 to give a product of 45. There is no such whole number, so we find the largest whole number that can be multiplied by 7 to give a product less than 45. Whatever is "left over" is indicated as a *remainder,* usually denoted by the letter R.

Example 4 $45 \div 7 = 6$ with a remainder of 3 or $45 \div 7 = 6$ R3.

To check we find the product of 6 and 7 and then add the remainder 3. That is, $(6 \times 7) + 3 = 42 + 3 = 45$.

Example 5 $58 \div 6 = 9$ R4 because $(6 \times 9) + 4 = 54 + 4 = 58$.

Notice that the remainder must always be smaller than the divisor.

▼ EXERCISE SET 1-7-1

The following division problems will require only the basic facts of multiplication.

1. $8 \div 2$ 2. $6 \div 3$ 3. $10 \div 5$ 4. $12 \div 2$

1. _____
2. _____
3. _____
4. _____

1-7 DIVISION OF WHOLE NUMBERS

5. 15 ÷ 3　　**6.** 18 ÷ 6　　**7.** 20 ÷ 4　　**8.** 32 ÷ 4

9. 21 ÷ 3　　**10.** 35 ÷ 5　　**11.** 27 ÷ 3　　**12.** 36 ÷ 4

13. 49 ÷ 7　　**14.** 56 ÷ 7　　**15.** 42 ÷ 6　　**16.** 72 ÷ 9

17. 63 ÷ 9　　**18.** 28 ÷ 4　　**19.** 72 ÷ 8　　**20.** 54 ÷ 6

21. 84 ÷ 9　　**22.** 42 ÷ 5　　**23.** 57 ÷ 6　　**24.** 68 ÷ 7

25. 17 ÷ 2　　**26.** 19 ÷ 2　　**27.** 23 ÷ 7　　**28.** 45 ÷ 6

29. 28 ÷ 3　　**30.** 49 ÷ 8　　**31.** 65 ÷ 7　　**32.** 78 ÷ 8

33. 73 ÷ 9　　**34.** 77 ÷ 8　　**35.** 61 ÷ 7　　**36.** 53 ÷ 6

37. 47 ÷ 5　　**38.** 33 ÷ 4　　**39.** 59 ÷ 6　　**40.** 65 ÷ 7

ANSWERS

5. _____
6. _____
7. _____
8. _____
9. _____
10. _____
11. _____
12. _____
13. _____
14. _____
15. _____
16. _____
17. _____
18. _____
19. _____
20. _____
21. _____
22. _____
23. _____
24. _____
25. _____
26. _____
27. _____
28. _____
29. _____
30. _____
31. _____
32. _____
33. _____
34. _____
35. _____
36. _____
37. _____
38. _____
39. _____
40. _____

Suppose that we did not know any of the multiplication facts. Would it still be possible to find the solution of 15 ÷ 5? Just as multiplication could be thought of as repeated addition, it is also possible to think of division as repeated subtraction. In other words, 15 ÷ 5 can be thought of as "How many times can 5 be subtracted from 15?"

$$\begin{array}{r}15\\-5\\\hline 10\\-5\\\hline 5\\-5\\\hline 0\end{array}$$

We therefore see that 5 can be subtracted from 15 three times. Thus $15 \div 5 = 3$.

Example 6 Solve $45 \div 7$ by repeated subtraction.

$$\begin{array}{r}45\\-7\\\hline 38\\-7\\\hline 31\\-7\\\hline 24\\-7\\\hline 17\\-7\\\hline 10\\-7\\\hline 3\end{array}$$

Since we can no longer subtract 7, we have 3 as a remainder. Therefore, $45 \div 7 = 6$ R3.

The **long-division algorithm** is based on the idea of repeated subtraction. We will show its use by an example.

Example 7 Solve $45 \div 7$ using the long-division algorithm.

We first write the problem using the long-division symbol as follows.

$$7\overline{)45}$$

We now estimate the largest number of times 7 will divide 45. We call this estimate the **trial quotient.** Example 6 has already given us a clue as to the best estimate, so we will use 6 as our trial quotient.

$$7\overline{)45}^{6}$$

We then multiply 6×7 to get 42, which is placed under the 45. We then subtract 42 from 45 and obtain 3, which is the remainder.

$$\begin{array}{r} 6 \\ 7{\overline{\smash{\big)}\,45}} \\ \underline{-42} \\ 3 \end{array}$$

Thus our quotient is 6 R3.

Notice that 6 × 7 = 42 is subtracted from 45. We have actually subtracted six 7s at once. We have a remainder of 3. If our remainder had been larger than 7, it would have indicated that the trial quotient was too small.

Suppose that we have guessed a trial quotient of 5. Then the remainder (10) would still contain a 7. Our trial quotient is therefore too small.

$$\begin{array}{r} 5 \\ 7{\overline{\smash{\big)}\,45}} \\ \underline{-35} \\ 10 \end{array}$$

WARNING

The remainder cannot be larger than the divisor.

If we guess too high for a trial quotient, it will not be possible to subtract. For instance, if we had guessed a trial quotient of 7, we could not subtract 49 from 45. Our trial quotient is too large.

$$\begin{array}{r} 7 \\ 7{\overline{\smash{\big)}\,45}} \\ -49 \end{array}$$

Example 8 Solve 96 ÷ 4.

First we divide 4 into 9 and multiply the trial quotient 2 by 4 and place the 8 under the 9. Subtracting, we obtain 1.

$$\begin{array}{r} 2 \\ 4{\overline{\smash{\big)}\,96}} \\ \underline{8} \\ 1 \end{array}$$

We now bring the 6 down with the 1 and divide 4 into 16. Multiplying the trial quotient 4 by 4, we obtain 16. Subtracting, we have zero as a remainder.

$$\begin{array}{r} 24 \\ 4{\overline{\smash{\big)}\,96}} \\ \underline{8} \\ 16 \\ \underline{16} \\ 0 \end{array}$$

Thus 96 ÷ 4 = 24.

> **WARNING**
>
> A very important part of the long-division algorithm is the proper placement of the trial quotient. Notice in example 8 that the 2 is placed over the 9 because we are dividing 4 into 9.

Example 9 Solve $112 \div 8$.

Since 8 cannot be divided into 1, we divide 8 into 11. We now multiply 1×8 and subtract.

$$\begin{array}{r} 1 \\ 8\overline{)112} \\ \underline{8} \\ 3 \end{array}$$

Bringing down the 2 gives our next division as $32 \div 8$. Our trial quotient for this division is 4 and since $4 \times 8 = 32$, we have zero as a remainder.

$$\begin{array}{r} 14 \\ 8\overline{)112} \\ \underline{8} \\ 32 \\ \underline{32} \\ 0 \end{array}$$

Thus $112 \div 8 = 14$.

Example 10 Find the quotient of $7\overline{)205}$.

7 will not divide into 2 so we must divide 7 into 20, giving 2 with a remainder of 6.

$$\begin{array}{r} 2 \\ 7\overline{)205} \\ \underline{14} \\ 6 \end{array}$$

We now bring down the 5 and divide 7 into 65. Our nearest trial quotient is 9. Subtracting, we have a remainder of 2.

$$\begin{array}{r} 29 \\ 7\overline{)205} \\ \underline{14} \\ 65 \\ \underline{63} \\ 2 \end{array}$$

The quotient is 29 R2.

Example 11 Find the quotient of $6\overline{)2{,}341}$.

$$\begin{array}{r} 390\ \text{R}1 \\ 6\overline{)2{,}341} \\ \underline{18} \\ 54 \\ \underline{54} \\ 1 \end{array}$$

> **WARNING**
>
> It is very important to note that each digit in the dividend to the right of the placement of the first trial quotient must have a digit over it in the answer. Thus, in example 11, when we cannot divide 6 into 1, we must place a zero over the 1.

Example 12 Find the quotient of $5\overline{)21{,}035}$.

$$\begin{array}{r} 4{,}207 \\ 5\overline{)21{,}035} \\ \underline{20} \\ 1\ 0 \\ \underline{1\ 0} \\ 3 \\ \underline{0} \\ 35 \\ \underline{35} \\ 0 \end{array}$$

Example 13 At snack time a kindergarten teacher opened a box of 35 cookies. She said "I'm going to give each of you the same number of cookies and have the remainder for myself." If she had 8 students, how many cookies did each one receive and how many were left for the teacher?

$$\begin{array}{r} 4\ \text{R}3 \\ 8\overline{)35} \\ \underline{32} \\ 3 \end{array}$$

Each student had 4 cookies and the teacher had 3.

Remember that we must always answer the question asked in a word problem. Note that this problem asks two questions and both must be answered.

EXERCISE SET 1-7-2

ANSWERS

Find the following quotients.

1. 3)36
2. 2)28
3. 4)48
4. 5)55

5. 2)46
6. 3)69
7. 3)93
8. 4)84

9. 3)45
10. 4)96
11. 7)84
12. 6)72

13. 7)91
14. 6)84
15. 3)75
16. 2)58

17. 5)79
18. 4)61
19. 2)93
20. 3)82

21. 6)215
22. 5)321
23. 4)628
24. 7)934

25. 2)509
26. 3)701
27. 5)803
28. 4)906

29. 7)450 **30.** 6)840 **31.** 3)720 **32.** 5)760

33. 4)1,682 **34.** 2)1,261 **35.** 5)1,254 **36.** 3)2,042

37. 6)1,816 **38.** 4)2,805 **39.** 8)1,670 **40.** 7)6,327

41. 5)15,535 **42.** 6)25,230 **43.** 4)24,016 **44.** 7)21,049

45. 4)20,206 **46.** 7)35,276 **47.** 8)34,453 **48.** 3)21,307

49. If $152 is to be divided equally among four people, how much does each person receive?

50. An eight-ounce container of dairy cream costs 48 cents. What is the price per ounce?

51. In a livestock yard 464 head of cattle are to be equally divided into eight pens. How many cattle should be placed in each pen?

52. If one pie serves eight people, how many pies are needed to serve 256 people?

ANSWERS

29. _____
30. _____
31. _____
32. _____
33. _____
34. _____
35. _____
36. _____
37. _____
38. _____
39. _____
40. _____
41. _____
42. _____
43. _____
44. _____
45. _____
46. _____
47. _____
48. _____
49. _____
50. _____
51. _____
52. _____

ANSWERS

53. _____

54. _____

55. _____

56. _____

57. _____

58. _____

59. _____

60. _____

53. If you owe $336 and agree to pay it in six monthly payments, how much do you pay each month?

54. If an RV van gets 9 miles to a gallon of fuel, how many gallons are needed to drive 207 miles?

55. A box of 32 chocolates is equally divided among five people. How many chocolates does each person receive and how many are left over?

56. How many eight-ounce glasses of milk can be obtained from 410 ounces? How many ounces are left over?

57. You drove 182 miles and used seven gallons of gasoline. How many miles per gallon did you get?

58. Four neighbors decide to split the cost equally for a satellite TV system. If the total cost of the system is $6,048, how much is each neighbor's share?

59. Forty hamburgers are divided equally among nine people. How many hamburgers does each person receive and how many are left over?

60. Three fishermen caught 74 fish. They divide them equally. How many fish are left over?

When dividing by a number consisting of more than one digit, the long-division process does not change. There is an increase in difficulty, however, in finding the proper trial quotient.

1-7 DIVISION OF WHOLE NUMBERS

Example 14 Find the quotient of $22\overline{)653}$.

Since 22 will not divide into 6, we must first divide 65 by 22. We try 2 as our trial quotient. In this case we can subtract and the remainder (21) is smaller than the divisor (22).

$$\begin{array}{r} 2 \\ 22\overline{)653} \\ \underline{44} \\ 21 \end{array}$$

We bring down the 3 and must now make a guess as to 213 divided by 22. If we try 9, we find the remainder to be 15, which is smaller than the divisor.

$$\begin{array}{r} 29 \text{ R}15 \\ 22\overline{)653} \\ \underline{44} \\ 213 \\ \underline{198} \\ 15 \end{array}$$

> **WARNING**
>
> Be careful to distinguish between the phrases "divided by" and "divided into." For example, $5\overline{)39}$ can be read as "39 divided *by* 5" or "5 divided *into* 39."

Example 15 Find the quotient of $514\overline{)29{,}298}$.

We must first find, from the left, a number into which 514 can divide. The first such number is 2,929, and our trial quotient is 5. Be careful to place the 5 over the 9.

$$\begin{array}{r} 5 \\ 514\overline{)29{,}298} \\ \underline{2570} \\ 359 \end{array}$$

The remainder is less than 514, so we bring down the next digit (8). This time we use 7 as our trial quotient.

$$\begin{array}{r} 57 \\ 514\overline{)29{,}298} \\ \underline{2570} \\ 3598 \\ \underline{3598} \\ 0 \end{array}$$

Example 16 Find the quotient of $44\overline{)88{,}132}$.

Our first trial quotient is 2.

$$\begin{array}{r} 2 \\ 44\overline{)88{,}132} \\ \underline{88} \\ 1 \end{array}$$

44 will not divide into 1 so we place a zero in the answer.

$$\begin{array}{r} 20 \\ 44\overline{)88{,}132} \\ \underline{88} \\ 1 \\ \underline{0} \\ 13 \end{array}$$

44 will not divide into 13 so we place a zero in the answer.

$$\begin{array}{r} 200 \\ 44\overline{)88{,}132} \\ \underline{88} \\ 1 \\ \underline{0} \\ 13 \\ \underline{0} \\ 132 \end{array}$$

Now 44 will divide into 132 and using a trial quotient of 3 gives us zero as a remainder.

$$\begin{array}{r} 2{,}003 \\ 44\overline{)88{,}132} \\ \underline{88} \\ 1 \\ \underline{0} \\ 13 \\ \underline{0} \\ 132 \\ \underline{132} \\ 0 \end{array}$$

▼ EXERCISE SET 1-7-3

Find the following quotients.

1. $13\overline{)468}$
2. $12\overline{)648}$
3. $16\overline{)368}$
4. $15\overline{)465}$

1-7 DIVISION OF WHOLE NUMBERS

5. 11)693   6. 17)799   7. 21)672   8. 24)624

9. 32)1,358   10. 41)960   11. 34)967   12. 62)1,395

13. 47)14,310   14. 39)23,477   15. 28)22,504   16. 55)22,290

17. 17)5,780   18. 21)14,490   19. 18)10,440   20. 16)11,680

21. 10)340   22. 10)690   23. 10)1,440   24. 10)2,350

25. 20)11,260   26. 30)21,660   27. 50)17,050   28. 40)11,360

29. 28)3,226   30. 34)7,525   31. 44)9,393   32. 46)7,295

ANSWERS

5. _____
6. _____
7. _____
8. _____
9. _____
10. _____
11. _____
12. _____
13. _____
14. _____
15. _____
16. _____
17. _____
18. _____
19. _____
20. _____
21. _____
22. _____
23. _____
24. _____
25. _____
26. _____
27. _____
28. _____
29. _____
30. _____
31. _____
32. _____

ANSWERS

33. _____
34. _____
35. _____
36. _____
37. _____
38. _____
39. _____
40. _____
41. _____
42. _____
43. _____
44. _____
45. _____
46. _____
47. _____
48. _____
49. _____
50. _____
51. _____
52. _____
53. _____
54. _____
55. _____
56. _____
57. _____
58. _____
59. _____
60. _____

33. $53\overline{)21{,}369}$ **34.** $61\overline{)12{,}525}$ **35.** $45\overline{)22{,}516}$ **36.** $72\overline{)14{,}470}$

37. $71\overline{)156{,}413}$ **38.** $68\overline{)137{,}700}$ **39.** $59\overline{)177{,}826}$ **40.** $81\overline{)170{,}343}$

41. $511\overline{)27{,}083}$ **42.** $314\overline{)20{,}096}$ **43.** $425\overline{)28{,}900}$ **44.** $416\overline{)31{,}200}$

45. $256\overline{)21{,}339}$ **46.** $443\overline{)33{,}769}$ **47.** $610\overline{)29{,}082}$ **48.** $480\overline{)25{,}439}$

49. $100\overline{)28{,}500}$ **50.** $100\overline{)58{,}200}$ **51.** $100\overline{)35{,}000}$ **52.** $100\overline{)71{,}000}$

53. $43\overline{)129{,}172}$ **54.** $52\overline{)104{,}364}$ **55.** $38\overline{)152{,}304}$ **56.** $28\overline{)168{,}028}$

57. $33\overline{)66{,}185}$ **58.** $47\overline{)94{,}218}$ **59.** $27\overline{)81{,}215}$ **60.** $32\overline{)64{,}298}$

61. Martha earns $36,144 in one year. How much does she earn each month?

62. Twenty-two homeowners will pay equal shares of a tax bill for $4,510. How much does each person owe?

63. A case of soda contains 24 cans. How many cases can be filled from 1,350 cans? How many cans are left over?

64. How many 32-ounce bottles can be filled from 7,720 ounces? How many ounces are left over?

65. Total car payments for three years are $6,228. How much is each monthly payment?

66. A machine shop can make 25 auto parts per day. How many days will it take to make 1,300 parts?

67. How many gallons of gasoline are needed to travel 1,716 miles if the car gets 22 miles per gallon?

68. If the tax on a home is $1,260 per year, how much is it per month?

69. A tennis club purchases tennis balls by the gross (144 balls in a gross). If the club needs 3,000 tennis balls, how many gross should they buy? (*Hint:* It is all right if there are a few tennis balls left over.)

70. If one package contains twelve hot dogs, how many packages are needed to obtain 260 hot dogs?

A N S W E R S

61. _____

62. _____

63. _____

64. _____

65. _____

66. _____

67. _____

68. _____

69. _____

70. _____

71. If a car traveled 768 miles and used 32 gallons of gasoline, how many miles to the gallon did the car get?

72. If a person earns $475 per week, how many weeks will it take that person to earn $16,000?

71. _____

72. _____

1–8 GROUPING SYMBOLS

OBJECTIVES

In this section you will learn to perform operations in the correct order indicated by grouping symbols.

Parentheses ( ), brackets [ ], and braces { } are all used as grouping symbols. A number expression enclosed in a grouping symbol is treated as if it were a single number.

Example 1 $12 - (3 + 4)$ indicates that the sum $(3 + 4)$ is subtracted from 12.

$$12 - (3 + 4) = 12 - 7$$
$$= 5$$

Example 2 $7(5 + 3)$ indicates that the sum $(5 + 3)$ is multiplied by 7.

$$7(5 + 3) = 7(8)$$
$$= 56$$

In general, operations within the grouping symbols are performed first.

Brackets and braces can be used instead of parentheses. $12 - [3 + 4]$ and $12 - \{3 + 4\}$ mean the same as $12 - (3 + 4)$. Parentheses are most commonly used when no other grouping symbols are involved.

EXERCISE SET 1–8–1

Find the value of each of the following.

1. $8 - (5 + 2)$ **2.** $14 - (8 + 3)$

3. $9 - (3 - 1)$ **4.** $20 - (13 - 9)$

5. $(8 - 2) + 4$ **6.** $(18 - 12) + 6$

1. _____
2. _____
3. _____
4. _____
5. _____
6. _____

7. $(10 - 7) - 1$

8. $(15 - 8) - 5$

9. $18 + (6 - 3)$

10. $21 - (11 + 4)$

11. $(15 + 4) - 16$

12. $34 - (16 + 15)$

13. $(15 - 6) + 29$

14. $58 - (24 + 16)$

15. $63 - (91 - 52)$

16. $49 - (38 - 19)$

17. $(12 - 5) + (14 - 6)$

18. $(16 - 7) - (14 - 6)$

19. $3(6 + 8)$

20. $5(4 + 7)$

21. $4(11 - 5)$

22. $7(24 - 19)$

23. $(13 - 6)(4 + 10)$

24. $(5 - 2)(5 + 2)$

25. $(18 - 2) - (10 + 5)$

26. $(10 + 14) - (18 + 6)$

27. $(23 - 15) - (14 - 9)$

28. $(34 - 28) + (72 - 66)$

29. $8 + 6(10 - 4)$

30. $45 - 3(7 + 4)$

ANSWERS

7. ____
8. ____
9. ____
10. ____
11. ____
12. ____
13. ____
14. ____
15. ____
16. ____
17. ____
18. ____
19. ____
20. ____
21. ____
22. ____
23. ____
24. ____
25. ____
26. ____
27. ____
28. ____
29. ____
30. ____

An expression may have a different value if grouping symbols are not used. We saw that

$$12 - (3 + 4) = 5,$$

but if parentheses are not used, the value of the expression becomes

$$12 - 3 + 4 = 13.$$

This is due to the following rule.

> **Rule** If an expression without grouping symbols contains only additions and subtractions, these operations are performed in order from left to right.

Example 3 $\quad 10 + 5 - 3 + 8 = 15 - 3 + 8$
$ = 12 + 8$
$ = 20$

Example 4 $\quad 6 - 4 + 11 - 4 = 2 + 11 - 4$
$ = 13 - 4$
$ = 9$

Example 5 $\quad 5 + 9 - 7 - 3 + 4 - 8 = 14 - 7 - 3 + 4 - 8$
$ = 7 - 3 + 4 - 8$
$ = 4 + 4 - 8$
$ = 8 - 8$
$ = 0$

If an expression contains operations other than just addition and subtraction, we use the following rule.

> **Rule** If no grouping symbols occur in an expression, multiplication and division are performed from left to right, and then addition and subtraction from left to right.

Example 6 $\quad 2 \times 3 + 5 = 6 + 5 \qquad$ (We first multiply 2×3.)
$ = 11$

Example 7 $\quad 4 + 6 \div 3 = 4 + 2 \qquad$ (We first divide 6 by 3.)
$ = 6$

Example 8 $\quad 3 \times 2 - 10 \div 2 = 6 - 10 \div 2 \qquad$ (We first multiply 3×2.)
$ = 6 - 5 \qquad$ (Second divide 10 by 2.)
$ = 1$

EXERCISE SET 1-8-2

Evaluate the following expressions.

1. $5 + 3 + 7$
2. $8 - 4 + 6$
3. $5 + 9 - 14$
4. $3 + 7 - 8$
5. $12 - 8 + 2 - 4$
6. $11 + 4 - 6 - 3$
7. $4 + 19 - 8 + 1$
8. $13 - 10 - 1 + 5$
9. $15 + 37 - 15 + 3$
10. $7 - 5 + 3 - 5$
11. $4 \times 3 + 6$
12. $8 \times 3 + 5$
13. $12 + 6 \times 2$
14. $15 - 2 \times 4$
15. $7 + 3 \times 5 - 2$
16. $3 + 4 \times 3 + 2$
17. $15 + 10 \div 5 - 3$
18. $6 + 21 \div 3 - 5$
19. $4 \div 2 + 8 \times 3$
20. $40 \div 2 - 5 \times 4$
21. $3 \times 5 - 9 \div 3 + 5$
22. $12 \div 2 - 2 \times 3 + 1$

ANSWERS

1. ___
2. ___
3. ___
4. ___
5. ___
6. ___
7. ___
8. ___
9. ___
10. ___
11. ___
12. ___
13. ___
14. ___
15. ___
16. ___
17. ___
18. ___
19. ___
20. ___
21. ___
22. ___

ANSWERS

23. _____
24. _____
25. _____
26. _____
27. _____
28. _____
29. _____
30. _____

23. $6 + 4 \times 5 - 3$

24. $6 + 4 \times (5 - 3)$

25. $(6 + 4) \times 5 - 3$

26. $(6 + 4) \times (5 - 3)$

27. $3 + 2 \times 8 \div 4 - 5$

28. $4 + 27 \div 3 \times 2 - 6$

29. $6 \times 5 \div 3 + 12$

30. $24 \div 6 \times 2 - 8$

Sometimes more than one set of grouping symbols is needed in an expression. When this occurs we use brackets or braces along with parentheses for clarification. For instance, $5 + [7 - (2 + 1)]$ could be written using only parentheses, but $5 + (7 - (2 + 1))$ is not as clear at first glance. Therefore, we alternate the symbols to avoid confusion. To evaluate such an expression we use the following rule.

> **Rule** When simplifying an expression containing grouping symbols within grouping symbols, remove the *innermost* set of symbols first.

Example 9 To evaluate $5 + [7 - (2 + 1)]$ we simplify the innermost set of symbols, which is $(2 + 1)$. Writing $(2 + 1)$ as 3 we now obtain

$$5 + [7 - (2 + 1)] = 5 + [7 - 3]$$
$$= 5 + 4$$
$$= 9.$$

Example 10 $[11 - (9 - 5)] - 4 = [11 - 4] - 4$
$$= 7 - 4$$
$$= 3$$

Example 11 $18 - \{5 + [4(5 - 2) - 7]\} = 18 - \{5 + [4(3) - 7]\}$
$$= 18 - \{5 + [12 - 7]\}$$
$$= 18 - \{5 + 5\}$$
$$= 18 - 10$$
$$= 8$$

EXERCISE SET 1-8-3

Evaluate each of the following expressions.

ANSWERS

1. $4 + [9 - (3 + 4)]$
2. $5 + [10 - (4 + 3)]$

3. $14 - [10 - (5 - 1)]$
4. $10 - [14 - (6 - 1)]$

5. $[(8 + 5) - 6] - 3$
6. $[(10 + 4) - 8] - 6$

7. $4[8 + (11 - 3)]$
8. $3[6 + (9 - 5)]$

9. $2 + [9 - (3 - 1)]$
10. $16 + [13 - (6 - 2)]$

11. $10 - [(6 - 4) - 1]$
12. $23 - [(12 - 3) - 5]$

13. $5 + 3[16 - (11 + 3)]$
14. $10 + 2[24 + (8 - 2)]$

1. _____

2. _____

3. _____

4. _____

5. _____

6. _____

7. _____

8. _____

9. _____

10. _____

11. _____

12. _____

13. _____

ANSWERS

15. $7 + [15 + 3(4 - 1)]$

16. $18 - [10 + 2(8 - 5)]$

15. _____

17. $[2(6 - 3) + 4] - 5$

18. $[3(7 - 3) + 2] - 9$

16. _____

17. _____

19. $6 + 2[8 + 3(6 - 3)]$

20. $19 - 3[18 - 4(7 - 4)]$

18. _____

19. _____

21. $[8 - (5 + 2)] + 3(15 - 6)$

22. $2(19 + 4) - [4(8 + 2) + 3]$

20. _____

23. $3[4 + (3 - 1)] + 5[12 - (8 + 2)]$

21. _____

22. _____

24. $6[4(10 - 7) - 3] - 4[16 - 2(3 + 1)]$

23. _____

24. _____

25. $2 + [10 - (3 + 6) + 5]$

26. $8 - [13 - (2 + 4) - 6]$

25. _____

26. _____

27. $2\{5 + 3[14 - (6 + 2)]\}$

28. $4\{[2(6 - 3) + 4] - 10\}$

29. $16 - \{15 - [2(5 + 2) - 4] + 6\}$

30. $2[4(7 - 3) + 1] - \{12 + 3[14 - 2(8 - 5)] - 5\}$

27. _____

28. _____

29. _____

30. _____

1-9 FACTORS AND PRIME FACTORIZATION

We now introduce a very important concept that will soon be used in the study of fractions.

> **Definition**   If a and b are whole numbers, a is said to be a **factor** of b if a divides b with no remainder.

OBJECTIVES

In this section you will learn to:
1. Find factors of a given number.
2. Find all primes less than a given number.
3. Find the prime factorization of a given number.

Example 1   The factors of 12 are 1, 2, 3, 4, 6, and 12.

Notice that each of these whole numbers divides 12 evenly (that is, with no remainder).

Example 2   Find all factors of 20.

When we are asked to find *all* factors of a number, we must have a system that will not leave out any numbers. The best way to arrive at all factors is to start with the number 1 and try each successive digit. In this way we will find the factors two at a time.

We first divide 20 by 1. It divides exactly 20 times. Therefore, 1 and 20 are factors.

Next we divide 20 by 2, obtaining a quotient of 10. So 2 and 10 are factors.

Notice that if 2 divides 20 ten times, then 10 divides 20 two times. Thus we find two factors at once.

If we try to divide 20 by 3 we have a remainder, so 3 is not a factor.

We next divide 20 by 4 obtaining 5. Thus 4 and 5 are factors.

All the factors of 20 are 1, 2, 4, 5, 10, 20.

Example 3 Find all factors of 48.

They are 1 and 48, 2 and 24, 3 and 16, 4 and 12, 6 and 8. When we are obtaining the factors two at a time, we can stop when the numbers meet. For instance, in this case if we tried digits above 6, we would next get 8 and 6, which we already have.

> **WARNING**
>
> A common error when listing all factors is to forget 1 and the number itself.

Example 4 Find all factors of 100.

They are 1 and 100, 2 and 50, 4 and 25, 5 and 20, and 10.

When we try 3, 6, 7, 8, 9, none of them will divide 100 evenly (that is, with no remainder).

EXERCISE SET 1-9-1

ANSWERS

List all factors of the following numbers.

1. 4   2. 6   3. 9   4. 8   5. 5

6. 7   7. 10   8. 14   9. 22   10. 18

11. 24   12. 33   13. 17   14. 19   15. 42

					ANSWERS
16. 46	**17.** 27	**18.** 51	**19.** 58	**20.** 60	16. _____
					17. _____
					18. _____
					19. _____
21. 69	**22.** 81	**23.** 88	**24.** 56	**25.** 76	20. _____
					21. _____
					22. _____
					23. _____
					24. _____
26. 78	**27.** 99	**28.** 108	**29.** 124	**30.** 111	25. _____
					26. _____
					27. _____
					28. _____
					29. _____
31. 220	**32.** 234				30. _____
					31. _____
					32. _____

Example 5 Find all factors of 7.

1 and 7 are the only factors.

Example 6 Find all factors of 23.

1 and 23 are the only factors.

These two examples indicate that some numbers have exactly two different factors. This is a special class of numbers called **primes**.

> **Definition** A **prime number** is a whole number that has exactly two different factors.

Some prime numbers are 2, 3, 5, 7, 11, 13, 17, 19, and 23. Euclid, a famous mathematician (about 300 B.C.), proved there is no limit to the number of primes. But as of this date, no one has ever found a workable formula that will give all the primes.

However, there is a method called the **Sieve of Eratosthenes** that will give all primes up to any given number we choose. This method is a simple device that divides all the numbers that can be evenly divided (up to whatever number we choose) and leaves any number that can be divided only by 1 and itself.

Here is a Sieve of Eratosthenes giving us the primes less than 50.

~~1~~ ② ③ ~~4~~ ⑤ ~~6~~ ⑦ ~~8~~ ~~9~~ ~~10~~
11 ~~12~~ 13 ~~14~~ ~~15~~ ~~16~~ 17 ~~18~~ 19 ~~20~~
~~21~~ ~~22~~ 23 ~~24~~ ~~25~~ ~~26~~ ~~27~~ ~~28~~ 29 ~~30~~
31 ~~32~~ ~~33~~ ~~34~~ ~~35~~ ~~36~~ 37 ~~38~~ ~~39~~ ~~40~~
41 ~~42~~ 43 ~~44~~ ~~45~~ ~~46~~ 47 ~~48~~ ~~49~~ ~~50~~

This sieve is obtained by following these steps.

1. Strike through the number 1 since it has only one factor.
2. Circle 2 and then strike out all numbers that 2 divides evenly. Such a number would be divisible by 1, 2, and itself. That would be too many factors to be prime.
3. Now circle 3 and strike out all numbers that 3 divides evenly.
4. Circle the next number not crossed out and strike out all numbers evenly divisible by that number.
5. Continue this process until you reach a number that multiplied by itself will give 50 or more.

When this process is completed, all numbers not marked out will be prime. In the above sieve we are left with the numbers 2, 3, 5, 7, 11, 13, 17, 19, 23, 29, 31, 37, 41, 43, and 47.

▼ EXERCISE SET 1-9-2

Construct a Sieve of Eratosthenes to find all primes less than 200. List these primes in the spaces provided and refer to them when they are needed in future problems.

```
  1    2    3    4    5    6    7    8    9   10
 11   12   13   14   15   16   17   18   19   20
 21   22   23   24   25   26   27   28   29   30
 31   32   33   34   35   36   37   38   39   40
 41   42   43   44   45   46   47   48   49   50
 51   52   53   54   55   56   57   58   59   60
 61   62   63   64   65   66   67   68   69   70
 71   72   73   74   75   76   77   78   79   80
 81   82   83   84   85   86   87   88   89   90
 91   92   93   94   95   96   97   98   99  100
101  102  103  104  105  106  107  108  109  110
111  112  113  114  115  116  117  118  119  120
121  122  123  124  125  126  127  128  129  130
131  132  133  134  135  136  137  138  139  140
141  142  143  144  145  146  147  148  149  150
151  152  153  154  155  156  157  158  159  160
161  162  163  164  165  166  167  168  169  170
171  172  173  174  175  176  177  178  179  180
181  182  183  184  185  186  187  188  189  190
191  192  193  194  195  196  197  198  199  200
```

ANSWERS

___ ___ ___ ___ ___ ___ ___ ___ ___ ___
___ ___ ___ ___ ___ ___ ___ ___ ___ ___
___ ___ ___ ___ ___ ___ ___ ___ ___ ___
___ ___ ___ ___ ___ ___ ___ ___ ___

We must now make a distinction between listing the factors of a number and factoring a number.

> **Definition** **Factoring** a number means to show the number as a product of two or more numbers.

Example 7 Factor 36.

Our answer could be given several different ways.

$$36 = 1 \times 36$$
$$36 = 6 \times 6$$
$$36 = 3 \times 12$$
$$36 = 2 \times 3 \times 6$$
$$36 = 3 \times 3 \times 4$$

and so on

> **Definition** The **prime factorization** of a number means expressing the number as a product of primes.

Example 8 Find the prime factorization of 36.

$$36 = 2 \times 2 \times 3 \times 3$$

Notice that each factor is a prime and the product is 36.

Using exponents, we write $36 = 2^2 \times 3^2$.

A very important statement that we will accept, but not attempt to prove, is the fundamental theorem of arithmetic. When you read this theorem you will see why we said "the" prime factorization of 36 rather than "a" prime factorization of 36.

> **Fundamental Theorem of Arithmetic** Any whole number that is not a prime can be factored into primes in exactly one way except for the order of the primes.

Example 9 $36 = 2 \times 2 \times 3 \times 3$ and no other product of primes can give us 36. We could say $36 = 2 \times 3 \times 2 \times 3$, but this is the same group of primes in a different order.

We will discuss two methods of prime factorization. The first is commonly called the **tree method.**

Example 10 Factor 48 into primes.

First find any two numbers whose product is 48. For instance, we could write 48 as 4×12. Now factor each of these factors if they are not already prime, drawing down any that are already prime.

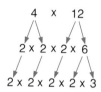

When all numbers are primes, we have the prime factorization. Thus $48 = 2^4 \times 3$.

▼ EXERCISE SET 1-9-3

Use the tree method to find the prime factorization of the following whole numbers.

1. 12
2. 18
3. 30
4. 70
5. 28

6. 42
7. 45
8. 75
9. 60
10. 56

11. 63 **12.** 28 **13.** 51 **14.** 57 **15.** 94

16. 65 **17.** 222 **18.** 231 **19.** 72 **20.** 108

21. 180 **22.** 450 **23.** 378 **24.** 2,310 **25.** 372

26. 328 **27.** 336 **28.** 240 **29.** 6,300 **30.** 8,316

ANSWERS

11. _____
12. _____
13. _____
14. _____
15. _____
16. _____
17. _____
18. _____
19. _____
20. _____
21. _____
22. _____
23. _____
24. _____
25. _____
26. _____
27. _____
28. _____
29. _____
30. _____

Another method of prime factorization that you may wish to use is repeated division by primes.

Example 11 Find the prime factorization of 48.

We first divide 48 by 2, placing our answer below the 48.

$$2 \underline{|48}$$
$$24$$

We divide by 2 again, and place the answer below the 24. Then by 2 again and again until we can no longer divide by 2. Then we divide by the next higher prime possible, and so on, until our final answer is 1. When this is done, the list of prime divisors is the prime factorization of 48.

$$2 \underline{|48}$$
$$2 \underline{|24}$$
$$2 \underline{|12}$$
$$2 \underline{|6}$$
$$3 \underline{|3}$$
$$1$$

Thus $48 = 2 \times 2 \times 2 \times 2 \times 3 = 2^4 \times 3$.

Example 12 Find the prime factorization of 108.

$$
\begin{array}{r|l}
2 & 108 \\
2 & 54 \\
3 & 27 \\
3 & 9 \\
3 & 3 \\
& 1
\end{array}
$$

Thus $108 = 2 \times 2 \times 3 \times 3 \times 3 = 2^2 \times 3^3$.

Always start with the smallest prime and work upward, dividing each as many times as possible before going to the next.

Example 13 Find the prime factorization of 882.

$$
\begin{array}{r|l}
2 & 882 \\
3 & 441 \\
3 & 147 \\
7 & 49 \\
7 & 7 \\
& 1
\end{array}
$$

Thus $882 = 2 \times 3 \times 3 \times 7 \times 7 = 2 \times 3^2 \times 7^2$.

Notice we first divided by 2, then 3, and since 5 would not divide evenly, we next went to 7. If 7 had not divided evenly, we would have tried 11, 13, and so on.

Example 14 Find the prime factorization of 2,244.

$$
\begin{array}{r|l}
2 & 2,244 \\
2 & 1,122 \\
3 & 561 \\
11 & 187 \\
17 & 17 \\
& 1
\end{array}
$$

Thus $2{,}244 = 2 \times 2 \times 3 \times 11 \times 17 = 2^2 \times 3 \times 11 \times 17$.

▼ EXERCISE SET 1-9-4

ANSWERS

1. _____
2. _____
3. _____
4. _____
5. _____

Use the repeated division by primes method to find the prime factorization of each of the following.

1. 21   **2.** 15   **3.** 12   **4.** 18   **5.** 16

1-9 FACTORS AND PRIME FACTORIZATION   75

6. 81   **7.** 60   **8.** 90   **9.** 56   **10.** 40

ANSWERS

6. _____
7. _____
8. _____
9. _____
10. _____

11. 45   **12.** 75   **13.** 70   **14.** 110   **15.** 77

11. _____
12. _____
13. _____
14. _____

16. 91   **17.** 196   **18.** 126   **19.** 390   **20.** 330

15. _____
16. _____
17. _____
18. _____
19. _____

21. 1,155   **22.** 2,695   **23.** 187   **24.** 209   **25.** 374

20. _____
21. _____
22. _____
23. _____
24. _____

26. 627   **27.** 1,615   **28.** 1,309   **29.** 2,728   **30.** 3,330

25. _____
26. _____
27. _____
28. _____
29. _____

31. 5,166   **32.** 9,975   **33.** 2,772   **34.** 1,980   **35.** 128

30. _____
31. _____
32. _____
33. _____
34. _____
35. _____

36. 432   **37.** 6,435   **38.** 6,622   **39.** 13,158   **40.** 32,085

36. _____
37. _____
38. _____
39. _____
40. _____

1-10 GREATEST COMMON FACTOR AND LEAST COMMON MULTIPLE

OBJECTIVES

In this section you will learn to:
1. Find the greatest common factor of two or more numbers.
2. Find the least common multiple of two or more numbers.

Another concept that will be needed in future work is the **greatest common factor** (abbreviated GCF). In the previous section we listed the factors of 36 and also the factors of 48. Look again at these factors.

36: 1, 2, 3, 4, 6, 9, 12, 18, 36

48: 1, 2, 3, 4, 6, 8, 12, 16, 24, 48

Now from these two lists pick out all numbers common to both. They are 1, 2, 3, 4, 6, and 12. The largest of these "common" factors is 12. So the greatest common factor, or GCF, of 36 and 48 is 12. We can, of course, find the GCF of any group of numbers.

Example 1 Find the GCF of 36, 42, and 90.

36: 1, 2, 3, 4, ⑥, 9, 12, 18, 36

42: 1, 2, 3, ⑥, 7, 14, 21, 42

90: 1, 2, 3, 5, ⑥, 9, 10, 15, 30, 45, 90

Looking at these lists we find that the GCF of 36, 42, and 90 is 6.

EXERCISE SET 1-10-1

Find the GCF of the following groups of numbers by listing the factors of each.

1. 12 and 18
2. 32 and 72
3. 42 and 49
4. 36 and 63
5. 54 and 144
6. 72, 80, and 96
7. 45, 63, and 105
8. 36, 48, and 64
9. 48, 64, and 72
10. 36, 54, 81, and 135

1–10 GREATEST COMMON FACTOR AND LEAST COMMON MULTIPLE

By now you are probably saying "there must be a better way." There is a more direct method of obtaining the GCF.

> **Definition** The **greatest common factor** of two or more numbers is the product of all prime factors common to the numbers.

Example 2 Find the GCF of 36, 42, and 90.

We first factor each number into primes.

$$36 = 2 \times 2 \times 3 \times 3$$
$$42 = 2 \times 3 \times 7$$
$$90 = 2 \times 3 \times 3 \times 5$$

The primes common to all three numbers are 2 and 3 so the GCF = $2 \times 3 = 6$.

Example 3 Find the GCF of 36 and 48.

$$36 = ②\times ②\times ③\times 3$$
$$48 = ②\times ②\times 2 \times 2 \times ③$$

The common primes are 2, 2, and 3 (note that 2 is a factor twice in each number). So the GCF = $2 \times 2 \times 3 = 12$.

> **Definition** If two whole numbers have no common prime factors, their GCF is 1 and the numbers are said to be **relatively prime.**

Example 4 Find the GCF of 21 and 55.

$$21 = 3 \times 7$$
$$55 = 5 \times 11$$

We see there are no common primes so in this case the GCF = 1. Thus 21 and 55 are relatively prime.

▼ EXERCISE SET 1–10–2

Use the prime factorization method to find the GCF of each of the following sets of numbers.

ANSWERS

1. 12 and 20
2. 15 and 18

1. _____

2. _____

ANSWERS

3. _____
4. _____
5. _____
6. _____
7. _____
8. _____
9. _____
10. _____
11. _____
12. _____
13. _____
14. _____
15. _____
16. _____
17. _____
18. _____
19. _____
20. _____
21. _____
22. _____
23. _____
24. _____
25. _____
26. _____

3. 28 and 30

4. 36 and 42

5. 8 and 15

6. 8 and 16

7. 42 and 84

8. 18 and 54

9. 54 and 72

10. 90 and 105

11. 35 and 88

12. 66 and 154

13. 84 and 105

14. 91 and 99

15. 108 and 126

16. 168 and 252

17. 117 and 315

18. 165 and 630

19. 36, 40, and 60

20. 24, 40, and 63

21. 72, 84, and 108

22. 30, 40, and 63

23. 84, 126, and 252

24. 99, 126, and 189

25. 42, 45, and 56

26. 24, 60, and 84

27. 72, 180, and 216   **28.** 180, 210, and 630

29. 252, 441, and 630   **30.** 270, 720, and 1,260

A N S W E R S

27. _____
28. _____
29. _____
30. _____

The concept of **least common multiple** (abbreviated LCM) of a set of whole numbers will also be of use in future work with fractions.

> **Definition** **Multiples** of a number *a* are the numbers obtained by multiplying the number *a* by the whole numbers 1, 2, 3, 4, 5, ...

From the definition we can see that the list of multiples of a number would never end since the list of whole numbers never ends.

Example 5 List several multiples of 3.

Such numbers would be 3, 6, 9, 12, 15, 18, 21, 24, . . . , where the dots indicate we could go on and on.

Now look at the following lists of some multiples of 3 and 5.

3: 3, 6, 9, 12, ⑮, 18, 21, 24, 27, ㉚, 33, 36, 39, 42, ㊺, . . .
5: 5, 10, ⑮, 20, 25, ㉚, 35, 40, ㊺, 50, 55, 60, 65, 70, . . .

Observing these lists we see that 3 and 5 have some multiples in common such as 15, 30, and 45. There would be more if our lists continued. Our interest, however, is really in the *least* of the multiples in both lists. In this case the *least common multiple* (LCM) is 15.

> **Definition** The **least common multiple** of a set of numbers is the smallest number that is a multiple of each number in the set.

We will now look at an example that will explain a method of finding the LCM of any set of numbers.

Example 6 Find the LCM of 12, 15, and 18.

First note that by definition this unknown number must contain 12 as a factor, 15 as a factor, and 18 as a factor. We now factor 12, 15, and 18 into primes.

$$12 = 2 \times 2 \times 3$$
$$15 = 3 \times 5$$
$$18 = 2 \times 3 \times 3$$

Now since we know that each of these numbers must be factors of the LCM and that no other prime factorization is possible, we can see that $2 \times 2 \times 3$ must be part of the LCM. This is also true for 3×5 and $2 \times 3 \times 3$. So we "build" the LCM in the following manner.

Write the first number as a product of primes: $2 \times 2 \times 3$. Now we look at the second number in prime factored form and ask ourselves if it is already included as a part of the first number, that is, $2 \times 2 \times 3$. We can see that it is not, but we need only a 5 to make it be there. So we now write $2 \times 2 \times 3 \times 5$ as a necessary part of our LCM.

We now look at the next number in prime factored form and see that $2 \times 3 \times 3$ is not already a part of the number listed thus far (in this case $2 \times 2 \times 3 \times 5$). We still need another 3 to ensure that $2 \times 3 \times 3$ or 18 is a part of the LCM. Including this additional 3 gives us $2 \times 2 \times 3 \times 5 \times 3 = 180$ as the LCM of 12, 15, and 18.

Example 7 Find the LCM of 21 and 30.

$$21 = 3 \times 7$$
$$30 = 2 \times 3 \times 5$$

We start with the first number, 3×7, and multiply it by 2×5, which are the other primes necessary to ensure that the second number is part of the LCM. Thus the LCM $= 2 \times 3 \times 5 \times 7 = 210$.

Example 8 Find the LCM of 4 and 14.

$$4 = 2 \times 2$$
$$14 = 2 \times 7$$
$$\text{LCM} = 2 \times 2 \times 7 = 28$$

Example 9 Find the LCM of 12, 90, and 105.

$$12 = 2 \times 2 \times 3$$
$$90 = 2 \times 3 \times 3 \times 5$$
$$105 = 3 \times 5 \times 7$$
$$\text{LCM} = 2 \times 2 \times 3 \times 3 \times 5 \times 7 = 1{,}260$$

EXERCISE SET 1-10-3

Find the LCM of the following sets of numbers.

1. 2 and 3
2. 3 and 5
3. 4 and 6
4. 6 and 9
5. 6 and 12
6. 6 and 15
7. 15 and 18
8. 18 and 24
9. 9 and 24
10. 12 and 16
11. 12 and 18
12. 10 and 14
13. 24 and 36
14. 25 and 40
15. 24 and 30
16. 42 and 70
17. 3, 4, and 6
18. 2, 8, and 12
19. 3, 6, and 8
20. 12, 18, and 20
21. 12, 20, and 24
22. 15, 24, and 36

ANSWERS

23. _____

24. _____

25. _____

26. _____

27. _____

28. _____

29. _____

30. _____

31. _____

32. _____

33. _____

34. _____

35. _____

36. _____

37. _____

38. _____

39. _____

40. _____

23. 9, 12, and 14

24. 8, 24, and 60

25. 27, 45, and 54

26. 18, 20, and 24

27. 18, 22, and 54

28. 9, 36, and 63

29. 14, 34, and 60

30. 16, 40, and 64

31. 12, 16, and 18

32. 6, 18, and 34

33. 3, 6, 9, and 15

34. 6, 8, 10, and 12

35. 8, 9, 12, and 18

36. 4, 8, 12, and 16

37. 12, 15, 18, and 24

38. 12, 18, 20, and 36

39. 9, 18, 27, and 36

40. 8, 12, 36, and 60

CHAPTER 1 SUMMARY

Digits

- All numbers are written using the ten digits 0, 1, 2, 3, 4, 5, 6, 7, 8, 9.
- Each digit has a digital value and a place value.
- Each group of three digits consists of hundreds, tens, ones. Some group names are thousands, millions, and billions.
- Our number system is a base ten system with each place value ten times as large as the place to its right.

Rounding

- In rounding to any place value if the number in the next place to the right is 5 or more, the number in the desired place is increased by one. If it is less than 5, the number in the desired place remains the same.

Addition

- Only like quantities can be added.
- Numbers can be added in any order and grouped in any manner. (commutative and associative properties of addition)
- Carrying in addition is founded on the base ten system.

Subtraction

- Subtraction is based on the addition facts.
- Only like quantities can be subtracted.
- Borrowing in subtraction is founded on the base ten system.

Multiplication

- Multiplication can be thought of as repeated addition.

Division

- Division can be thought of as repeated subtraction.

Grouping Symbols

- Parentheses, brackets, and braces are used as grouping symbols.
- In general, operations within grouping symbols are performed first.
- If an expression without grouping symbols contains only additions and subtractions, these operations are performed in order from left to right.
- If no grouping symbols occur in an expression, multiplication and division are performed from left to right, and then addition and subtraction from left to right.
- When simplifying an expression containing grouping symbols within grouping symbols, remove the innermost set of symbols first.

Primes

- A prime number has exactly two different factors.
- A number can be factored into primes in only one way.

Greatest Common Factor

- The greatest common factor (GCF) of two or more numbers is the product of all prime factors common to the numbers.

Least Common Multiple

- The least common multiple (LCM) of a set of numbers is the smallest number that is a multiple of each number in the set.

CHAPTER 1 REVIEW

NAME:

CLASS / SECTION: DATE:

ANSWERS

1. What is the place value of the digit 4 in the number 10,546?

2. What is the place value of the digit 2 in the number 12,385?

3. What is the place value of the digit 2 in the number 618,324,705?

4. What is the place value of the digit 7 in the number 704,251,638?

5. Write the number 10,325 in words.

6. Write the number 21,364 in words.

7. Write the number 724,803,241 in words.

8. Write the number 105,063,004 in words.

9. Write "five hundred twenty-three thousand, two hundred one" as a number.

10. Write "four million, eighty-six thousand, twenty-one" as a number.

11. Write the number 827 in expanded form.

12. Write the number 468 in expanded form.

13. Write the number 40,396 in expanded form.

14. Write the number 21,508 in expanded form.

15. Round 346 to the nearest ten.

16. Round 524 to the nearest ten.

17. Round 1,735 to the nearest ten.

18. Round 3,685 to the nearest ten.

CHAPTER 1 REVIEW

19. Round 21,346 to the nearest hundred.

20. Round 55,663 to the nearest hundred.

21. Round 45,568 to the nearest thousand.

22. Round 38,519 to the nearest thousand.

23. Round 461,528 to the nearest ten thousand.

24. Round 256,104 to the nearest ten thousand.

25. Add: 8
            9
            1
            7
            3

26. Add: 3
            8
            7
            4
            2

27. Add: 34
            63

28. Add: 28
            51

Add in problems 29–40.

29. 68
        55

30. 39
        61

31. 28
        43
        39

32. 53
        27
        49

33. 216
        389

34. 546
        375

35. 329
        671

36. 485
        515

37. 304
        685
        293

38. 416
        398
        502

39. 2,428
        1,373
          454
        4,162

40. 3,129
        4,361
          283
        1,256

41. Find the sum of 2,064; 348; 38; and 8,269.

42. Find the sum of 1,906; 86; 9,271; and 568.

ANSWERS

19. _____
20. _____
21. _____
22. _____
23. _____
24. _____
25. _____
26. _____
27. _____
28. _____
29. _____
30. _____
31. _____
32. _____
33. _____
34. _____
35. _____
36. _____
37. _____
38. _____
39. _____
40. _____
41. _____
42. _____

ANSWERS

43. _____
44. _____
45. _____
46. _____
47. _____
48. _____
49. _____
50. _____
51. _____
52. _____
53. _____
54. _____
55. _____
56. _____
57. _____
58. _____
59. _____
60. _____
61. _____
62. _____
63. _____
64. _____
65. _____
66. _____
67. _____
68. _____
69. _____
70. _____

Subtract in problems 43–50.

43. 587
    -354

44. 674
    -541

45. 324
    -65

46. 236
    -57

47. 653
    -456

48. 575
    -276

49. $4{,}604$
    $-1{,}426$

50. $7{,}507$
    $-4{,}319$

51. Find the difference of 35,027 and 12,848.

52. Find the difference of 15,304 and 7,219.

53. What property allows us to say that 5×8 is the same as 8×5?

54. What property allows us to write $7 \times (20 + 5) = (7 \times 20) + (7 \times 5)$?

Multiply in problems 55–70.

55. 83
    $\times6$

56. 75
    $\times8$

57. 34
    $\times 23$

58. 27
    $\times 34$

59. 325
    $\times34$

60. 519
    $\times27$

61. 228
    $\times10$

62. 349
    $\times10$

63. 426
    $\times 153$

64. 365
    $\times 218$

65. 548
    $\times 100$

66. 634
    $\times 100$

67. $1{,}345$
    $\times268$

68. $5{,}418$
    $\times325$

69. $2{,}041$
    $\times304$

70. $1{,}605$
    $\times704$

CHAPTER 1 REVIEW

Divide in problems 71–82.

71. 6)162 **72.** 3)138 **73.** 4)1,564 **74.** 7)3,311

75. 5)213 **76.** 8)284 **77.** 14)784 **78.** 12)756

79. 21)531 **80.** 24)785 **81.** 43)129,215 **82.** 53)212,318

Evaluate the expressions in problems 83–90.

83. $15 - (6 + 5)$

84. $31 - (15 - 9)$

85. $(16 + 4) - (18 - 13)$

86. $(8 + 3)(14 - 6)$

87. $5 + 3[4 + (8 - 5)]$

88. $12 - 2[17 - (6 + 5)]$

89. $2\{6[(5 - 2) - 2] + 4\}$

90. $25 - \{15 - [3(4 - 2) - 6]\}$

91. List all factors of 38.

92. List all factors of 39.

93. List all factors of 84.

94. List all factors of 96.

ANSWERS

71. _____
72. _____
73. _____
74. _____
75. _____
76. _____
77. _____
78. _____
79. _____
80. _____
81. _____
82. _____
83. _____
84. _____
85. _____
86. _____
87. _____
88. _____
89. _____
90. _____
91. _____
92. _____
93. _____
94. _____

ANSWERS

95. _____

96. _____

97. _____

98. _____

99. _____

100. _____

101. _____

102. _____

103. _____

104. _____

105. _____

106. _____

107. _____

108. _____

109. _____

110. _____

111. _____

112. _____

95. Find the prime factorization of 132.

96. Find the prime factorization of 84.

97. Find the prime factorization of 1,260.

98. Find the prime factorization of 4,158.

99. Find the greatest common factor of 22 and 26.

100. Find the greatest common factor of 12 and 44.

101. Find the greatest common factor of 56 and 60.

102. Find the greatest common factor of 210 and 462.

103. Find the greatest common factor of 60, 132, and 210.

104. Find the greatest common factor of 84, 120, and 396.

105. Find the least common multiple of 4 and 5.

106. Find the least common multiple of 3 and 8.

107. Find the least common multiple of 6 and 8.

108. Find the least common multiple of 12 and 15.

109. Find the least common multiple of 9, 15, and 20.

110. Find the least common multiple of 24, 27, and 36.

111. Find the least common multiple of 6, 8, 9, and 12.

112. Find the least common multiple of 8, 12, 16, and 24.

CHAPTER 1 REVIEW

In problems 113 and 114 write the number in the sentence in words.

113. The Amazon River is 3,915 miles long.

114. One kilogram of seawater contains 10,561 milligrams of sodium.

In problems 115 and 116 write the number in the sentence using digits.

115. The cruise ship Queen Elizabeth II can carry one thousand, eight hundred fifteen passengers.

116. In 1985 the population of Alaska was estimated at five hundred twenty-one thousand people.

Solve each of the following.

117. You receive three checks for $218, $46, and $109 and wish to deposit them in your bank account. What is the total deposit?

118. A person borrows $875 and agrees to pay back $1,003 at the end of one year. How much interest does the person pay?

119. If your monthly salary was $1,245, how much would you make in a year?

120. Real estate taxes on a building amount to $168 per month. How much is that per year?

121. If a Boeing 727 aircraft can carry 189 passengers, how many planes would it take to carry 2,400 passengers?

122. A man purchases a car for $10,295. He pays $4,000 down. The finance charge on the balance is $2,921. How much are the monthly payments if it is paid off in 36 months?

ANSWERS

113. _____

114. _____

115. _____

116. _____

117. _____

118. _____

119. _____

120. _____

121. _____

122. _____

SCORE: _____

CHAPTER 1 TEST

NAME: _____
CLASS / SECTION: _____ DATE: _____

ANSWERS

1. _____
2. _____
3. _____
4. _____
5. _____
6. _____
7. _____
8. _____
9. _____
10. _____
11. _____
12. _____
13. _____
14. _____
15. _____
16. _____
17. _____
18. _____
19. _____
20. _____

SCORE: _____

1. What is the place value of the digit 5 in the number 25,604?

2. Write the number 52,619 in words.

3. Write the number 3,428 in expanded form.

4. Round 375 to the nearest ten.

5. Round 12,496 to the nearest thousand.

6. Add: 475
        528

7. Add: 268
         723
         356

8. Find the sum of 3,196; 38; 506; and 1,624.

9. Subtract: 843
             −548

10. Subtract: 608
              −199

11. Find the difference of 10,135 and 6,128.

12. Multiply: 364
                28

13. Multiply: 1,235
                 406

14. Divide: $8\overline{)512}$

15. Divide: $23\overline{)1,097}$

16. Divide: $13\overline{)39,221}$

17. Evaluate:
    $3\{4(6 − 4) + 5[2(7 − 2)]\}$

18. Find the prime factorization of 408.

19. Find the greatest common factor of 126 and 140.

20. Find the least common multiple of 12 and 15.

CHAPTER 2 PRETEST

NAME: _____

CLASS / SECTION: _____ DATE: _____

ANSWERS

Answer as many of the following problems as you can before starting this chapter. When you finish the chapter, take the test at the end and compare your scores to see how much you have learned.

1. _____

2. _____

3. _____

4. _____

5. _____

6. _____

7. _____

8. _____

9. _____

10. _____

11. _____

12. _____

13. _____

14. _____

15. _____

16. _____

17. _____

18. _____

19. _____

20. _____

SCORE: _____

1. Reduce $\dfrac{24}{84}$.

2. Change $9\dfrac{11}{12}$ to an improper fraction.

3. Find the missing numerator: $\dfrac{3}{16} = \dfrac{?}{64}$

4. Find the LCD of $\dfrac{1}{5}, \dfrac{2}{3}, \dfrac{7}{9}$.

Perform the indicated operation. Give all answers in reduced form.

5. $\dfrac{1}{3} \times \dfrac{2}{5}$

6. $\dfrac{3}{7} \div \dfrac{5}{6}$

7. $\dfrac{2}{11} + \dfrac{7}{11}$

8. $\dfrac{3}{8} - \dfrac{1}{8}$

9. $3\dfrac{3}{4} \times 2\dfrac{2}{5}$

10. $\dfrac{4}{7} + \dfrac{3}{8}$

11. $\dfrac{9}{11} \div 6$

12. $\dfrac{3}{5} - \dfrac{1}{4}$

13. $\dfrac{7}{12} + \dfrac{5}{9} + \dfrac{3}{4}$

14. $\dfrac{3}{8} \times \dfrac{4}{11}$

15. $5\dfrac{3}{4} - 2\dfrac{1}{3}$

16. $\dfrac{4}{9} \div \dfrac{14}{15}$

17. $\dfrac{6}{7} \times \dfrac{14}{27}$

18. $3\dfrac{2}{3} + 1\dfrac{7}{12} + 6\dfrac{8}{9}$

19. $3\dfrac{3}{4} \div 6\dfrac{2}{3}$

20. $13\dfrac{3}{8} - 7\dfrac{5}{6}$

CHAPTER 2

Fractions and Mixed Numbers

Harness horse A paced the first quarter-mile in $29\frac{2}{5}$ seconds. Horse B paced it in $28\frac{4}{5}$ seconds. How much faster was horse B?

In the preceding chapter we studied the four basic operations on whole numbers. In this chapter we will be concerned with another set of numbers called **fractions**. Whole numbers and fractions together are sometimes called the **numbers of arithmetic**. As you will see, fractions are composed of whole numbers and the operations you learned in the previous chapter form the foundation for operations on this new set of numbers.

 2-1 SIMPLIFYING FRACTIONS

OBJECTIVES

In this section you will learn to:
1. Identify the numerator and denominator of a fraction.
2. Classify a fraction as proper or improper.
3. Use the fundamental principle of fractions to reduce a fraction.

Definition   A **common fraction** is the indicated quotient of two whole numbers.

Examples of common fractions are $\frac{3}{5}$, $\frac{1}{2}$, $\frac{5}{8}$, $\frac{7}{3}$, and $\frac{2}{2}$. These fractions could be written as $3 \div 5$, $1 \div 2$, $5 \div 8$, and so on, but the usual method is to write one number over the other with a division bar between them. Some special names make it easier to talk about fractions.

Division or fractional bar → $\frac{a}{b}$ ← Top number is called the **numerator**
← Bottom number is called the **denominator**

Fractions can be classified as *proper* or *improper*.

If the numerator is smaller than the denominator, we call the indicated division a **proper fraction.**

If the numerator is equal to, or larger than, the denominator, we call the indicated division an **improper fraction.**

A fraction is read as a division.

Example 1 $\frac{3}{4}$ is read as "three divided by four" or "three-fourths."

Example 2 $\frac{1}{2}$ is read as "one divided by two" or "one-half."

Example 3 $\frac{2}{3}$ is read as "two divided by three" or "two-thirds."

Example 4 $\frac{12}{7}$ is read as "twelve divided by seven" or "twelve-sevenths."

2-1 SIMPLIFYING FRACTIONS

Fractions that are equal in value may be written in many ways. For instance, if you have ten dimes and give away five of them, or if you have four quarters and give away two of them, or if you have twenty nickels and give away ten of them, you have given away one-half of a dollar in each case. In other words, $\frac{5}{10}$, $\frac{2}{4}$, $\frac{10}{20}$, and $\frac{1}{2}$ all represent the same value. This illustrates the following very important principle.

Fundamental Principle of Fractions If $\frac{a}{b}$ is a fraction and c is any number except zero, then $\frac{a}{b} = \frac{a \times c}{b \times c}$.

In other words, the principle states that we can multiply or divide both the numerator and denominator of a fraction by the same nonzero number and the value of the fraction remains the same.

Example 5 $\quad \frac{5}{8} = \frac{5 \times 2}{8 \times 2} = \frac{10}{16} \quad$ We multiply numerator and denominator by 2.

Example 6 $\quad \frac{5}{8} = \frac{5 \times 3}{8 \times 3} = \frac{15}{24} \quad$ We multiply numerator and denominator by 3.

Example 7 $\quad \frac{10}{15} = \frac{2 \times 5}{3 \times 5} = \frac{2}{3} \quad$ We divide numerator and denominator by 5.

We now wish to use the fundamental principle of fractions to **simplify** or **reduce** fractions.

Definition A fraction $\frac{a}{b}$ is in **simplified** or **reduced** form if a and b have no factor in common except the number 1.

Example 8 Each of the fractions $\frac{2}{3}$, $\frac{4}{7}$, and $\frac{5}{8}$ is in simplified form because 1 is the only factor common to the numerator and denominator.

Example 9 $\frac{4}{6}$, $\frac{9}{12}$, and $\frac{7}{14}$ are *not* in simplified form. Can you find a common factor (other than 1) of the numerator and denominator in each fraction?

Rule To *reduce* or *simplify* a fraction, factor both the numerator and denominator into primes and then divide out all common factors using the fundamental principle of fractions.

Example 10 Reduce $\frac{12}{18}$.

First factor the numerator and denominator into primes.

$$\frac{12}{18} = \frac{2 \times 2 \times 3}{2 \times 3 \times 3}$$

Notice that we have a factor of 2 common to the numerator and denominator. There is also a common factor of 3. Now dividing them out, we have

$$\frac{\cancel{2} \times 2 \times \cancel{3}}{\cancel{2} \times 3 \times \cancel{3}} = \frac{1 \times 2 \times 1}{1 \times 3 \times 1} = \frac{2}{3}.$$

So $\frac{12}{18} = \frac{2}{3}$ in reduced or simplified form. Notice that when we divide 2 into 2 and 3 into 3, we are left with a factor of 1. This is not necessary to write unless 1 is the only number in the numerator.

Example 11 Reduce $\frac{7}{14}$.

$$\frac{7}{14} = \frac{\cancel{7} \times 1}{\cancel{7} \times 2} = \frac{1}{2}$$

Example 12 Simplify $\frac{24}{68}$.

$$\frac{24}{68} = \frac{\cancel{2} \times \cancel{2} \times 2 \times 3}{\cancel{2} \times \cancel{2} \times 17} = \frac{6}{17}$$

Example 13 Simplify $\frac{15}{28}$.

$$\frac{15}{28} = \frac{3 \times 5}{2 \times 2 \times 7}$$

There are no common factors other than 1, so the fraction is already in simplified form.

Simplifying or reducing fractions is important because fractional answers should be given in simplest form.

EXERCISE SET 2-1-1

Reduce the following fractions.

1. $\dfrac{4}{6}$
2. $\dfrac{6}{9}$
3. $\dfrac{9}{12}$

4. $\dfrac{6}{8}$
5. $\dfrac{4}{10}$
6. $\dfrac{6}{15}$

7. $\dfrac{6}{10}$
8. $\dfrac{8}{10}$
9. $\dfrac{15}{18}$

10. $\dfrac{10}{12}$
11. $\dfrac{14}{21}$
12. $\dfrac{6}{14}$

13. $\dfrac{15}{35}$
14. $\dfrac{12}{15}$
15. $\dfrac{8}{12}$

16. $\dfrac{12}{18}$
17. $\dfrac{3}{6}$
18. $\dfrac{3}{9}$

19. $\dfrac{5}{20}$
20. $\dfrac{5}{10}$
21. $\dfrac{7}{21}$

22. $\dfrac{5}{15}$
23. $\dfrac{2}{18}$
24. $\dfrac{9}{18}$

25. $\dfrac{4}{16}$
26. $\dfrac{9}{36}$
27. $\dfrac{6}{12}$

ANSWERS

1. ____
2. ____
3. ____
4. ____
5. ____
6. ____
7. ____
8. ____
9. ____
10. ____
11. ____
12. ____
13. ____
14. ____
15. ____
16. ____
17. ____
18. ____
19. ____
20. ____
21. ____
22. ____
23. ____
24. ____
25. ____
26. ____
27. ____

28. $\dfrac{8}{24}$

29. $\dfrac{12}{16}$

30. $\dfrac{8}{24}$

31. $\dfrac{12}{18}$

32. $\dfrac{24}{28}$

33. $\dfrac{36}{42}$

34. $\dfrac{16}{72}$

35. $\dfrac{24}{64}$

36. $\dfrac{15}{32}$

37. $\dfrac{10}{27}$

38. $\dfrac{45}{54}$

39. $\dfrac{32}{88}$

40. $\dfrac{12}{78}$

41. $\dfrac{28}{39}$

42. $\dfrac{20}{50}$

43. $\dfrac{20}{65}$

44. $\dfrac{27}{72}$

45. $\dfrac{21}{36}$

46. $\dfrac{48}{66}$

47. $\dfrac{28}{63}$

48. $\dfrac{24}{56}$

49. $\dfrac{36}{54}$

50. $\dfrac{36}{91}$

51. $\dfrac{24}{112}$

52. $\dfrac{72}{90}$

53. $\dfrac{63}{68}$

54. $\dfrac{36}{96}$

55. $\dfrac{108}{144}$ **56.** $\dfrac{64}{144}$ **57.** $\dfrac{108}{198}$

58. $\dfrac{168}{273}$ **59.** $\dfrac{196}{336}$ **60.** $\dfrac{176}{288}$

A N S W E R S

55. _____
56. _____
57. _____
58. _____
59. _____
60. _____

2-2 MULTIPLICATION OF FRACTIONS

Definition If $\dfrac{a}{b}$ and $\dfrac{c}{d}$ are fractions, then their product is $\dfrac{a \times c}{b \times d}$.

O B J E C T I V E S

In this section you will learn to multiply fractions and give the product in reduced form.

This definition stated simply is *to multiply two fractions, multiply numerators to get the numerator of the product and denominators to get the denominator of the product.*

Example 1 $\dfrac{2}{3} \times \dfrac{5}{7} = \dfrac{2 \times 5}{3 \times 7} = \dfrac{10}{21}$

Example 2 $\dfrac{5}{2} \times \dfrac{17}{21} = \dfrac{5 \times 17}{2 \times 21} = \dfrac{85}{42}$

Always giving the product in reduced form sometimes requires more than just multiplying as in the preceding two examples. Consider the following.

Example 3 $\dfrac{2}{5} \times \dfrac{3}{4} = \dfrac{2 \times 3}{5 \times 4} = \dfrac{6}{20}$

We note that the fraction $\dfrac{6}{20}$ can be reduced. So we factor and obtain

$$\dfrac{6}{20} = \dfrac{\cancel{2} \times 3}{\cancel{2} \times 2 \times 5} = \dfrac{3}{10}.$$

In example 3 we multiplied 2×3 to get 6 and then factored 6 to get 2×3. This, of course, is "wasted motion" or unnecessary work. To avoid this we actually reduce the answer before we find it. We use the following rule, which is a direct use of the fundamental principle of fractions.

> **Rule** In an indicated multiplication of two or more fractions, any factor of any numerator can be divided out with a like factor of any denominator.

Example 4 $\quad \dfrac{2}{3} \times \dfrac{6}{7} = \dfrac{2}{\cancel{3}} \times \dfrac{2 \times \cancel{3}}{7} = \dfrac{4}{7}$

A more efficient way of writing this example is

$$\dfrac{2}{\cancel{3}} \times \dfrac{\overset{2}{\cancel{6}}}{7} = \dfrac{4}{7}.$$

Notice that we simply divided the factor 3 into 6 and wrote the quotient above it.

Example 5 $\quad \dfrac{3}{7} \times \dfrac{21}{12} = \dfrac{\cancel{3}}{\cancel{7}} \times \dfrac{\overset{3}{\cancel{21}}}{\underset{4}{\cancel{12}}} = \dfrac{3}{4}$

The number of fractions being multiplied does not change the rule or method.

Example 6 $\quad \dfrac{2}{3} \times \dfrac{9}{11} \times \dfrac{5}{4} = \dfrac{\cancel{2}}{\cancel{3}} \times \dfrac{\overset{3}{\cancel{9}}}{11} \times \dfrac{5}{\underset{2}{\cancel{4}}} = \dfrac{15}{22}$

Example 7 $\quad \dfrac{5}{8} \times \dfrac{3}{4} \times \dfrac{4}{5} = \dfrac{\cancel{5}}{8} \times \dfrac{3}{\cancel{4}} \times \dfrac{\cancel{4}}{\cancel{5}} = \dfrac{3}{8}$

It is important to recognize that the set of whole numbers can be written as fractions. For instance, the whole number 5 can be written as $\dfrac{5}{1}, \dfrac{10}{2}, \dfrac{15}{3}$, and so on. So multiplication of a fraction by a whole number follows the same rule.

Example 8 Find $\dfrac{2}{3}$ of 6.

In word problems *of* usually means multiply.

$$\dfrac{2}{3} \times 6 = \dfrac{2}{\cancel{3}} \times \dfrac{\overset{2}{\cancel{6}}}{1} = \dfrac{4}{1} = 4$$

Example 9 $\quad \dfrac{5}{8} \times 4 \times \dfrac{3}{5} = \dfrac{\cancel{5}}{\underset{2}{\cancel{8}}} \times \dfrac{\cancel{4}}{1} \times \dfrac{3}{\cancel{5}} = \dfrac{3}{2}$

Example 10 Bob and his friend earned $24 mowing a lawn and each took one-half of the total earned. Bob gave one-third of his half to his sister for her birthday. How much did his sister get?

Here, as in many word problems, you actually have a problem within a problem. Before you can answer the question "How much did his sister get?", you must answer "How much did Bob earn?"

The amount Bob earned is

$$\frac{1}{\cancel{2}} \times \frac{\cancel{24}^{12}}{1} = \$12.$$

Then the amount he gave his sister is

$$\frac{1}{\cancel{3}} \times \frac{\cancel{12}^{4}}{1} = \$4.$$

WARNING

When working word problems, make sure you read the problem carefully and be sure you have answered the question asked.

Example 11 A man decided to give one-tenth of his income to charity. He also designated that one-third of the amount for charity will go to the Red Cross. What part of his income goes to the Red Cross?

$\frac{1}{10}$ of his income goes to charity and $\frac{1}{3}$ of that or $\frac{1}{3} \times \frac{1}{10}$

$= \frac{1}{30}$ goes to the Red Cross.

EXERCISE SET 2-2-1

Multiply the following.

1. $\frac{1}{2} \times \frac{3}{5}$

2. $\frac{1}{3} \times \frac{2}{5}$

ANSWERS

3. $\dfrac{1}{4} \times \dfrac{1}{2}$

4. $\dfrac{1}{3} \times \dfrac{1}{6}$

3. _____

4. _____

5. $\dfrac{2}{3} \times \dfrac{5}{7}$

6. $\dfrac{3}{5} \times \dfrac{2}{11}$

5. _____

6. _____

7. $\dfrac{3}{4} \times \dfrac{3}{5}$

8. $\dfrac{2}{3} \times \dfrac{2}{5}$

7. _____

8. _____

9. $\dfrac{2}{3} \times \dfrac{1}{3}$

10. $\dfrac{4}{5} \times \dfrac{2}{5}$

9. _____

10. _____

11. $\dfrac{1}{3} \times \dfrac{3}{4}$

12. $\dfrac{1}{2} \times \dfrac{2}{3}$

11. _____

12. _____

13. $\dfrac{2}{5} \times \dfrac{5}{7}$

14. $\dfrac{3}{8} \times \dfrac{1}{3}$

13. _____

14. _____

15. $\dfrac{5}{6} \times \dfrac{6}{11}$

16. $\dfrac{3}{4} \times \dfrac{4}{7}$

17. $\dfrac{3}{8} \times \dfrac{5}{8}$

18. $\dfrac{2}{5} \times \dfrac{3}{5}$

19. $\dfrac{4}{5} \times \dfrac{3}{4}$

20. $\dfrac{7}{8} \times \dfrac{3}{7}$

21. $\dfrac{1}{2} \times \dfrac{4}{5}$

22. $\dfrac{2}{3} \times \dfrac{6}{7}$

23. $\dfrac{14}{3} \times \dfrac{5}{7}$

24. $\dfrac{12}{13} \times \dfrac{1}{4}$

25. $\dfrac{2}{5} \times \dfrac{10}{11}$

26. $\dfrac{3}{4} \times \dfrac{8}{13}$

15. _____

16. _____

17. _____

18. _____

19. _____

20. _____

21. _____

22. _____

23. _____

24. _____

25. _____

26. _____

27. $\dfrac{2}{5} \times \dfrac{3}{6}$

28. $\dfrac{3}{8} \times \dfrac{5}{6}$

29. $\dfrac{1}{2} \times \dfrac{1}{4} \times \dfrac{3}{5}$

30. $\dfrac{1}{3} \times \dfrac{2}{5} \times \dfrac{1}{7}$

31. $\dfrac{1}{3} \times \dfrac{9}{5} \times \dfrac{1}{2}$

32. $\dfrac{2}{3} \times \dfrac{1}{10} \times \dfrac{1}{5}$

33. $\dfrac{3}{4} \times \dfrac{2}{3}$

34. $\dfrac{2}{5} \times \dfrac{5}{8}$

35. $\dfrac{4}{10} \times \dfrac{2}{5}$

36. $\dfrac{3}{4} \times \dfrac{12}{15}$

37. $\dfrac{2}{3} \times \dfrac{3}{4} \times \dfrac{1}{9}$

38. $\dfrac{1}{4} \times \dfrac{2}{3} \times \dfrac{6}{8}$

27. _____

28. _____

29. _____

30. _____

31. _____

32. _____

33. _____

34. _____

35. _____

36. _____

37. _____

38. _____

39. $\dfrac{4}{7} \times \dfrac{21}{22}$

40. $\dfrac{7}{8} \times \dfrac{18}{21}$

41. $\dfrac{3}{16} \times \dfrac{6}{9}$

42. $\dfrac{4}{18} \times \dfrac{6}{20}$

43. $\dfrac{21}{32} \times \dfrac{18}{28} \times \dfrac{16}{40}$

44. $\dfrac{14}{36} \times \dfrac{9}{24} \times \dfrac{6}{21}$

45. $\dfrac{4}{9} \times \dfrac{30}{50}$

46. $\dfrac{16}{21} \times \dfrac{28}{38}$

47. $\dfrac{15}{28} \times \dfrac{14}{25}$

48. $\dfrac{15}{17} \times \dfrac{51}{54}$

49. $\dfrac{21}{14} \times \dfrac{10}{15}$

50. $\dfrac{30}{18} \times \dfrac{9}{15}$

ANSWERS

39. _____

40. _____

41. _____

42. _____

43. _____

44. _____

45. _____

46. _____

47. _____

48. _____

49. _____

50. _____

ANSWERS

51. _____

52. _____

53. _____

54. _____

55. _____

56. _____

57. _____

58. _____

59. _____

60. _____

61. _____

62. _____

51. $\dfrac{2}{3} \times 6$

52. $\dfrac{1}{2} \times 4$

53. $\dfrac{1}{4} \times 12$

54. $\dfrac{3}{4} \times 8$

55. $\dfrac{3}{5} \times 15$

56. $\dfrac{2}{3} \times 18$

57. $\dfrac{1}{2} \times 6 \times \dfrac{5}{3}$

58. $\dfrac{3}{5} \times 10 \times \dfrac{3}{8}$

59. $\dfrac{3}{4} \times 12 \times \dfrac{1}{9}$

60. $\dfrac{5}{12} \times 16 \times \dfrac{3}{20}$

61. Find $\dfrac{3}{4}$ of 80.

62. Find $\dfrac{2}{5}$ of 35.

63. Find $\frac{3}{7}$ of 21.

64. Find $\frac{1}{8}$ of 96.

65. Find $\frac{3}{5}$ of $\frac{15}{21}$.

66. Find $\frac{2}{3}$ of $\frac{9}{14}$.

67. Find $\frac{12}{13}$ of $\frac{39}{36}$.

68. Find $\frac{4}{11}$ of $\frac{121}{122}$.

69. Find $\frac{1}{12}$ of 156.

70. Find $\frac{3}{7}$ of $\frac{28}{36}$.

71. A bottle of a certain perfume contains $\frac{1}{2}$ ounce. How many ounces are there in 36 bottles?

72. A can of vegetables weighs $\frac{2}{3}$ pound. What is the weight of 48 cans?

ANSWERS

63. _____
64. _____
65. _____
66. _____
67. _____
68. _____
69. _____
70. _____
71. _____
72. _____

ANSWERS

73. A 2-liter bottle of Coke is $\frac{1}{4}$ full. How much is in the bottle?

74. A man owned $\frac{1}{4}$ interest in a business. If he sold $\frac{1}{2}$ of his share, how much of the business does he now own?

73. _____

74. _____

75. A certain recipe requires $\frac{2}{3}$ cup of sugar, $\frac{1}{4}$ teaspoon of vanilla, and $\frac{1}{2}$ cup of flour. How much of each ingredient is needed for one-half the recipe?

76. A rectangular field measures $\frac{2}{3}$ mile by $\frac{3}{8}$ mile. What is the area of the field?

75. _____

76. _____

77. Janet earns $225 per week. She spends $\frac{1}{3}$ of her earnings for rent. How much does she spend for rent each week?

78. The scale on a map reads that 1 inch represents 400 miles. How many miles are represented by $\frac{3}{8}$ inch?

77. _____

78. _____

79. _____

79. Mike's pickup truck can carry $\frac{3}{4}$ ton of gravel. If he carries six loads and is paid $20 per ton, how much does he earn?

80. A man gave $\frac{1}{10}$ of his income to charity. He gave $\frac{1}{3}$ of the amount for charity to the United Fund. If his income was $36,000, how much did he give to the United Fund?

80. _____

2-3 DIVISION OF FRACTIONS

To establish the rule for dividing fractions we must first consider the reciprocal of a number. Observe the following multiplications.

$$\frac{2}{3} \times \frac{3}{2} = \frac{\cancel{2}^{1}}{\cancel{3}} \times \frac{\cancel{3}^{1}}{\cancel{2}} = 1$$

$$5 \times \frac{1}{5} = \frac{\cancel{5}}{1} \times \frac{1}{\cancel{5}} = 1$$

$$\frac{7}{8} \times \frac{8}{7} = \frac{\cancel{7}^{1}}{\cancel{8}} \times \frac{\cancel{8}^{1}}{\cancel{7}} = 1$$

In each case the product is 1 so the two numbers are said to be **reciprocals** of each other.

> **Definition**   If a and b are numbers and if $a \times b = 1$, then a and b are **reciprocals** of each other. Every number, except zero, has a reciprocal.

OBJECTIVES

In this section you will learn to:
1. Simplify complex fractions.
2. Divide fractions.

We sometimes call the reciprocal the **invert.**

Example 1   The reciprocal or invert of $\frac{7}{8}$ is $\frac{8}{7}$.

Example 2   The reciprocal or invert of $\frac{3}{4}$ is $\frac{4}{3}$.

Example 3   The reciprocal or invert of the whole number 5 is $\frac{1}{5}$.

Now consider this division problem.

Example 4   Find the quotient $\frac{2}{3} \div \frac{5}{8}$.

Since the fraction bar always means division, we can write

$$\frac{2}{3} \div \frac{5}{8} \text{ as } \frac{\frac{2}{3}}{\frac{5}{8}}.$$

Now we have what is known as a **complex fraction,** that is, a fraction containing fractions. To simplify the complex fraction we will use the fundamental principle of fractions and multiply both the numerator and denominator of the complex fraction by a number that will yield 1 in the denominator. In other words, we multiply by the reciprocal of the denominator of the complex fraction. In this case we multiply by the reciprocal of $\frac{5}{8}$, which is $\frac{8}{5}$.

$$\frac{\frac{2}{3}}{\frac{5}{8}} = \frac{\frac{2}{3} \times \frac{8}{5}}{\frac{5}{8} \times \frac{8}{5}} = \frac{\frac{2}{3} \times \frac{8}{5}}{1} = \frac{2}{3} \times \frac{8}{5}$$

So $\frac{2}{3} \div \frac{5}{8} = \frac{2}{3} \times \frac{8}{5} = \frac{16}{15}$.

Example 4 is an illustration of the following rule.

Rule If $\frac{a}{b}$ and $\frac{c}{d}$ are fractions, then $\frac{a}{b} \div \frac{c}{d} = \frac{a}{b} \times \frac{d}{c}$.

In simpler words, if we wish to divide one fraction by another, we change the problem to multiplication by inverting the divisor. We often shorten the statement to "invert and multiply."

WARNING

Make sure to invert the correct fraction, which is the divisor (that is, the fraction doing the dividing).

Example 5 Divide: $\frac{5}{8} \div \frac{1}{2}$

$$\frac{5}{8} \div \frac{1}{2} = \frac{5}{\underset{4}{\cancel{8}}} \times \frac{\cancel{2}}{1} = \frac{5}{4}$$

Example 6 Find the quotient: $\dfrac{2}{3} \div \dfrac{5}{6}$

$$\dfrac{2}{3} \div \dfrac{5}{6} = \dfrac{2}{\cancel{3}} \times \dfrac{\cancel{6}^{2}}{5} = \dfrac{4}{5}$$

Example 7 Divide: $7 \div \dfrac{3}{4}$

$$7 \div \dfrac{3}{4} = \dfrac{7}{1} \times \dfrac{4}{3} = \dfrac{28}{3}$$

Example 8 Find the quotient: $\dfrac{4}{5} \div 6$

$$\dfrac{4}{5} \div 6 = \dfrac{\cancel{4}^{2}}{5} \times \dfrac{1}{\cancel{6}_{3}} = \dfrac{2}{15}$$

Example 9 Find the quotient: $\dfrac{2}{9} \div \dfrac{2}{3}$

$$\dfrac{2}{9} \div \dfrac{2}{3} = \dfrac{\cancel{2}^{1}}{\cancel{9}_{3}} \times \dfrac{\cancel{3}^{1}}{\cancel{2}_{1}} = \dfrac{1}{3}$$

Example 10 How many $\dfrac{1}{4}$-pound candy bars are there in a 3-pound box?

The question is how many times will $\dfrac{1}{4}$ divide into 3?

$$3 \div \dfrac{1}{4} = \dfrac{3}{1} \times \dfrac{4}{1} = 12$$

EXERCISE SET 2-3-1

Simplify the following complex fractions.

ANSWERS

1. $\dfrac{\;\dfrac{2}{3}\;}{\dfrac{3}{4}}$

2. $\dfrac{\;\dfrac{2}{3}\;}{\dfrac{5}{7}}$

1. _____

2. _____

3. $\dfrac{\dfrac{1}{2}}{\dfrac{2}{3}}$

4. $\dfrac{\dfrac{1}{3}}{\dfrac{3}{4}}$

5. $\dfrac{5}{\dfrac{1}{4}}$

6. $\dfrac{\dfrac{3}{2}}{3}$

7. $\dfrac{\dfrac{1}{4}}{5}$

8. $\dfrac{\dfrac{2}{3}}{3}$

9. $\dfrac{\dfrac{3}{5}}{\dfrac{9}{10}}$

10. $\dfrac{\dfrac{2}{7}}{\dfrac{6}{35}}$

3. _____

4. _____

5. _____

6. _____

7. _____

8. _____

9. _____

10. _____

2-3 DIVISION OF FRACTIONS   113

11. $\dfrac{\frac{5}{6}}{\frac{15}{18}}$

12. $\dfrac{\frac{3}{7}}{\frac{15}{28}}$

ANSWERS

11. _____

12. _____

Find the following quotients.

13. $\dfrac{1}{2} \div \dfrac{3}{4}$

14. $\dfrac{1}{3} \div \dfrac{5}{6}$

13. _____

14. _____

15. $\dfrac{3}{4} \div \dfrac{1}{2}$

16. $\dfrac{5}{6} \div \dfrac{1}{3}$

15. _____

16. _____

17. $\dfrac{2}{3} \div \dfrac{4}{9}$

18. $\dfrac{3}{4} \div \dfrac{9}{16}$

17. _____

18. _____

19. _____

20. _____

21. _____

22. _____

23. _____

24. _____

25. _____

26. _____

19. $\dfrac{4}{9} \div \dfrac{2}{3}$

20. $\dfrac{9}{16} \div \dfrac{3}{4}$

21. $\dfrac{5}{9} \div \dfrac{2}{5}$

22. $\dfrac{1}{3} \div \dfrac{3}{5}$

23. $\dfrac{4}{3} \div \dfrac{8}{6}$

24. $\dfrac{4}{5} \div \dfrac{3}{10}$

25. $\dfrac{5}{6} \div \dfrac{6}{7}$

26. $\dfrac{3}{8} \div \dfrac{8}{5}$

27. $\dfrac{6}{11} \div \dfrac{4}{11}$

28. $\dfrac{4}{9} \div \dfrac{2}{3}$

27. _____

29. $\dfrac{5}{8} \div \dfrac{7}{8}$

30. $\dfrac{4}{7} \div \dfrac{3}{7}$

28. _____

29. _____

31. $8 \div \dfrac{4}{5}$

32. $6 \div \dfrac{2}{3}$

30. _____

31. _____

33. $\dfrac{3}{4} \div \dfrac{7}{8}$

34. $\dfrac{5}{12} \div \dfrac{2}{9}$

32. _____

33. _____

34. _____

ANSWERS

35. $\dfrac{6}{7} \div \dfrac{7}{8}$

36. $\dfrac{2}{9} \div \dfrac{9}{10}$

35. _____

37. $5 \div \dfrac{2}{5}$

38. $9 \div \dfrac{2}{3}$

36. _____

37. _____

39. $\dfrac{2}{5} \div 5$

40. $\dfrac{2}{3} \div 9$

38. _____

39. _____

41. $\dfrac{3}{4} \div \dfrac{3}{8}$

42. $\dfrac{5}{9} \div \dfrac{5}{3}$

40. _____

41. _____

42. _____

43. $\dfrac{13}{28} \div 52$

44. $\dfrac{11}{20} \div 44$

45. $16 \div \dfrac{8}{9}$

46. $30 \div \dfrac{6}{5}$

47. $\dfrac{3}{8} \div \dfrac{5}{6}$

48. $3 \div \dfrac{6}{7}$

49. $\dfrac{7}{24} \div \dfrac{21}{18}$

50. $\dfrac{6}{25} \div \dfrac{3}{10}$

43. _____

44. _____

45. _____

46. _____

47. _____

48. _____

49. _____

50. _____

51. $\dfrac{4}{5} \div 8$

52. $\dfrac{12}{13} \div 4$

53. $\dfrac{7}{18} \div \dfrac{7}{12}$

54. $\dfrac{8}{15} \div \dfrac{6}{35}$

55. $\dfrac{5}{28} \div \dfrac{25}{42}$

56. $\dfrac{7}{11} \div \dfrac{14}{33}$

57. $\dfrac{11}{60} \div \dfrac{44}{50}$

58. $\dfrac{12}{21} \div \dfrac{16}{35}$

51. _____

52. _____

53. _____

54. _____

55. _____

56. _____

57. _____

58. _____

59. $\dfrac{17}{35} \div \dfrac{51}{56}$

60. $\dfrac{11}{38} \div \dfrac{22}{19}$

61. $\dfrac{33}{7} \div \dfrac{11}{21}$

62. $\dfrac{24}{27} \div \dfrac{8}{9}$

63. $\dfrac{18}{7} \div \dfrac{9}{21}$

64. $\dfrac{6}{11} \div \dfrac{2}{3}$

65. $\dfrac{15}{21} \div \dfrac{3}{14}$

66. $\dfrac{4}{15} \div \dfrac{9}{20}$

59. _____

60. _____

61. _____

62. _____

63. _____

64. _____

65. _____

66. _____

67. $\dfrac{4}{25} \div \dfrac{12}{35}$

68. $\dfrac{32}{15} \div \dfrac{16}{25}$

67. _____

68. _____

69. $\dfrac{5}{21} \div \dfrac{15}{49}$

70. $\dfrac{14}{15} \div \dfrac{35}{36}$

69. _____

70. _____

71. Each student is served $\dfrac{1}{2}$ pint of milk. How many students can be served from 50 pints of milk?

72. How many "quarter-pounder" hamburgers can be made from 12 pounds of hamburger?

71. _____

72. _____

73. Jean has a 10-yard spool of ribbon from which she wants to make some bows. If it takes $\dfrac{2}{3}$ yard of ribbon to make each bow, how many bows can she make?

74. Bob wants to buy twelve liters of wine for a party but the store only has $\dfrac{3}{4}$-liter bottles. How many bottles will it take to get twelve liters?

73. _____

74. _____

75. A furlong is $\frac{1}{8}$ mile. How many furlongs are there in $\frac{3}{4}$ mile?

76. A certain tablet contains $\frac{5}{8}$ grain of medicine. How many tablets can be made from 40 grains?

77. A person with a $\frac{3}{4}$-gallon pail must fill a 24-gallon tank. How many full pails will it take?

78. A $\frac{3}{4}$-acre tract of land is to be divided into six equal sections. How much of an acre will each section be?

79. A truck can carry $\frac{3}{4}$ ton of gravel. How many loads will it take to carry 12 tons?

80. A car travels $\frac{7}{10}$ of a mile per minute. How long does it take to travel 14 miles?

ANSWERS

75. _____

76. _____

77. _____

78. _____

79. _____

80. _____

2-4 ADDITION AND SUBTRACTION OF FRACTIONS

OBJECTIVES

In this section you will learn to:
1. Add and subtract like fractions.
2. Find the LCD of two or more fractions.
3. Change a fraction to an equivalent fraction.
4. Add and subtract any two fractions.

Definition Two fractions are said to be **like fractions** if and only if they have the same (common) denominator.

$\frac{2}{3}$ and $\frac{5}{3}$ are like fractions.

$\frac{7}{8}$ and $\frac{5}{8}$ are like fractions.

$\frac{2}{3}$ and $\frac{2}{5}$ are *not* like fractions.

In chapter 1 we stated that in all of mathematics only like quantities can be added or subtracted. This means then that only like fractions can be added or subtracted.

Rule To add *like* fractions add the numerators and place this sum over the common denominator.

Rule To subtract *like* fractions subtract the numerators and place this difference over the common denominator.

Example 1 $\frac{2}{3} + \frac{5}{3} = \frac{7}{3}$

Example 2 $\frac{7}{3} - \frac{5}{3} = \frac{2}{3}$

Example 3 $\frac{1}{2} + \frac{5}{2} = \frac{6}{2} = 3$

Note that $\frac{6}{2}$ reduces to 3.

Example 4 $\frac{11}{5} - \frac{1}{5} = \frac{10}{5} = 2$

Example 5 $\dfrac{5}{8} + \dfrac{7}{8} = \dfrac{12}{8} = \dfrac{3}{2}$

Notice that the sum or difference of two fractions may be reduced even though neither of the fractions can be reduced. All fractional answers should be in reduced form, if possible.

EXERCISE SET 2-4-1

Perform the indicated operation.

1. $\dfrac{1}{5} + \dfrac{2}{5}$
2. $\dfrac{3}{5} - \dfrac{1}{5}$
3. $\dfrac{5}{8} + \dfrac{1}{8}$
4. $\dfrac{3}{7} + \dfrac{4}{7}$

5. $\dfrac{8}{9} - \dfrac{2}{9}$
6. $\dfrac{11}{12} - \dfrac{5}{12}$
7. $\dfrac{4}{5} + \dfrac{6}{5}$
8. $\dfrac{22}{13} - \dfrac{9}{13}$

9. $\dfrac{38}{15} + \dfrac{17}{15}$
10. $\dfrac{34}{18} - \dfrac{19}{18}$
11. $\dfrac{11}{12} + \dfrac{7}{12}$
12. $\dfrac{31}{36} - \dfrac{13}{36}$

13. $\dfrac{31}{18} - \dfrac{15}{18}$
14. $\dfrac{27}{15} + \dfrac{18}{15}$
15. $\dfrac{11}{40} + \dfrac{15}{40}$
16. $\dfrac{31}{19} - \dfrac{12}{19}$

17. $\dfrac{52}{15} - \dfrac{12}{15}$
18. $\dfrac{39}{41} + \dfrac{43}{41}$
19. $\dfrac{31}{38} + \dfrac{1}{38}$
20. $\dfrac{15}{32} - \dfrac{3}{32}$

ANSWERS

1. _____
2. _____
3. _____
4. _____
5. _____
6. _____
7. _____
8. _____
9. _____
10. _____
11. _____
12. _____
13. _____
14. _____
15. _____
16. _____
17. _____
18. _____
19. _____
20. _____

The preceding examples and exercises all involved the sum or difference of like fractions. But what about the sum of two fractions such as $\dfrac{1}{3}$ and $\dfrac{1}{2}$? Obviously these two fractions cannot be added in their present form because they are not alike.

Recalling the fundamental principle of fractions, we know that $\dfrac{1}{3}$ can be written in many equivalent forms.

$$\frac{1}{3} = \frac{2}{6} = \frac{3}{9} = \frac{4}{12} = \frac{5}{15}, \text{ and so on.}$$

Also

$$\frac{1}{2} = \frac{2}{4} = \frac{3}{6} = \frac{4}{8} = \frac{5}{10} = \frac{6}{12}, \text{ and so on.}$$

In the list of ways to write $\frac{1}{3}$ and $\frac{1}{2}$ we find some ways in which the fractions are expressed as like fractions. Notice that $\frac{1}{3} = \frac{2}{6}$ and $\frac{1}{2} = \frac{3}{6}$. Also $\frac{1}{3} = \frac{4}{12}$ and $\frac{1}{2} = \frac{6}{12}$. In these forms they can be added (or subtracted) and this is our clue to adding and subtracting unlike fractions.

$$\frac{1}{3} + \frac{1}{2} = \frac{2}{6} + \frac{3}{6} = \frac{5}{6}$$

or

$$\frac{1}{3} + \frac{1}{2} = \frac{4}{12} + \frac{6}{12} = \frac{10}{12} = \frac{5}{6}$$

Notice that if we do not find the least common denominator that makes $\frac{1}{3}$ and $\frac{1}{2}$ like fractions, it will be necessary to reduce the sum. This will always be true, so it is to our advantage to find the least number that is a common denominator. In other words, we want the **least common denominator.**

> **Definition** The **least common denominator** (LCD) of two or more fractions is the least common multiple of their denominators.

If you need to review LCM refer to chapter 1.

Example 6 Find the LCD of $\frac{5}{12}$ and $\frac{7}{18}$.

Remember, to find the LCM of 12 and 18 we factor each number into primes.

$$12 = 2 \times 2 \times 3$$
$$18 = 2 \times 3 \times 3$$

$$\text{LCM} = 2 \times 2 \times 3 \times 3 = 36$$

So the LCD of $\frac{5}{12}$ and $\frac{7}{18}$ is 36.

EXERCISE SET 2-4-2

Find the LCD of the following sets of fractions.

1. $\dfrac{1}{2}$ and $\dfrac{1}{3}$

2. $\dfrac{1}{3}$ and $\dfrac{1}{5}$

3. $\dfrac{1}{2}$ and $\dfrac{5}{6}$

4. $\dfrac{2}{3}$ and $\dfrac{1}{6}$

5. $\dfrac{1}{6}$ and $\dfrac{4}{15}$

6. $\dfrac{4}{9}$ and $\dfrac{5}{12}$

7. $\dfrac{5}{9}$ and $\dfrac{1}{6}$

8. $\dfrac{3}{8}$ and $\dfrac{7}{12}$

9. $\dfrac{3}{16}$ and $\dfrac{11}{24}$

10. $\dfrac{7}{20}$ and $\dfrac{11}{36}$

11. $\dfrac{5}{12}$ and $\dfrac{13}{42}$

12. $\dfrac{3}{14}$ and $\dfrac{16}{21}$

ANSWERS

1. _____
2. _____
3. _____
4. _____
5. _____
6. _____
7. _____
8. _____
9. _____
10. _____
11. _____
12. _____

13. _____

14. _____

15. _____

16. _____

17. _____

18. _____

19. _____

20. _____

13. $\dfrac{5}{36}$ and $\dfrac{7}{24}$

14. $\dfrac{5}{12}$ and $\dfrac{3}{10}$

15. $\dfrac{1}{2}$, $\dfrac{5}{6}$, and $\dfrac{1}{9}$

16. $\dfrac{1}{2}$, $\dfrac{3}{8}$, and $\dfrac{5}{6}$

17. $\dfrac{1}{6}$, $\dfrac{3}{8}$, and $\dfrac{4}{9}$

18. $\dfrac{1}{3}$, $\dfrac{5}{6}$, and $\dfrac{1}{10}$

19. $\dfrac{5}{12}$, $\dfrac{9}{16}$, and $\dfrac{7}{18}$

20. $\dfrac{5}{12}$, $\dfrac{4}{15}$, and $\dfrac{1}{36}$

The fundamental principle of fractions can be used to change a fraction to an equivalent fraction with a different denominator.

Example 7 Change $\dfrac{3}{4}$ to a fraction with a denominator of 12.

$$\dfrac{3}{4} = \dfrac{?}{12}$$

We know that we can multiply the numerator and denominator by the same nonzero number. Our task then is to determine what number multiplied by 4 (the original denominator) will give 12 (the new denominator). In other words, we divide 4 into 12 to find the number 3. Multiplying the numerator and denominator by 3, we have

$$\dfrac{3}{4} = \dfrac{3 \times 3}{4 \times 3} = \dfrac{9}{12}.$$

Example 8 Change $\frac{3}{4}$ to a fraction with a denominator of 20.

$$\frac{3}{4} = \frac{?}{20}$$

We divide 4 into 20 and see that 4 must be multiplied by 5 to obtain the new denominator. So the numerator 3 must also be multiplied by 5.

$$\frac{3}{4} = \frac{3 \times 5}{4 \times 5} = \frac{15}{20}$$

Example 9 $\frac{7}{8} = \frac{?}{24}$

$$\frac{7}{8} = \frac{7 \times 3}{8 \times 3} = \frac{21}{24}$$

Example 10 $\frac{5}{3} = \frac{?}{15}$

$$\frac{5}{3} = \frac{5 \times 5}{3 \times 5} = \frac{25}{15}$$

▼ EXERCISE SET 2-4-3

Find the missing numerator.

1. $\frac{1}{2} = \frac{?}{6}$

2. $\frac{1}{2} = \frac{?}{8}$

3. $\frac{1}{3} = \frac{?}{12}$

4. $\frac{1}{5} = \frac{?}{10}$

5. $\frac{1}{6} = \frac{?}{18}$

6. $\frac{1}{7} = \frac{?}{35}$

7. $\frac{2}{3} = \frac{?}{24}$

8. $\frac{5}{6} = \frac{?}{24}$

ANSWERS

1. _____
2. _____
3. _____
4. _____
5. _____
6. _____
7. _____
8. _____

9. $\dfrac{3}{8} = \dfrac{?}{32}$

10. $\dfrac{2}{3} = \dfrac{?}{42}$

11. $\dfrac{4}{5} = \dfrac{?}{20}$

12. $\dfrac{3}{8} = \dfrac{?}{24}$

13. $\dfrac{2}{7} = \dfrac{?}{63}$

14. $\dfrac{7}{9} = \dfrac{?}{72}$

15. $\dfrac{3}{11} = \dfrac{?}{44}$

16. $\dfrac{3}{5} = \dfrac{?}{30}$

17. $\dfrac{13}{6} = \dfrac{?}{30}$

18. $\dfrac{4}{7} = \dfrac{?}{91}$

19. $\dfrac{4}{9} = \dfrac{?}{108}$

20. $\dfrac{7}{6} = \dfrac{?}{114}$

Since we now know how to find the LCD of any two or more fractions and also how to change a fraction to one with a new denominator, we can state a rule for adding any two or more fractions.

Rule To add any two or more fractions find the LCD and change each fraction to an equivalent fraction with that denominator. The fractions are now like fractions and can be added by using the rule for adding like fractions.

Example 11 Add: $\dfrac{3}{5} + \dfrac{2}{3}$

The LCD of 5 and 3 is 15.

$$\dfrac{3}{5} + \dfrac{2}{3} = \dfrac{9}{15} + \dfrac{10}{15} = \dfrac{19}{15}$$

Example 12 Add: $\dfrac{5}{12} + \dfrac{1}{4}$

The LCD of 12 and 4 is 12.

$$\dfrac{5}{12} + \dfrac{1}{4} = \dfrac{5}{12} + \dfrac{3}{12} = \dfrac{8}{12} = \dfrac{2}{3}$$

Example 13 Add: $\dfrac{1}{2} + \dfrac{2}{3} + \dfrac{5}{8}$

The LCD of 2, 3, and 8 is 24.

$$\dfrac{1}{2} + \dfrac{2}{3} + \dfrac{5}{8} = \dfrac{12}{24} + \dfrac{16}{24} + \dfrac{15}{24} = \dfrac{43}{24}$$

> **Rule** To subtract two fractions find the LCD and change each fraction to an equivalent fraction with that denominator. The fractions are now like fractions and can be subtracted using the rule for subtracting like fractions.

Example 14 Subtract: $\dfrac{5}{8} - \dfrac{1}{2}$

The LCD of 8 and 2 is 8.

$$\dfrac{5}{8} - \dfrac{1}{2} = \dfrac{5}{8} - \dfrac{4}{8} = \dfrac{1}{8}$$

Example 15 Subtract: $\dfrac{5}{6} - \dfrac{11}{15}$

The LCD of 6 and 15 is 30.

$$\dfrac{5}{6} - \dfrac{11}{15} = \dfrac{25}{30} - \dfrac{22}{30} = \dfrac{3}{30} = \dfrac{1}{10}$$

Example 16 Farmer Brown's milk bucket contained $\frac{5}{6}$ gallon of milk. His cow kicked the bucket and spilled $\frac{1}{2}$ gallon of milk before he could upright the pail. How much milk was left in the bucket?

We must subtract the amount spilled from the original amount.

Subtract: $\frac{5}{6} - \frac{1}{2}$

The LCD of 6 and 2 is 6.
Thus

$$\frac{5}{6} - \frac{1}{2} = \frac{5}{6} - \frac{3}{6} = \frac{2}{6} = \frac{1}{3}.$$

The amount left in the bucket was $\frac{1}{3}$ gallon.

EXERCISE SET 2-4-4

ANSWERS

Perform the indicated operation.

1. $\frac{1}{3} + \frac{1}{5}$

2. $\frac{1}{2} + \frac{1}{7}$

3. $\frac{2}{3} + \frac{3}{4}$

4. $\frac{3}{5} + \frac{2}{3}$

5. $\frac{5}{8} - \frac{1}{3}$

6. $\frac{4}{7} - \frac{2}{5}$

7. $\dfrac{3}{5} + \dfrac{3}{10}$

8. $\dfrac{2}{3} + \dfrac{5}{6}$

9. $\dfrac{2}{3} - \dfrac{1}{6}$

10. $\dfrac{5}{8} - \dfrac{1}{4}$

11. $\dfrac{1}{15} + \dfrac{5}{12}$

12. $\dfrac{1}{6} + \dfrac{4}{15}$

13. $\dfrac{7}{9} - \dfrac{3}{4}$

14. $\dfrac{11}{12} - \dfrac{5}{18}$

15. $\dfrac{3}{8} + \dfrac{7}{20}$

16. $\dfrac{5}{6} + \dfrac{2}{15}$

17. $\dfrac{5}{6} - \dfrac{5}{12}$

18. $\dfrac{3}{7} - \dfrac{3}{14}$

ANSWERS

7. _____

8. _____

9. _____

10. _____

11. _____

12. _____

13. _____

14. _____

15. _____

16. _____

17. _____

18. _____

ANSWERS

19. $\dfrac{1}{6} + \dfrac{11}{15}$

20. $\dfrac{2}{9} + \dfrac{5}{6}$

19. _____

20. _____

21. $\dfrac{4}{9} - \dfrac{5}{12}$

22. $\dfrac{5}{6} - \dfrac{7}{9}$

21. _____

22. _____

23. $\dfrac{5}{14} - \dfrac{4}{21}$

24. $\dfrac{5}{12} - \dfrac{3}{10}$

23. _____

24. _____

25. $\dfrac{1}{2} + \dfrac{1}{4} + \dfrac{1}{8}$

26. $\dfrac{1}{3} + \dfrac{1}{6} + \dfrac{1}{2}$

25. _____

26. _____

27. $\dfrac{3}{4} + \dfrac{2}{5} + \dfrac{3}{10}$

28. $\dfrac{3}{4} + \dfrac{5}{6} + \dfrac{3}{8}$

27. _____

28. _____

29. $\dfrac{1}{6} + \dfrac{3}{8} + \dfrac{4}{9}$

30. $\dfrac{1}{3} + \dfrac{5}{6} + \dfrac{1}{10}$

29. _____

30. _____

2-4 ADDITION AND SUBTRACTION OF FRACTIONS

31. $\dfrac{3}{4} - \dfrac{5}{12}$

32. $\dfrac{8}{15} - \dfrac{1}{3}$

31. _____

33. $\dfrac{5}{12} + \dfrac{3}{16} + \dfrac{7}{32}$

34. $\dfrac{2}{9} + \dfrac{5}{12} + \dfrac{3}{8}$

32. _____

33. _____

35. $\dfrac{11}{18} - \dfrac{4}{15}$

36. $\dfrac{11}{14} - \dfrac{13}{42}$

34. _____

35. _____

37. $\dfrac{3}{10} + \dfrac{2}{15} + \dfrac{1}{21}$

38. $\dfrac{5}{12} + \dfrac{7}{18} + \dfrac{3}{32}$

36. _____

37. _____

39. $\dfrac{2}{3} + \dfrac{5}{12} - \dfrac{7}{8}$

40. $\dfrac{3}{4} + \dfrac{5}{6} - \dfrac{4}{9}$

38. _____

39. _____

41. $\dfrac{2}{5} - \dfrac{1}{4} \div \dfrac{5}{6}$

42. $\left(\dfrac{2}{5} - \dfrac{1}{4}\right) \div \dfrac{5}{6}$

40. _____

41. _____

42. _____

43. $\dfrac{3}{4} + \dfrac{2}{3} \times \dfrac{1}{4}$

44. $\dfrac{3}{8} \div \dfrac{1}{4} \times \dfrac{2}{3}$

45. $\dfrac{1}{2} - \dfrac{1}{3}\left(\dfrac{3}{5} \div \dfrac{1}{2}\right)$

46. $\dfrac{3}{4} - \dfrac{5}{6} \div \left(\dfrac{2}{3} + 1\right)$

47. $\dfrac{1}{8} \div \dfrac{1}{4} + \dfrac{9}{8} \times \dfrac{1}{3} - \dfrac{3}{4}$

48. $\dfrac{7}{3} \times \dfrac{1}{2} - \dfrac{1}{4} \div \dfrac{5}{8} + \dfrac{3}{2}$

49. One-third of a cup of water was added to a container that had $\dfrac{3}{8}$ cup of water in it. What is the total amount of water in the container?

50. If $\dfrac{1}{4}$ ounce of a liquid is added to $\dfrac{5}{8}$ ounce of the same liquid, what is the total amount of liquid?

51. A hamburger that weighed $\dfrac{1}{4}$ pound before cooking was found to weigh $\dfrac{3}{16}$ pound after cooking. How much weight was lost?

52. A $\dfrac{1}{8}$-pound slice is removed from a $\dfrac{3}{4}$-pound block of butter. How much butter is left?

53. Al starts to walk to Julia's house, which is $\frac{7}{10}$ mile away. After he has walked $\frac{2}{5}$ of a mile, what distance must still be walked?

54. Margaret owns $\frac{3}{4}$ acre on which she wants to place an office building and a parking lot. The zoning board declares that she cannot use $\frac{1}{6}$ acre of her property for that purpose. How much can she use?

55. A man owns $\frac{7}{8}$ acre of land. He purchases an adjacent $\frac{2}{3}$-acre parcel. How much land does he now own?

56. If a person spends $\frac{1}{3}$ of the day working and $\frac{1}{4}$ of the day sleeping, how much of the day is left for other activities?

57. A company that sells gravel receives orders for $\frac{3}{4}$ ton, $\frac{5}{8}$ ton, and $\frac{1}{2}$ ton. What is the total tonnage of the three orders?

58. A woman gives $\frac{1}{10}$ of her salary to charity, $\frac{1}{4}$ is spent for rent, and $\frac{1}{6}$ for food and clothing. What total portion of her salary is used for these three purposes?

ANSWERS

53. _____

54. _____

55. _____

56. _____

57. _____

58. _____

59.

59. $\frac{1}{8}$ cup of water is added to an empty container. $\frac{1}{2}$ cup is added to that. Finally, $\frac{1}{4}$ cup of water is added. If the container has a capacity of one cup, how much water must be added to fill the container?

60. A one-gallon tank is full of water. $\frac{1}{4}$ gallon is drawn off. Another $\frac{2}{3}$ gallon is removed. Then $\frac{3}{4}$ gallon is added to the tank. Finally, $\frac{1}{2}$ gallon is drawn off. How much water remains in the tank?

2-5 MIXED NUMBERS

OBJECTIVES

In this section you will learn to:
1. Change a mixed number to an improper fraction and vice versa.
2. Multiply, divide, add, and subtract mixed numbers.

We have learned the four basic operations on whole numbers and fractions. We now turn our attention to a combination of the two called **mixed numbers**.

Example 1 $4\frac{2}{3}$, read as "four and two-thirds," is a way of writing $4 + \frac{2}{3}$.

Example 2 $5\frac{1}{2}$, read as "five and one-half," is a way of writing $5 + \frac{1}{2}$.

These are examples of mixed numbers. Every improper fraction can be changed to a mixed number and every mixed number can be changed to an improper fraction.

Example 3 Change $4\frac{2}{3}$ to an improper fraction.

No new rule is needed here since $4\frac{2}{3}$ means $4 + \frac{2}{3}$. The LCD is 3, so

$$4\frac{2}{3} = 4 + \frac{2}{3} = \frac{12}{3} + \frac{2}{3} = \frac{14}{3}.$$

Example 4 Change $5\frac{1}{2}$ to an improper fraction.

$$5\frac{1}{2} = 5 + \frac{1}{2} = \frac{10}{2} + \frac{1}{2} = \frac{11}{2}$$

Since the denominator of the fractional part of the mixed number will always be the LCD, we can use a short cut.

Example 5 Change $5\frac{1}{2}$ to an improper fraction.

The shortcut method involves multiplying the denominator (2) by the whole number (5) and then adding the product to the numerator (1). This gives a result of 11 in this case. Place this number over the original denominator (2) to obtain $5\frac{1}{2} = \frac{11}{2}$.

Example 6 Change $7\frac{5}{8}$ to an improper fraction.

Multiply 8×7, then add 5 to get $56 + 5 = 61$. So

$$7\frac{5}{8} = \frac{61}{8}.$$

Example 7 $2\frac{3}{4} = \frac{4 \times 2 + 3}{4} = \frac{11}{4}$

Example 8 $9\frac{1}{6} = \frac{6 \times 9 + 1}{6} = \frac{55}{6}$

Changing an improper fraction to a mixed number uses division to obtain the whole number part and the remainder to obtain the fractional part.

Example 9 Change $\frac{39}{4}$ to a mixed number.

We divide 4 into 39. This gives 9 with a remainder of 3. In other words,

$$\frac{39}{4} = \frac{36}{4} + \frac{3}{4} = 9 + \frac{3}{4} = 9\frac{3}{4}.$$

Example 10 Change $\frac{7}{3}$ to a mixed number.

$$7 \div 3 = 2 \text{ R}1; \text{ so } \frac{7}{3} = 2\frac{1}{3}.$$

Example 11 Change $\dfrac{38}{3}$ to a mixed number.

$$38 \div 3 = 12 \text{ R}2; \text{ so } \dfrac{38}{3} = 12\dfrac{2}{3}.$$

EXERCISE SET 2-5-1

ANSWERS

Change each of the following to an improper fraction.

1. $4\dfrac{1}{2}$
2. $2\dfrac{1}{8}$
3. $3\dfrac{2}{3}$
4. $6\dfrac{3}{4}$
5. $6\dfrac{5}{8}$
6. $9\dfrac{2}{3}$
7. $7\dfrac{5}{6}$
8. $8\dfrac{4}{9}$
9. $12\dfrac{1}{4}$
10. $13\dfrac{8}{9}$

Change each of the following to a mixed number.

11. $\dfrac{5}{2}$
12. $\dfrac{8}{3}$
13. $\dfrac{9}{2}$
14. $\dfrac{12}{5}$

15. $\dfrac{18}{7}$ **16.** $\dfrac{25}{6}$

17. $\dfrac{34}{5}$ **18.** $\dfrac{22}{3}$

19. $\dfrac{53}{4}$ **20.** $\dfrac{68}{7}$

ANSWERS

15. _____

16. _____

17. _____

18. _____

19. _____

20. _____

Rule To multiply or divide mixed numbers change the mixed numbers into improper fractions and proceed by the rule for multiplying or dividing fractions.

Example 12 Multiply: $4\dfrac{1}{2} \times 3\dfrac{1}{3}$

$$4\dfrac{1}{2} \times 3\dfrac{1}{3} = \dfrac{\overset{3}{\cancel{9}}}{\cancel{2}} \times \dfrac{\overset{5}{\cancel{10}}}{\cancel{3}} = 15$$

Example 13 Multiply: $7\dfrac{1}{8} \times 4\dfrac{1}{5}$

$$7\dfrac{1}{8} \times 4\dfrac{1}{5} = \dfrac{57}{8} \times \dfrac{21}{5} = \dfrac{1{,}197}{40} = 29\dfrac{37}{40}$$

Example 14 Multiply: $5\dfrac{1}{3} \times 3\dfrac{5}{8}$

$$5\dfrac{1}{3} \times 3\dfrac{5}{8} = \dfrac{\overset{2}{\cancel{16}}}{3} \times \dfrac{29}{\cancel{8}} = \dfrac{58}{3} = 19\dfrac{1}{3}$$

Example 15 Divide: $3\frac{1}{2} \div 2\frac{3}{4}$

$$3\frac{1}{2} \div 2\frac{3}{4} = \frac{7}{2} \div \frac{11}{4} = \frac{7}{\cancel{2}} \times \frac{\cancel{4}^{2}}{11} = \frac{14}{11} = 1\frac{3}{11}$$

Example 16 Divide: $19\frac{1}{2} \div 2$

$$19\frac{1}{2} \div 2 = \frac{39}{2} \div \frac{2}{1} = \frac{39}{2} \times \frac{1}{2} = \frac{39}{4} = 9\frac{3}{4}$$

EXERCISE SET 2-5-2

Perform the indiated operation. Give answers in mixed number form.

1. $2\frac{1}{2} \times 4\frac{1}{3}$

2. $3\frac{1}{2} \times 5\frac{1}{3}$

3. $3\frac{3}{8} \times 2\frac{1}{3}$

4. $4\frac{1}{2} \div 3\frac{1}{3}$

5. $5\frac{1}{8} \div 2\frac{3}{4}$

6. $7\frac{1}{2} \div 3\frac{5}{8}$

7. $8\dfrac{1}{3} \times 5\dfrac{2}{5}$

8. $4\dfrac{2}{5} \div 1\dfrac{4}{7}$

9. $10\dfrac{2}{3} \div 4$

10. $3\dfrac{4}{7} \times 4\dfrac{1}{5}$

11. $8 \times 3\dfrac{3}{4}$

12. $7 \div 4\dfrac{2}{3}$

13. $3\dfrac{1}{5} \times 4\dfrac{2}{7}$

14. $4\dfrac{5}{6} \times 1\dfrac{3}{5}$

15. $3\dfrac{4}{7} \times 1\dfrac{5}{8}$

16. $5 \div 1\dfrac{1}{4}$

7. _____

8. _____

9. _____

10. _____

11. _____

12. _____

13. _____

14. _____

15. _____

16. _____

17. $13\frac{1}{3} \div 3\frac{3}{4}$

18. $3\frac{3}{4} \times 2\frac{2}{5}$

19. $3\frac{3}{5} \times 4\frac{2}{3}$

20. $4\frac{3}{8} \div 5$

21. $3\frac{3}{5} \div 2\frac{4}{7}$

22. $5\frac{3}{8} \div 2\frac{4}{5}$

23. $2\frac{1}{3} \div 1\frac{3}{4} \times 3\frac{4}{5}$

24. $4\frac{2}{3} \times 3\frac{1}{4} \div 2\frac{5}{8}$

25. A man works $6\frac{1}{2}$ hours a day for five days. What is the total number of hours worked?

26. If a car averages $28\frac{2}{5}$ miles per gallon of gasoline, how far will it travel on $7\frac{1}{2}$ gallons?

27. A car traveled $76\frac{4}{5}$ miles on $2\frac{1}{4}$ gallons of gasoline. How many miles per gallon did it get?

28. What is the area of a room that is $16\frac{1}{2}$ feet long and $12\frac{2}{3}$ feet wide?

27. _____

29. If $5\frac{1}{3}$ pounds of hamburger is divided into four equal portions, what is the weight of each portion?

30. Five overnight hikers decide to divide $23\frac{3}{4}$ pounds of equipment so that each of them carries the same weight. How much is each person's share?

28. _____

29. _____

30. _____

Addition of mixed numbers again requires us to come back to the basic fact that only like quantities can be added. So we have the following rule.

Rule To add mixed numbers add the fractional parts to the fractional parts and the whole number parts to the whole number parts.

Example 17 Add: $5\frac{3}{8} + 2\frac{1}{2}$

We will add $\frac{3}{8}$ to $\frac{1}{2}$, and 5 to 2.

$$5\frac{3}{8} + 2\frac{1}{2} = 5\frac{3}{8} + 2\frac{4}{8} = 7\frac{7}{8}$$

Example 18 Add: $2\frac{1}{2} + 3\frac{5}{8} + 7\frac{1}{3}$

$$2\frac{1}{2} + 3\frac{5}{8} + 7\frac{1}{3} = 2\frac{12}{24} + 3\frac{15}{24} + 7\frac{8}{24} = 12\frac{35}{24}$$

This answer is a mixed number with an improper fraction. This is not an acceptable way to leave the answer, so we will simplify it.

$$12\frac{35}{24} = 12 + \frac{35}{24} = 12 + 1 + \frac{11}{24} = 13\frac{11}{24}$$

Example 19 Add: $1\frac{7}{8} + 9\frac{3}{4} + 6\frac{4}{5}$

$$1\frac{7}{8} + 9\frac{3}{4} + 6\frac{4}{5} = 1\frac{35}{40} + 9\frac{30}{40} + 6\frac{32}{40} = 16\frac{97}{40}$$
$$= 16 + 2 + \frac{17}{40} = 18\frac{17}{40}$$

▼ EXERCISE SET 2-5-3

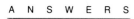

Add the following.

1. $5\frac{1}{3} + 2\frac{1}{3}$

2. $3\frac{5}{8} + 9\frac{1}{8}$

3. $2\frac{1}{4} + 3\frac{1}{2}$

4. $3\frac{1}{5} + 4\frac{2}{3}$

1. _____

2. _____

3. _____

4. _____

5. $1\frac{2}{5} + 5\frac{1}{3}$

6. $7\frac{2}{3} + 5\frac{3}{4}$

7. $5\frac{3}{8} + 6\frac{3}{4}$

8. $12\frac{1}{5} + 2\frac{7}{8}$

9. $9\frac{3}{4} + 7\frac{5}{6}$

10. $14\frac{3}{5} + 11\frac{7}{10}$

11. $2\frac{1}{2} + 1\frac{1}{3} + 5\frac{1}{8}$

12. $3\frac{1}{4} + 2\frac{1}{6} + 1\frac{1}{8}$

5. _____

6. _____

7. _____

8. _____

9. _____

10. _____

11. _____

12. _____

ANSWERS

13. $4\dfrac{2}{3} + 3\dfrac{1}{7} + 2\dfrac{2}{9}$

14. $6\dfrac{3}{4} + 4\dfrac{1}{12} + 2\dfrac{4}{9}$

13. _____

15. $3\dfrac{3}{4} + 5\dfrac{9}{10} + 7\dfrac{5}{6}$

16. $4\dfrac{7}{8} + 10\dfrac{11}{12} + 8\dfrac{1}{6}$

14. _____

15. _____

17. If $6\dfrac{2}{3}$ cups of water are poured into a container that has $8\dfrac{3}{4}$ cups of water in it, what is the total number of cups of water in the container?

18. A woman decides to carpet her living room and hallway. She needs $14\dfrac{3}{4}$ square yards for the living room and $9\dfrac{1}{2}$ square yards for the hallway. What is the total number of square yards needed for both?

16. _____

17. _____

18. _____

19. Bill drove for $5\frac{3}{4}$ hours on Monday, $6\frac{1}{3}$ hours on Tuesday, and $8\frac{1}{2}$ hours on Wednesday. What was the total number of hours that he drove for the three days?

20. A man worked $6\frac{1}{2}$ hours on Monday, $9\frac{1}{3}$ hours on Tuesday, $5\frac{2}{3}$ hours on Wednesday, $13\frac{1}{4}$ hours on Thursday, and $4\frac{3}{4}$ hours on Friday. What was the total number of hours he worked for the five days?

A N S W E R S

19. _____

20. _____

Subtraction of mixed numbers can involve the idea of borrowing that we used in dealing with whole numbers.

Example 20 Subtract: $5\frac{1}{4} - 1\frac{3}{4}$

Notice that if we attempt to subtract the fractional parts we run into the fact that we cannot take $\frac{3}{4}$ from $\frac{1}{4}$. We can write $5\frac{1}{4}$ as $5 + \frac{1}{4} = (4 + 1) + \frac{1}{4} = 4 + \left(\frac{4}{4} + \frac{1}{4}\right) = 4\frac{5}{4}$.
Then
$$5\frac{1}{4} - 1\frac{3}{4} = 4\frac{5}{4} - 1\frac{3}{4} = 3\frac{2}{4} = 3\frac{1}{2}.$$

Example 21 Subtract: $25\frac{3}{8} - 5\frac{1}{4}$

The LCD of 8 and 4 is 8.
$$25\frac{3}{8} - 5\frac{1}{4} = 25\frac{3}{8} - 5\frac{2}{8} = 20\frac{1}{8}$$

Example 22 Subtract: $15\frac{2}{3} - 4\frac{7}{8}$

$$15\frac{2}{3} - 4\frac{7}{8} = 15\frac{16}{24} - 4\frac{21}{24} = 14\frac{40}{24} - 4\frac{21}{24} = 10\frac{19}{24}$$

EXERCISE SET 2-5-4

Subtract the following.

1. $3\dfrac{3}{4} - 1\dfrac{1}{4}$

2. $4\dfrac{5}{8} - 3\dfrac{3}{8}$

3. $4\dfrac{1}{3} - 1\dfrac{2}{3}$

4. $5\dfrac{1}{5} - 2\dfrac{4}{5}$

5. $8\dfrac{1}{4} - 3\dfrac{3}{4}$

6. $12\dfrac{3}{8} - 7\dfrac{5}{8}$

7. $5\dfrac{1}{4} - 2\dfrac{1}{8}$

8. $7\dfrac{1}{2} - 5\dfrac{1}{4}$

ANSWERS

1. _____
2. _____
3. _____
4. _____
5. _____
6. _____
7. _____
8. _____

2-5 MIXED NUMBERS   **149**

9. $6\dfrac{3}{4} - 5\dfrac{2}{3}$

10. $4\dfrac{5}{8} - 3\dfrac{4}{5}$

9. _____

11. $16\dfrac{3}{7} - 5\dfrac{4}{5}$

12. $14\dfrac{5}{8} - 6\dfrac{7}{9}$

10. _____

11. _____

13. $5 - 3\dfrac{2}{3}$

14. $9 - 6\dfrac{3}{4}$

12. _____

13. _____

15. $18\dfrac{5}{12} - 9\dfrac{5}{6}$

16. $24\dfrac{3}{8} - 17\dfrac{7}{12}$

14. _____

15. _____

16. _____

17. A container has $30\frac{1}{4}$ ounces of water in it. If $7\frac{2}{3}$ ounces are poured out, how many ounces remain?

18. If a certain trip takes $8\frac{1}{3}$ hours, how much time remains after you have traveled $6\frac{3}{4}$ hours?

17. _____

19. How much yard goods remain after $3\frac{3}{4}$ yards are cut from a bolt containing $11\frac{1}{2}$ yards?

20. A man owned $82\frac{1}{2}$ acres of land. If he sold $24\frac{2}{3}$ acres, how many acres does he have left?

18. _____

19. _____

21. A woman worked $8\frac{1}{2}$ hours on Monday, $6\frac{3}{4}$ hours on Tuesday, and $10\frac{1}{3}$ hours on Wednesday. How many more hours must she work to total 40 hours for the week?

22. A spool of ribbon contains 100 yards. $23\frac{1}{2}$ yards are used by one person. Another person takes enough to make five bows that require $2\frac{1}{3}$ yards for each bow. How many yards of ribbon remain on the spool?

20. _____

21. _____

22. _____

CHAPTER 2 SUMMARY

Fractions

- A common fraction is the indicated quotient of two whole numbers.
- Fractions are classified as proper and improper.

Simplification

- The fundamental principle of fractions is used to simplify fractions. A fraction is in simplified form when the numerator and denominator have no factor in common except the number 1.
- Reduce a fraction by factoring the numerator and denominator into primes and dividing out all common factors.

Multiplication

- To multiply two or more fractions multiply the numerators to obtain the numerator of the product, and multiply the denominators to obtain the denominator of the product. Common factors should be divided out before we multiply. The product should always be given in reduced form.
- If the product of two numbers is 1, then the numbers are reciprocals of each other.

Division

- To divide two fractions we invert the divisor and multiply.

Addition and Subtraction

- To add (subtract) like fractions add (subtract) the numerators and place the result over the common denominator.
- To add or subtract unlike fractions find the least common denominator (LCD) and change each fraction to an equivalent fraction with that denominator. Then apply the rule for adding or subtracting like fractions.

Mixed Numbers

- When multiplying or dividing mixed numbers, change them to improper fractions and use the rule for multiplying or dividing fractions.
- To add (subtract) mixed numbers add (subtract) the fractional parts to the fractional parts and the whole number parts to the whole number parts. In subtracting mixed numbers it may be necessary to borrow.

CHAPTER 2 REVIEW

ANSWERS

Reduce the fractions in problems 1–8.

1. $\dfrac{12}{28}$
2. $\dfrac{8}{20}$
3. $\dfrac{18}{48}$
4. $\dfrac{45}{105}$

5. $\dfrac{54}{63}$
6. $\dfrac{60}{700}$
7. $\dfrac{36}{126}$
8. $\dfrac{189}{198}$

Find the missing numerator in problems 9–12.

9. $\dfrac{3}{4} = \dfrac{?}{20}$
10. $\dfrac{4}{5} = \dfrac{?}{60}$

11. $\dfrac{11}{12} = \dfrac{?}{72}$
12. $\dfrac{7}{18} = \dfrac{?}{126}$

13. Change $9\dfrac{6}{7}$ to an improper fraction.
14. Change $11\dfrac{3}{8}$ to an improper fraction.

15. Change $\dfrac{61}{8}$ to a mixed number.
16. Change $\dfrac{45}{4}$ to a mixed number.

152

Perform the indicated operation in problems 17–86. Give all answers in reduced form. If answers are improper fractions, give them as mixed numbers.

17. $\dfrac{3}{4} \times \dfrac{10}{27}$

18. $\dfrac{4}{5} \times \dfrac{15}{16}$

19. $\dfrac{3}{5} \div \dfrac{9}{10}$

20. $\dfrac{4}{9} \div \dfrac{16}{27}$

21. $\dfrac{3}{4} + \dfrac{1}{6}$

22. $\dfrac{1}{9} + \dfrac{5}{6}$

23. $\dfrac{5}{8} - \dfrac{3}{8}$

24. $\dfrac{7}{10} - \dfrac{3}{10}$

25. $\dfrac{5}{13} + \dfrac{7}{13}$

26. $\dfrac{4}{11} + \dfrac{6}{11}$

ANSWERS

17. _____

18. _____

19. _____

20. _____

21. _____

22. _____

23. _____

24. _____

25. _____

26. _____

27. _____

28. _____

29. _____

30. _____

31. _____

32. _____

33. _____

34. _____

35. _____

36. _____

37. _____

38. _____

27. $\dfrac{5}{9} \times \dfrac{18}{35}$

28. $\dfrac{2}{3} \times \dfrac{9}{10}$

29. $\dfrac{3}{8} \div 9$

30. $\dfrac{4}{7} \div 10$

31. $\dfrac{4}{5} \times \dfrac{13}{16}$

32. $\dfrac{5}{6} \times \dfrac{18}{23}$

33. $\dfrac{7}{10} + \dfrac{9}{10}$

34. $\dfrac{7}{8} + \dfrac{5}{8}$

35. $\dfrac{5}{7} - \dfrac{2}{3}$

36. $\dfrac{5}{6} - \dfrac{3}{5}$

37. $\dfrac{2}{3} + \dfrac{5}{6}$

38. $\dfrac{5}{7} + \dfrac{9}{14}$

39. $\dfrac{5}{6} \times 24$

40. $\dfrac{3}{8} \times 32$

41. $\dfrac{7}{8} \div \dfrac{3}{16}$

42. $\dfrac{3}{5} \div \dfrac{6}{7}$

43. $4\dfrac{5}{12} - 2\dfrac{3}{16}$

44. $5\dfrac{1}{6} - 3\dfrac{7}{15}$

45. $\dfrac{5}{12} + \dfrac{7}{18}$

46. $\dfrac{4}{9} + \dfrac{9}{12}$

47. $7 \times \dfrac{13}{14}$

48. $5 \times \dfrac{9}{10}$

49. $4 \div 2\dfrac{2}{3}$

50. $7 \div 4\dfrac{2}{3}$

39. _____

40. _____

41. _____

42. _____

43. _____

44. _____

45. _____

46. _____

47. _____

48. _____

49. _____

50. _____

51. $3\frac{5}{6} + 2\frac{2}{3}$

52. $5\frac{1}{3} + 3\frac{3}{4}$

53. $\frac{7}{9} - \frac{5}{12}$

54. $\frac{7}{10} - \frac{8}{15}$

55. $2\frac{1}{2} \times 4$

56. $3\frac{1}{8} \times 5$

57. $6 \div \frac{2}{3}$

58. $12 \div \frac{3}{4}$

59. $\frac{5}{16} + \frac{7}{10}$

60. $\frac{8}{9} + \frac{11}{12}$

61. $9\frac{1}{6} - 2\frac{5}{8}$

62. $4\frac{3}{8} - 1\frac{5}{12}$

CHAPTER 2 REVIEW

63. $1\frac{4}{5} \times 2\frac{2}{3}$

64. $4\frac{1}{2} \times 3\frac{3}{5}$

63. _____

65. $\frac{5}{9} \div \frac{3}{5}$

66. $\frac{3}{4} \div \frac{8}{15}$

64. _____

65. _____

67. $\frac{1}{4} + \frac{3}{5} + \frac{7}{10}$

68. $\frac{3}{8} + \frac{7}{12} + \frac{5}{16}$

66. _____

67. _____

69. $3\frac{5}{12} - 2\frac{5}{8}$

70. $7\frac{2}{9} - 6\frac{1}{12}$

68. _____

69. _____

71. $4\frac{3}{8} \div \frac{21}{26}$

72. $3\frac{2}{3} \div \frac{22}{27}$

70. _____

71. _____

73. $\frac{7}{9} \times \frac{1}{2} \times \frac{4}{21}$

74. $\frac{3}{8} \times \frac{4}{9} \times \frac{3}{5}$

72. _____

73. _____

74. _____

75. $7 - \dfrac{5}{8}$

76. $9 - \dfrac{3}{5}$

77. $2\dfrac{1}{8} + 1\dfrac{5}{6} + 5\dfrac{4}{9}$

78. $3\dfrac{3}{4} + 2\dfrac{7}{18} + 1\dfrac{5}{24}$

79. $6\dfrac{1}{2} \div 3\dfrac{1}{4}$

80. $5\dfrac{2}{3} \div 4\dfrac{1}{4}$

81. $8\dfrac{5}{14} - 1\dfrac{8}{21}$

82. $4\dfrac{1}{16} - 1\dfrac{1}{18}$

83. $2\dfrac{1}{2} \times 4\dfrac{1}{6} \times 1\dfrac{1}{5}$

84. $5\dfrac{1}{3} \times 3\dfrac{1}{4} \times 2\dfrac{5}{8}$

75. _____

76. _____

77. _____

78. _____

79. _____

80. _____

81. _____

82. _____

83. _____

84. _____

85. $1\dfrac{6}{15} \div \dfrac{7}{18}$

86. $2\dfrac{11}{12} \div 5\dfrac{1}{4}$

87. A family spends $\dfrac{1}{4}$ of its income for food. If the weekly income is $600, how much is spent for food per week?

88. A girl pours 24 pails of water into an empty tank. If the pail holds $\dfrac{2}{3}$ gallon, how many gallons are in the tank?

89. A pump can deliver $\dfrac{3}{8}$ liter of water per second. How long will it take to pump 48 liters?

90. How many cartons of milk, each containing $\dfrac{1}{8}$ gallon, can be filled from 20 gallons?

91. A man works $8\dfrac{2}{3}$ hours a day for five days. What is the total number of hours worked?

92. Jean drove $7\dfrac{2}{3}$ hours on Monday, $8\dfrac{1}{2}$ hours on Tuesday, and $4\dfrac{1}{4}$ hours on Wednesday. What was the total number of hours she drove for the three days?

ANSWERS

85. _____

86. _____

87. _____

88. _____

89. _____

90. _____

91. _____

92. _____

93. A piece of ribbon $5\frac{1}{3}$ yards long is cut from a spool containing $20\frac{3}{4}$ yards. How many yards remain on the spool?

94. If $2\frac{3}{4}$ liters of water are drained out of a car radiator containing $6\frac{1}{2}$ liters, how many liters are left in the radiator?

95. Harness horse A paced the first quarter-mile in $29\frac{2}{5}$ seconds. Horse B paced it in $28\frac{4}{5}$ seconds. How much faster was horse B?

96. A certain stock was selling in the morning at $39\frac{1}{2}$. In the afternoon it was selling at $42\frac{3}{8}$. How much had it gained?

93. _____

94. _____

95. _____

96. _____

SCORE: _____

CHAPTER 2 TEST

NAME: _____
CLASS / SECTION: _____   DATE: _____

1. Reduce $\dfrac{54}{90}$.

2. Change $11\dfrac{3}{5}$ to an improper fraction.

3. Find the missing numerator: $\dfrac{5}{19} = \dfrac{?}{57}$

4. Find the LCD of $\dfrac{2}{3}, \dfrac{5}{8}, \dfrac{3}{4}$.

Perform the indicated operation. Give all answers in reduced form.

5. $\dfrac{2}{3} \times \dfrac{1}{7}$

6. $\dfrac{3}{5} \div \dfrac{10}{11}$

7. $\dfrac{3}{13} + \dfrac{8}{13}$

8. $\dfrac{5}{9} - \dfrac{2}{9}$

9. $5\dfrac{1}{3} \times 2\dfrac{1}{4}$

10. $\dfrac{2}{5} + \dfrac{4}{9}$

11. $\dfrac{15}{16} \div 10$

12. $\dfrac{5}{6} - \dfrac{4}{7}$

ANSWERS

1. _____
2. _____
3. _____
4. _____
5. _____
6. _____
7. _____
8. _____
9. _____
10. _____
11. _____
12. _____

ANSWERS

13. $\dfrac{11}{12} + \dfrac{7}{18} + \dfrac{2}{3}$

14. $\dfrac{2}{9} \times \dfrac{3}{5}$

13. _____

14. _____

15. $3\dfrac{2}{3} - 1\dfrac{1}{8}$

16. $\dfrac{6}{35} \div \dfrac{9}{14}$

15. _____

16. _____

17. $\dfrac{5}{8} \times \dfrac{12}{25}$

18. $4\dfrac{5}{6} + 5\dfrac{7}{8} + 2\dfrac{11}{12}$

17. _____

18. _____

19. $4\dfrac{2}{3} \div 5\dfrac{3}{5}$

20. $21\dfrac{1}{6} - 4\dfrac{5}{9}$

19. _____

20. _____

SCORE: _____

CHAPTERS 1-2 CUMULATIVE TEST

1. Write the number 5,309 in words.

2. Write the number 2,543 in expanded form.

3. Round 16,752 to the nearest hundred.

4. Add: 143
         651
         407

5. Subtract: 1,603
           −   927

6. Multiply: 258
           ×   36

7. Divide: $32\overline{)4{,}156}$

8. Evaluate: $3[2(8-5)-4] - 2[25 - 3(7+1)]$

9. Find the prime factorization of 252.

10. Find the greatest common factor of 180 and 450.

11. Reduce $\frac{30}{78}$.

12. Find the LCD of $\frac{1}{8}, \frac{5}{16}, \frac{7}{24}$.

Perform the indicated operations. Give all answers in reduced form.

13. $\frac{2}{3} \times \frac{1}{8}$

14. $\frac{2}{3} \div \frac{3}{4}$

ANSWERS

1. _____
2. _____
3. _____
4. _____
5. _____
6. _____
7. _____
8. _____
9. _____
10. _____
11. _____
12. _____
13. _____
14. _____

163

15. $\dfrac{2}{3} - \dfrac{1}{2}$

16. $\dfrac{2}{5} + \dfrac{3}{10}$

15. _____

16. _____

17. $2\dfrac{1}{3} \times 3\dfrac{3}{5}$

18. $\dfrac{11}{12} + \dfrac{3}{8} - \dfrac{5}{6}$

17. _____

18. _____

19. $5\dfrac{1}{4} - 2\dfrac{3}{8}$

20. $5\dfrac{1}{4} \div 1\dfrac{2}{5}$

19. _____

20. _____

SCORE: _____

CHAPTER 3 PRETEST

NAME: _____
CLASS / SECTION: _____ DATE: _____

ANSWERS

Answer as many of the following problems as you can before starting this chapter. When you finish the chapter, take the test at the end and compare your scores to see how much you have learned.

1. _____
2. _____
3. _____
4. _____
5. _____
6. _____
7. _____
8. _____
9. _____
10. _____
11. _____
12. _____
13. _____
14. _____
15. _____
16. _____
17. _____
18. _____
19. _____
20. _____
21. _____
22. _____
23. _____
24. _____

SCORE: _____

1. Write 38.752 in expanded form.

2. Write 406.98 in words.

3. Round 5.6937 to the nearest hundredth.

4. Change .075 to a common fraction in simplest form.

5. Change $6\frac{3}{8}$ to a decimal rounded to three decimal places.

Perform the indicated operation. In division problems round the quotient to three decimal places.

6. $\;\;7.48$
   $+16.27$

7. $\;\;23.142$
   $-\;\;8.355$

8. $7\overline{)25.2}$

9. $.5\overline{)5.67}$

10. $\;\;3.16$
    $\times\;\;.5$

11. $\;\;6.321$
    $\;\;15.4$
    $+\;\;7.589$

12. $\;\;32.24$
    -17.955

13. $15\overline{)183.4}$

14. $.6\overline{)6.876}$

15. $\;\;135.8$
    $\times\;\;1.07$

16. $\;\;61.05$
    $\;\;417.876$
    $+\;\;23.144$

17. $\;\;40.003$
    -16.454

18. $24\overline{)121.44}$

19. $\;\;34.04$
    $\times\;\;.067$

20. $11.2\overline{).0784}$

21. Write 5.14×10^{-5} as a decimal number.

22. Write 45,200,000,000 in scientific notation.

23. Simplify: $\sqrt{18}$

24. Simplify: $\sqrt{176}$

CHAPTER 3

Decimals, Scientific Notation, and Square Roots

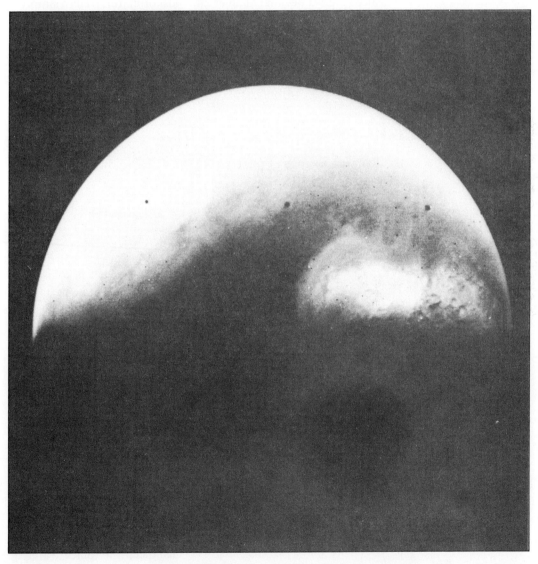

The planet Mars is approximately 49,000,000 miles from Earth. Express this distance in scientific notation.

CHAPTER 3 DECIMALS, SCIENTIFIC NOTATION, AND SQUARE ROOTS

In this chapter we are going to study a set of numbers called *decimals*. Concepts established in both chapters 1 and 2 form a foundation for this new set of numbers. In chapter 1 we noted that our number system is a base 10 system. In expanded notation we saw that each place value is ten times the place value to its right. That same idea will be extended to include the set of decimal numbers. Also, in chapter 2, we dealt with fractions. Since decimals are another way of expressing fractions, many of the same concepts will apply.

3-1 EXPANDED FORM AND READING AND WRITING DECIMALS

OBJECTIVES

In this section you will learn to:
1. Write decimals in expanded notation.
2. Write a decimal in words and vice versa.

If a **decimal point** precedes a digit or set of digits, a fraction is indicated. Thus a number containing a decimal point is often referred to as a **decimal fraction**. It is also called a **decimal number** and most often referred to simply as a **decimal**. Each place after the decimal point has a specific value.

Example 1 .7 is read as "seven tenths." (This could be written as the common fraction $\frac{7}{10}$.)

Example 2 .65 is read as "sixty-five hundredths." (This could be written as the common fraction $\frac{65}{100}$.)

Example 3 3.6 is a mixed number and is read as "three and six tenths." (This could be written as $3\frac{6}{10}$.)

The fact that each place value is ten times the place value to its right is still true in decimal notation. Observe the following names of some place values.

Hundreds	Tens	Ones		Tenths	Hundredths	Thousandths
10 × 10	10	1	.	$\frac{1}{10}$	$\frac{1}{10 \times 10}$	$\frac{1}{10 \times 10 \times 10}$

Notice that each place value when multiplied by ten gives the place value of the next place to the left. For instance, the place value $\frac{1}{10 \times 10}$ or $\frac{1}{100}$, if multiplied by 10, gives $10 \times \frac{1}{100} = \frac{1}{10}$, which is the next place value to the left. The expanded form of a decimal number or mixed number in decimal form uses this fact.

Example 4 Write 7.35 in expanded form.

$$7.35 = 7 \times 1 + 3 \times \frac{1}{10} + 5 \times \frac{1}{100}$$

Example 5 Write 25.301 in expanded form.

$$25.301 = 2 \times 10 + 5 \times 1 + 3 \times \frac{1}{10} + 0 \times \frac{1}{100} + 1 \times \frac{1}{1,000}$$

In reading or writing decimal notation we must remember that the word *and* is used only for the decimal point. Also, we must remember that the names of places to the right of the decimal point represent fractions and always must end in "ths."

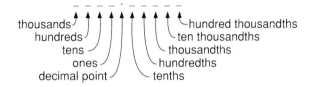

Example 6 27.36 is read as "twenty-seven and thirty-six hundredths."

Example 7 .003 is read as "three thousandths."

Example 8 .0125 is read as "one hundred twenty-five ten thousandths."

Example 9 200.017 is read as "two hundred and seventeen thousandths."

Example 10 Express "twenty-seven and fifteen hundredths" in decimal notation.

27 is a whole number so it is placed before the decimal point. The 5 in 15 must be placed in the hundredths place. The decimal form is 27.15.

Example 11 Express "two hundred seven thousandths" in decimal notation.

The 7 in 207 must be placed in the thousandths place. Therefore, the decimal form is .207.

Example 12 Express "seventeen thousandths" in decimal notation.

The 7 in 17 must be placed in the thousandths place. Thus the decimal form is .017.

EXERCISE SET 3-1-1

ANSWERS

Write the following decimal numbers in expanded form.

1. .3
2. .24
3. .04
4. .135
5. .309
6. .004
7. .1053
8. 6.39
9. 38.124
10. 104.1005

Write the following decimal numbers in words.

11. .6
12. .32
13. .129
14. .04
15. .802
16. 3.5
17. 27.03
18. 41.276
19. 16.0004
20. 100.0105

Express the following in decimal notation.

21. Four tenths

22. One hundred five thousandths

23. Three hundred sixty-one thousandths

24. Twenty-one and seven hundredths

25. One hundred twenty-five and two hundred four thousandths

26. Five hundred three and seventy-five hundredths

27. One hundred and seven thousandths

28. Three hundred and five ten thousandths

29. If you wrote a check for $105.73, how would you write it in words?

30. If you wrote a check for $1,004.06, how would you write it in words?

ANSWERS

21. _____
22. _____
23. _____
24. _____
25. _____
26. _____
27. _____
28. _____
29. _____
30. _____

3-2 ROUNDING DECIMALS

In chapter 1 we established a rule for rounding whole numbers. That same rule applies to rounding decimals to any desired place. We will repeat that rule here for your convenience.

Rule To round a number to any place value if the number in the next place to the right is 5 or more, the number in the desired place value is increased by one. If it is less than 5, the number in the desired place remains the same. All numbers to the right of the desired place are changed to zero.

OBJECTIVES

In this section you will learn to round a given number to any desired place.

Example 1 Round 2.3764 to the nearest hundredth.

Since 6 is in the next place to the right of the hundredths place, 2.3764 rounds to 2.38 to the nearest hundredth. It is understood that all numbers to the right of 8 are zero and we do not write them.

Example 2 Round 2.3764 to the nearest tenth.

Since 7 is in the next place to the right of the tenths place, 2.3764 rounds to 2.4 to the nearest tenth.

Example 3 Round 52.3764 to the nearest whole number.

Since 3 is in the next place to the right of the ones place, 52.3764 rounds to 52 to the nearest whole number.

Example 4 Round 13.15972 to the nearest thousandth.

Since 7 is in the next place to the right of the thousandths place, 13.15972 rounds to 13.160 to the nearest thousandth. In this instance we need to write the zero after the 6 to indicate that we have rounded to the nearest thousandth.

▼ EXERCISE SET 3-2-1

ANSWERS

1. Round .4813 to the nearest tenth.
2. Round .4813 to the nearest hundredth.
3. Round .4813 to the nearest whole number.
4. Round .4813 to the nearest thousandth.
5. Round .8372 to the nearest whole number.
6. Round .8372 to the nearest thousandth.
7. Round .8372 to the nearest tenth.
8. Round .8372 to the nearest hundredth.
9. Round .5163 to the nearest thousandth.
10. Round .5163 to the nearest tenth.

11. Round .5163 to the nearest hundredth.

12. Round .5163 to the nearest whole number.

13. Round 9.0539 to the nearest whole number.

14. Round 9.0539 to the nearest thousandth.

15. Round 9.0539 to the nearest tenth.

16. Round 9.0539 to the nearest hundredth.

17. Round 24.5534 to the nearest tenth.

18. Round 24.5534 to the nearest hundredth.

19. Round 24.5534 to the nearest whole number.

20. Round 24.5534 to the nearest thousandth.

21. Round 61.9605 to the nearest whole number.

22. Round 61.9605 to the nearest thousandth.

23. Round 61.9605 to the nearest tenth.

24. Round 61.9605 to the nearest hundredth.

ANSWERS

11. _____
12. _____
13. _____
14. _____
15. _____
16. _____
17. _____
18. _____
19. _____
20. _____
21. _____
22. _____
23. _____
24. _____

3–3 ADDING AND SUBTRACTING DECIMAL NUMBERS

We refer again to the basic fact that only like quantities can be added or subtracted. Since this is true, we must add tenths to tenths, hundredths to hundredths, and so on. This can be accomplished by following this simple rule.

OBJECTIVES

In this section you will learn to add and subtract decimal numbers.

> **Rule** When adding or subtracting decimal numbers in column form, the decimal points must be placed in the same column.

Example 1 Find the sum of 2.35; 7.01; 16.352; 7.5216; and 6.2.

Placing these numbers in column form, we have

$$\begin{array}{r} 2.35 \\ 7.01 \\ 16.352 \\ 7.5216 \\ \underline{6.2} \\ 39.4336. \end{array}$$

Add in the usual manner. Notice we are keeping tenths under tenths, hundredths under hundredths, and so on. Thus keeping the decimal points in the same column is like getting a common denominator.

If it helps to keep columns of numbers straight, zeros can be written in the missing places.

Example 2 Find the sum of 16.35; 14.003; and 12.1.

Placing these numbers in column form, we have

$$\begin{array}{r} 16.35 \\ 14.003 \\ \underline{12.1} \\ 42.453 \end{array} \quad \text{or} \quad \begin{array}{r} 16.350 \\ 14.003 \\ \underline{12.100} \\ 42.453. \end{array}$$

Decimal numbers may be added or subtracted without placing them in column form, but it is not easy to make sure that only like place values are combined. So it is strongly suggested that the numbers be placed in columns.

Example 3 Find $257.38 - 41.62$.

$$\begin{array}{r} 257.38 \\ -41.62 \\ \hline 215.76 \end{array}$$

Example 4 Find $15.05 - 2.634$.

Since we are subtracting a number having three decimal places, the minuend must also have three decimal places. We therefore write 15.05 as 15.050.

$$\begin{array}{r} 15.050 \\ -2.634 \\ \hline 12.416 \end{array}$$

The most common use of decimals is one we encounter almost every day: money. $3.75 is read as "three and seventy-five hundredths dollars" or "three dollars and seventy-five cents." $.37 is read as "thirty-seven hundredths dollars" or "thirty-seven cents."

3-3 ADDING AND SUBTRACTING DECIMAL NUMBERS

Example 5 Jim bought a pair of shoes for $37.95, a shirt for $14.25, and pants for $32.50 (all taxes included). He gave the salesperson a one hundred dollar bill. How much change should he get?

We first add the prices of the individual items to get the total cost of his purchase.

$$\begin{array}{r} \$37.95 \\ 14.25 \\ \underline{32.50} \\ \$84.70 \end{array}$$

Then we subtract the total cost from $100.00 to obtain the amount of change he should receive.

$$\begin{array}{r} \$100.00 \\ -\underline{84.70} \\ \$15.30 \end{array}$$

▼ EXERCISE SET 3-3-1

Add.

1. 4.13	**2.** 8.07	**3.** 6.27	**4.** 9.63	**5.** 2.143
2.765	2.98	13.085	3.37	4.531
				1.214

6. 2.17	**7.** 2.053	**8.** 18.3	**9.** 14.826	**10.** 4.0068
.6343	11.67	2.62	7.0154	17.365
4.516	7.46	27.398	13.27	6.406
			35.8	14.95

Place in column form and add.

11. 21.67 + 3.75 + 2.61

12. 15.274 + 11.37 + 21.6

13. 6.38 + 24.2015 + 13.98

14. 28.076 + 18.3 + 7.65

15. 41.6725 + 16.087 + 33.47 + 12.09

ANSWERS

1. _____
2. _____
3. _____
4. _____
5. _____
6. _____
7. _____
8. _____
9. _____
10. _____
11. _____
12. _____
13. _____
14. _____
15. _____

16. $25.8074 + 21.695 + 7.23 + 10.09$

Subtract.

17. 16.75 − 4.62	**18.** 12.32 − 9.45	**19.** 21.04 − 6.21	**20.** 18.53 −12.47
21. 16.427 −11.348	**22.** 13.214 − 9.382	**23.** 21.374 −14.536	**24.** 7.0034 −4.2605
25. 12.09 − 8.352	**26.** 17.34 −11.005		

Place in column form and subtract.

27. $34.685 - 9.205$ **28.** $18.724 - 13.26$ **29.** $12.1 - 6.34$

30. $35.03 - 13.4761$ **31.** $12 - 9.704$ **32.** $7 - 5.632$

33. If you have $46.04 in your bank account and deposit a check for $25.96, what is the new balance?

34. If you bought a book for $12.95 and a calculator for $29.75, what is the total cost for both?

35. Mary bought a book for $4.95. What is the change from a twenty dollar bill?

36. Ellen's checkbook balance was $73.25. She wrote a check for $5.63. What is her new balance?

3-4 MULTIPLYING DECIMAL NUMBERS   177

ANSWERS

37. Jim went grocery shopping and bought a roast for $17.63, milk for $2.27, bread for $1.39, and chicken for $5.49. What change did he receive from $40?

38. Sally received four checks for $23.58, $6.45, $15.25, and $16.03. She deposits all four checks in her bank account. What is the total deposit?

37. _____

39. The regular price of a sport coat is $129.00. It is on sale at a discount of $25.80. What is the sale price?

40. You are given a discount of $4.95 on an item costing $21.50. How much change will you receive from $20?

38. _____

39. _____

41. A man's bank balance is $283.42. He writes checks for $27.95 and $31.08. He also receives a check for $38.27 that he deposits in his account. What is his new balance?

42. A woman's grocery bill is $63.04. She has discount coupons for $3.50, $.65, and $2.25. How much change will she receive from $100.00?

40. _____

41. _____

42. _____

3-4 MULTIPLYING DECIMAL NUMBERS

The next operation with decimal numbers that we will examine is multiplication. Of course, the rules for this operation will differ from the rules for addition and subtraction. The first thing we notice is that in multiplication the decimal points do not have to be kept in the same column. We will look at some examples to establish the rule needed for multiplication.

OBJECTIVES

In this section you will learn to multiply any two decimal numbers.

Example 1   Multiply: 3.75 × 2.6

We first write the problem in column form as if no decimal point existed.

$$\begin{array}{r} 3.75 \\ \times\ 2.6 \\ \hline 2250 \\ 750 \\ \hline 9750 \end{array}$$

Now the question is, "Where do we place the decimal point?" To answer this let's take a closer look at the problem. We have 3.75, which is three and seventy-five hundredths. In the last chapter this would have been written as $3\frac{75}{100}$. Also, we have 2.6, or two and six tenths, which we can write as $2\frac{6}{10}$. Now if we look at $3\frac{75}{100} \times 2\frac{6}{10}$ as our problem, we can work it using the rules from the preceding chapter.

$$3\frac{75}{100} \times 2\frac{6}{10} = \frac{375}{100} \times \frac{26}{10}$$
$$= \frac{375 \times 26}{100 \times 10} = \frac{9{,}750}{1{,}000} = 9\frac{750}{1{,}000}$$

In decimal form this would be 9.750.

If we look at the original numbers being multiplied, we see hundredths (3.75) being multiplied by tenths (2.6), so it is reasonable that the product should be thousandths.

If we multiply tenths by tenths, we get hundredths. If we multiply hundredths by hundredths, we get ten thousandths, and so on.

Example 2   Multiply: $.6 \times .7$

$$.6 \times .7 = \frac{6}{10} \times \frac{7}{10} = \frac{42}{100} = .42$$

Example 3   Multiply: $.24 \times .12$

$$.24 \times .12 = \frac{24}{100} \times \frac{12}{100} = \frac{288}{10{,}000} = .0288$$

Rule   When multiplying decimal numbers, the number of places to the right of the decimal point in the product must equal the total number of places to the right of the decimal point in both numbers to be multiplied.

Example 4   Multiply: 4.38×7.1

4.38 has two decimal places and 7.1 has one. Therefore the product must have three.

```
     4.38
  ×  7.1
     438
   30 66
   31.098
```

3-4 MULTIPLYING DECIMAL NUMBERS

Example 5 Multiply: .761 × .32

The product must have 3 + 2 = 5 decimal places.

$$\begin{array}{r}.761\\ \times\ .32\\ \hline 1522\\ 2283\ \\ \hline .24352\end{array}$$

Example 6 Multiply: .73 × .045

Again the product must have five decimal places.

$$\begin{array}{r}.73\\ \times .045\\ \hline 365\\ 292\ \\ \hline .03285\end{array}$$

Notice here that to have five places to the right of the decimal point it was necessary to add a zero.

Example 7 If chocolate bars cost $.42 each, what is the cost of eight of them?

$$\begin{array}{r}\$\ .42\\ \times\ \ 8\\ \hline \$3.36\end{array}$$

Notice in example 7 that a whole number times a decimal number still uses the same rule of the total number of places to the right of the decimal point. The decimal point can be considered to be to the right of the whole number. Thus 8 could be written as 8.0.

▼ EXERCISE SET 3-4-1

Multiply.

1. 3.2 × .4
2. 8.5 × .2
3. .8 × .4
4. .24 × .03
5. 1.32 × .7

6. 4.34 × 2.6
7. 7.05 × .24
8. 12.62 × .07
9. 31.6 × .412
10. 3.172 × .43

ANSWERS

1. _____
2. _____
3. _____
4. _____
5. _____
6. _____
7. _____
8. _____
9. _____
10. _____

ANSWERS

11. _____

12. _____

13. _____

14. _____

15. _____

16. _____

17. _____

18. _____

19. _____

20. _____

21. _____

22. _____

23. _____

24. _____

25. _____

26. _____

27. _____

28. _____

29. _____

30. _____

31. _____

32. _____

33. _____

34. _____

11. 5.27
    ×   9

12. .028
    × .13

13. .124
    × .36

14. .004
    × .09

15. 3.146
    ×   10

16. 7.139
    × 100

17. 13.217
    × 1,000

18. 14.62
    ×   .1

19. 68.13
    × .01

20. 45.62
    × .001

21. Find the product of 2.41 and 3.7.

22. Find the product of 3.29 and 2.6.

23. Find the product of 25.14 and .28.

24. Find the product of 12.89 and .35.

25. Find the product of 3.14 and 6.12.

26. Find the product of 4.27 and 3.18.

27. Find the product of .305 and .24.

28. Find the product of .704 and .45.

29. Find the product of 17.138 and 13.

30. Find the product of 15.231 and 25.

31. Find the product of 12.005 and 2.06.

32. Find the product of 21.015 and 4.02.

33. What is the cost of six pencils if they are priced at $.55 each?

34. If the price of a candy bar is $.45, find the cost of seven of them.

35. Find the cost of five cans of coffee if the price per can is $3.58.

36. What is the cost of twelve melons if they are priced at $1.29 each?

37. If steak is priced at $5.98 per pound, what is the cost of a steak that weighs 1.74 pounds? (Round answer to the nearest cent.)

38. What is the cost of 3.28 pounds of hamburger if the price is $2.99 per pound? (Round answer to two decimal places.)

39. What is the area of a room that is 10.25 feet wide and 15.5 feet long? (Round answer to two decimal places.)

40. Find the area of a lot 100.8 feet by 75.6 feet. (Round answer to one decimal place.)

41. If ten gallons of gasoline were purchased at 95.9 cents per gallon, how much change was received from a twenty dollar bill? (Be careful with the decimal point!)

42. A shopper bought five pounds of ground beef at $1.99 per pound and three cans of green beans at $.75 per can. How much change was received from $20?

ANSWERS

35. _____

36. _____

37. _____

38. _____

39. _____

40. _____

41. _____

42. _____

A short method of multiplying a decimal number by a power of ten is given below.

> **Rule** To multiply a decimal number by a power of ten move the decimal point to the *right* a number of places equal to the number of zeros in the power of ten.

Example 8 Find the product of 3.142(100).

We are multiplying by 100, so we move the decimal point *two* places to the right.

$$3.142(100) = 314.2$$

Example 9 Find the product of 12.63(1,000).

We move the decimal point *three* places to the right.

$$12.63(1,000) = 12,630$$

▼ EXERCISE SET 3-4-2

A N S W E R S

Find the following products.

1. 3.5(10)
2. 16.4(10)
3. 16.13(10)

4. 12.531(10)
5. 2.417(100)
6. 8.261(100)

7. 301.2(100)
8. 72.5(100)
9. 4.21(1,000)

10. 9.16(1,000)
11. 8.259(10)
12. 5.117(10)

13. .319(100)
14. .173(100)
15. .041(100)

16. .061(100)
17. 21.351(10,000)
18. 6.492(10,000)

19. .014(10)
20. .0003(100)

3-5 DIVIDING DECIMAL NUMBERS BY WHOLE NUMBERS

The operation of division will basically follow the long-division algorithm for whole numbers. We will first divide decimal numbers by whole numbers.

OBJECTIVES

In this section you will learn to:
1. Divide a decimal number by a whole number.
2. Round the quotient to any desired place.
3. Find the average of a set of numbers.

Example 1 Divide .36 by 4.

.36 (thirty-six hundredths) can be written as the common fraction $\frac{36}{100}$. If we divide this by 4, we have

$$\frac{36}{100} \div 4 = \frac{\overset{9}{\cancel{36}}}{100} \times \frac{1}{\cancel{4}} = \frac{9}{100}.$$

So $.36 \div 4 = .09$.

Writing this in long-division form, we have

$$4\overline{).36.}\overset{.09}{}$$

with .09 labeled as quotient, .36 as dividend, and 4 as divisor.

Example 2 Divide .7851 by 3.

This problem written using common fractions is

$$\frac{7,851}{10,000} \div 3 = \frac{\overset{2,617}{\cancel{7,851}}}{10,000} \times \frac{1}{\cancel{3}} = \frac{2,617}{10,000}.$$

So $.7851 \div 3 = .2617$.

If we write the division in long-division form, we have

$$3\overline{).7851.}\overset{.2617}{}$$

Rule To divide a decimal number by a whole number use the long-division algorithm and place the decimal in the quotient directly above the decimal in the dividend.

Example 3 Divide 7.435 by 5 using long division.

$$
\begin{array}{r}
1.487 \\
5{\overline{\smash{\big)}\,7.435}} \\
\underline{5} \\
24 \\
\underline{20} \\
43 \\
\underline{40} \\
35 \\
\underline{35} \\
0
\end{array}
$$

Example 4 Divide .00384 by 4.

$$
\begin{array}{r}
.00096 \\
4{\overline{\smash{\big)}\,.00384}} \\
\underline{36} \\
24 \\
\underline{24} \\
0
\end{array}
$$

Each digit to the right of the decimal point in the dividend must have a corresponding digit in the quotient.

Sometimes it is necessary to increase the number of decimal places to more than appear in the problem. One situation in which this occurs is when the quotient does not come out exactly. Adding more decimal places can sometimes give us an answer with no remainder. At other times we do it to round our answer.

Example 5 Divide 22.4 by 5.

$$
\begin{array}{r}
4.4 \\
5{\overline{\smash{\big)}\,22.4}} \\
\underline{20} \\
24 \\
\underline{20} \\
4
\end{array}
$$

We write a zero in the next place value in the dividend and continue dividing.

$$
\begin{array}{r}
4.48 \\
5{\overline{\smash{\big)}\,22.40}} \\
\underline{20} \\
24 \\
\underline{20} \\
40 \\
\underline{40} \\
0
\end{array}
$$

We now look at a combination of division and rounding that we studied in section 3–2.

Example 6 Find 5.325 divided by 8. Round the quotient to three decimal places.

$$\begin{array}{r} .6656 \\ 8{\overline{\smash{\big)}\,5.3250}} \\ \underline{4\ 8} \\ 52 \\ \underline{48} \\ 45 \\ \underline{40} \\ 50 \\ \underline{48} \\ 2 \end{array}$$ = .666 (rounded to three places)

Notice that the instructions stated to round to three decimal places. Why then divide to four places? The answer to the question is simply that, if we are going to round to three places, we must have one more place to the right of the third place to apply the rounding rule.

Rule If a division is to be rounded to a certain number of decimal places, the quotient must be carried to one more place to correctly round to the desired place.

Example 7 Find 7 divided by 6. Round the quotient to two decimal places.

We first note that the decimal point is always understood to be to the right of any whole number. Therefore we may write 7 as 7.000 to give us three decimal places so we can round to two places.

$$\begin{array}{r} 1.166 \\ 6{\overline{\smash{\big)}\,7.000}} \\ \underline{6} \\ 1\ 0 \\ \underline{6} \\ 40 \\ \underline{36} \\ 40 \\ \underline{36} \\ 4 \end{array}$$ = 1.17 (rounded to two places)

EXERCISE SET 3-5-1

ANSWERS

Divide each of the following so that there is no remainder.

1. $4\overline{)2.8}$
2. $5\overline{)8.5}$
3. $9\overline{)31.5}$

4. $6\overline{)8.04}$
5. $8\overline{)56.32}$
6. $12\overline{)1.68}$

7. $6\overline{).102}$
8. $7\overline{).0126}$
9. $8\overline{)128.32}$

10. $6\overline{)60.54}$
11. $21\overline{)84.273}$
12. $34\overline{)170.918}$

1. _____
2. _____
3. _____
4. _____
5. _____
6. _____
7. _____
8. _____
9. _____
10. _____
11. _____
12. _____

13. 10)14.83 **14.** 10)28.64 **15.** 100)31.265

16. 100)73.468

Round each of the following to two decimal places.

17. 6)29 **18.** 3)1.4 **19.** 8)5.3

20. 7)15.31 **21.** 9)44.6 **22.** 12)39.2

ANSWERS

13. _____

14. _____

15. _____

16. _____

17. _____

18. _____

19. _____

20. _____

21. _____

22. _____

ANSWERS

23. $16\overline{)54.7}$

24. $11\overline{)23.07}$

25. $48\overline{).304}$

23. _____

24. _____

26. $34\overline{).285}$

25. _____

26. _____

27. _____

Round each of the following to three decimal places.

27. $7\overline{)24.32}$

28. $8\overline{)20.15}$

29. $13\overline{)40}$

28. _____

29. _____

30. _____

30. $17\overline{)50}$

31. $103\overline{)294}$

32. $108\overline{)531}$

31. _____

32. _____

33. _____

33. $24\overline{).72}$

34. $23\overline{).56}$

34. _____

A short method of dividing a decimal number by a power of ten is given below.

> **Rule** To divide a decimal number by a power of ten move the decimal point to the *left* a number of places equal to the number of zeros in the power of ten.

Example 8 Divide: 231.5 ÷ 10

We move the decimal point one place to the left.

$$231.5 \div 10 = 23.15$$

Example 9 Divide: 1.32 ÷ 1,000

We move the decimal point three places to the left.

$$1.32 \div 1,000 = .00132$$

▼ EXERCISE SET 3-5-2

Divide the following.

1. 18.3 ÷ 10
2. 143.2 ÷ 10
3. 412.6 ÷ 100

4. 17.31 ÷ 100
5. 1.36 ÷ 10
6. 1.05 ÷ 10

7. 12.94 ÷ 100
8. 14.52 ÷ 100
9. 8.31 ÷ 100

10. .014 ÷ 10
11. 29.3 ÷ 1,000
12. 513.7 ÷ 1,000

13. .001 ÷ 10
14. .035 ÷ 100
15. 1.003 ÷ 100

16. .139 ÷ 10
17. 349.6 ÷ 10,000
18. 81.63 ÷ 10,000

ANSWERS

1. _____
2. _____
3. _____
4. _____
5. _____
6. _____
7. _____
8. _____
9. _____
10. _____
11. _____
12. _____
13. _____
14. _____
15. _____
16. _____
17. _____
18. _____

ANSWERS

19. .0204 ÷ 10   20. .0194 ÷ 100

19. _____
20. _____

One very important use of division by whole numbers is in finding an average. An *average* is found by dividing the sum of a set of numbers by the number of numbers in the set.

Example 10 A math professor gave a test to 14 students with the following grade results: 56, 73, 62, 91, 87, 64, 72, 84, 97, 77, 72, 40, 83, and 90. What was the average grade rounded to one decimal place?

We first find the sum of all the grades, obtaining a result of 1,048. Next we divide the sum by the number of grades.

$$\begin{array}{r} 74.85 \\ 14{\overline{\smash{\big)}\,1{,}048.00}} \\ \underline{98} \\ 68 \\ \underline{56} \\ 12\,0 \\ \underline{11\,2} \\ 80 \\ \underline{70} \\ 10 \end{array} = 74.9 \text{ (rounded to one place)}$$

Example 11 One September the rainfall in Miami, Florida was 23 inches for the month. What was the average daily rainfall to the nearest tenth of an inch?

$$\begin{array}{r} .76 \\ 30{\overline{\smash{\big)}\,23.00}} \\ \underline{21\,0} \\ 2\,00 \\ \underline{1\,80} \\ 20 \end{array} = .8 \text{ inch per day}$$

EXERCISE SET 3-5-3

1. Sam took five tests in a math class during the semester. His scores were 73, 79, 85, 94, and 88. What was his average rounded to one decimal place?

2. The average annual rainfall for Atlanta is 48.66 inches. What is the average per month rounded to two decimal places?

1. _____

2. _____

3-5 DIVIDING DECIMAL NUMBERS BY WHOLE NUMBERS

3. During one season, Larry Bird of the Boston Celtics scored 1,745 points in 82 games. What was his average number of points per game rounded to one decimal place?

4. During one season, the New York Islanders hockey team played 80 games and scored 91 points. How many points per game did they average rounded to two decimal places?

5. Thirty-four people donated a total of $211.00. What was the average donation per person rounded to the nearest cent?

6. Twenty-three people invested a total of $2,506.00 in a stock. What was the average investment per person rounded to the nearest cent?

7. A certain company has three employees. Employee A earns $18,595.00 per year, employee B earns $20,400.00, and employee C earns $17,620.00. What is the average earnings for an employee of the company rounded to the nearest cent?

8. Barbara earned $135.00 the first week, $194.50 the second, and $203.45 the third. What was her average earnings per week rounded to the nearest cent?

9. The ages of seven people in a group are 18, 23, 18, 34, 55, 23, and 18. What is the average age rounded to one decimal place?

10. The weights of eight people are 120, 230, 145, 105, 114, 176, 168, and 125 pounds. What is the average weight per person rounded to two decimal places?

11. A man drove 452.7 miles on Monday, 341.2 miles on Tuesday, 103.8 miles on Wednesday, 316.4 miles on Thursday, and 265.0 miles on Friday. What was his average mileage per day for the five days rounded to the nearest tenth?

12. Helen worked 8.4 hours on Monday, 7.0 hours on Tuesday, 5.5 hours on Wednesday, 7.6 hours on Thursday, and 7.8 hours on Friday. What was her average number of hours per day rounded to the nearest tenth?

ANSWERS

3. _____

4. _____

5. _____

6. _____

7. _____

8. _____

9. _____

10. _____

11. _____

12. _____

3-6 DIVIDING BY DECIMAL NUMBERS

OBJECTIVES

In this section you will learn to divide any whole number or decimal number by a decimal number.

We will now examine how to divide by a decimal. Consider the following problem.

Example 1 Divide: $7.834 \div .02$

$7.834 \div .02$ can be written as $\dfrac{7.834}{.02}$ since the fractional bar always means division. The fundamental principle of fractions allows us to multiply both the numerator and denominator by the same nonzero number. Since we already know how to divide by a whole number, we will multiply both the numerator and denominator by a number that will make our problem one of dividing by a whole number.

$$\frac{7.834}{.02} \times \frac{100}{100} = \frac{783.4}{2}$$

Thus we can always change division by a decimal to division by a whole number. Writing this in long-division form, we have

$$\begin{array}{r} 391.7 \\ 2\overline{)783.4} \\ \underline{6} \\ 18 \\ \underline{18} \\ 3 \\ \underline{2} \\ 1\,4 \\ \underline{1\,4} \\ 0. \end{array}$$

So the fact is that we never actually divide by a decimal number because we always make our divisor a whole number.

Example 2 Divide: $17.28 \div 1.2$

If we multiply both of these numbers by 10, we have

$$\frac{17.28}{1.2} \times \frac{10}{10} = \frac{172.8}{12}$$

Placing this in long-division form, we have

$$\begin{array}{r} 14.4 \\ 12\overline{)172.8} \\ \underline{12} \\ 52 \\ \underline{48} \\ 4\,8 \\ \underline{4\,8} \\ 0. \end{array}$$

The number used to multiply in each case will be 10; 100; 1,000; 10,000; and so on. Multiplying by these numbers simply moves the decimal point to the right a number of spaces equal to the number of zeros.

> **Rule** To divide a decimal number or whole number by a decimal number move the decimal point of the divisor all the way to the right until the divisor is a whole number. Then move the decimal point in the dividend the same number of places to the right.

Example 3 Divide: $.03\overline{)52.152}$

We move the decimal point two places in the divisor and two places in the dividend. We now have

$$
\begin{array}{r}
1{,}738.4 \\
3\overline{)5{,}215.2} \\
\underline{3} \\
2\,2 \\
\underline{2\,1} \\
1\,1 \\
\underline{9} \\
2\,5 \\
\underline{2\,4} \\
1\,2 \\
\underline{1\,2} \\
0.
\end{array}
$$

Example 4 Divide: $3.725\overline{)78}$ (Round the answer to one decimal place.)

We move the decimal point three places in both the divisor and dividend. Zeros are added to the dividend to give enough places to move the decimal point. We then have

$$
\begin{array}{r}
20.93 = 20.9 \text{ (rounded to one place).} \\
3{,}725\overline{)78{,}000.00} \\
\underline{74\,50} \\
3\,500\,0 \\
\underline{3\,352\,5} \\
147\,50 \\
\underline{111\,75} \\
35\,75
\end{array}
$$

Example 5 Divide: $.3\overline{)5}$ (Round the answer to two decimal places.)

Moving the decimal point one place in both the divisor and dividend, we have

$$
\begin{array}{r}
16.666 \\
3\overline{)50.000} \\
\underline{3} \\
20 \\
\underline{18} \\
2\,0 \\
\underline{1\,8} \\
20 \\
\underline{18} \\
20 \\
\underline{18} \\
2
\end{array}
$$

= 16.67 (rounded to two places).

EXERCISE SET 3-6-1

ANSWERS

Divide the following.

1. $.2\overline{).7}$
2. $.4\overline{)1.24}$
3. $.02\overline{).502}$

1. _____

2. _____

3. _____

4. $.07\overline{).742}$
5. $.21\overline{).4494}$
6. $.32\overline{)1.184}$

4. _____

5. _____

6. _____

7. $3.6\overline{)29.592}$
8. $1.28\overline{)5.2096}$
9. $.015\overline{).00021}$

7. _____

8. _____

9. _____

3-6 DIVIDING BY DECIMAL NUMBERS

10. .218)‾.00109 11. .1)‾3.15 12. .01)‾16.2

ANSWERS

10. _____

11. _____

Round each of the following to one decimal place.

13. .7)‾23.4 14. .03)‾8.612 15. 1.6)‾17.24

12. _____

13. _____

14. _____

16. 3.8)‾88 17. 4.3)‾90 18. 1.61)‾4.3

15. _____

16. _____

17. _____

Round each of the following to two decimal places.

18. _____

19. .7)‾8 20. 2.6)‾18.32 21. 4.2)‾22.41

19. _____

20. _____

21. _____

22. 2.137)‾45 23. 5.013)‾21.43 24. 3.028)‾17.52

22. _____

23. _____

24. _____

25. If 5.5 pounds of apples cost $3.80, what is the cost per pound rounded to the nearest cent?

26. If a 6.34-pound roast cost $25.23, what is the cost per pound rounded to the nearest cent?

27. If a car traveled 347.5 miles on 14.6 gallons of gasoline, how many miles did it get per gallon rounded to the nearest tenth?

28. After traveling 382.3 miles with your car you fill the tank and find it used 11.8 gallons of gasoline. How many miles per gallon did the car get rounded to the nearest tenth?

29. Mark Spitz swam 100 meters in 51.20 seconds. How many meters per second did he average? Round your answer to two decimal places.

30. Anne Henning skated 500 meters in 43.3 seconds. How many meters per second did she average? Round your answer to the nearest tenth.

3–7 INTERCHANGING FRACTIONS AND DECIMALS

In this and the previous chapter we learned the four basic operations on fractions and decimals. It is sometimes desirable to change fractions to decimals or decimals to fractions.

First we will discuss changing decimals to fractions. The process comes from the fact that the number of places to the right of the decimal point gives us the denominator of the fraction.

1. One place means tenths, so the denominator is 10.
2. Two places means hundredths, so the denominator is 100.
3. Three places means thousandths, so the denominator is 1,000. And so on.

OBJECTIVES

In this section you will learn to:
1. Change a decimal to a fraction.
2. Change a fraction to a decimal.

Example 1 Change .375 to a common fraction.

.375 has three decimal places and is read as "three hundred seventy-five thousandths." Writing this as a common fraction, we have

$$.375 = \frac{375}{1,000}.$$

Now we must refer back to the fact that we always reduce a fractional answer to simplest form. So

$$.375 = \frac{375}{1,000} = \frac{3 \times \cancel{5} \times \cancel{5} \times \cancel{5}}{2 \times 2 \times 2 \times \cancel{5} \times \cancel{5} \times \cancel{5}} = \frac{3}{8}.$$

Rule To change a decimal to a common fraction rewrite the decimal as a fraction with the proper denominator and reduce to simplest form.

Example 2 Change .22 to a common fraction.

.22 or twenty-two hundredths is written as

$$.22 = \frac{22}{100} = \frac{11}{50}.$$

If digits other than zero appear to the left of the decimal point, we will have a mixed number, but we only need deal with the decimal part.

Example 3 Change 3.75 to a mixed number.

Three and seventy-five hundredths is written as

$$3.75 = 3\frac{75}{100} = 3\frac{3}{4}.$$

▼ EXERCISE SET 3-7-1

Change each of the following to a common fraction or mixed number.

1. .7 2. .11 3. .4 4. .16 5. .34

6. .55 7. 2.9 8. 4.8 9. .124 10. .025

11. .238 12. .144 13. .416 14. .732 15. 7.104

16. 3.496 17. 11.775 18. 19.016 19. .1152 20. 18.1104

The task of changing a fraction to a decimal depends on the fact that the fractional bar indicates division. For instance, $\frac{3}{4}$ can be written as $4\overline{)3}$.

Example 4 Change $\frac{3}{4}$ to a decimal.

Writing this in long-division form, we have

$$\begin{array}{r} .75 \\ 4\overline{)3.00} \\ \underline{2\ 8} \\ 20 \\ \underline{20} \\ 0. \end{array}$$

The answer in example 4 did not have a remainder. If there had been a remainder, we could have continued dividing.

The decimal form of any given fraction will either eventually terminate or repeat indefinitely.

Example 5 Change $\frac{2}{3}$ to a decimal.

$$\begin{array}{r} .666\ldots \\ 3\overline{)2.000} \\ \underline{1\ 8} \\ 20 \\ \underline{18} \\ 20 \\ \underline{18} \\ 2 \end{array}$$

This is an example of a repeating decimal. In this case the 6 keeps repeating forever. We sometimes place a bar over the repeating digit or digits to indicate that they repeat indefinitely. We could thus write .666 . . . as $.\overline{6}$ to indicate the repetition.

If we are going to round an answer in decimal form, we need to know how many places are needed. Remember that we must divide to one more place than the desired place.

Example 6 Change $\frac{7}{9}$ to a decimal rounded to three places.

$$\begin{array}{r} .7777 = .778 \text{ (rounded to three places)} \\ 9\overline{)7.0000} \\ \underline{6\ 3} \\ 70 \\ \underline{63} \\ 70 \\ \underline{63} \\ 70 \\ \underline{63} \\ 7 \end{array}$$

An improper fraction is treated the same way.

Example 7 Change $\frac{12}{7}$ to a decimal rounded to two places.

$$\begin{array}{r} 1.714 = 1.71 \text{ (rounded to two places)} \\ 7\overline{)12.000} \\ \underline{7} \\ 5\ 0 \\ \underline{4\ 9} \\ 10 \\ \underline{7} \\ 30 \\ \underline{28} \\ 2 \end{array}$$

Example 8 Change $2\dfrac{5}{8}$ to decimal form rounded to three places.

First change the mixed number to an improper fraction.

$$2\dfrac{5}{8} = \dfrac{21}{8}$$

Then

```
        2.625
     8)21.000
       16
        5 0
        4 8
          20
          16
           40
           40
            0.
```

EXERCISE SET 3-7-2

Change each of the following to a decimal number rounded to three decimal places.

1. $\dfrac{1}{2}$ 2. $\dfrac{1}{4}$ 3. $\dfrac{4}{5}$ 4. $\dfrac{1}{5}$ 5. $\dfrac{1}{8}$

6. $\dfrac{3}{8}$ 7. $\dfrac{1}{6}$ 8. $\dfrac{5}{6}$ 9. $\dfrac{5}{9}$ 10. $\dfrac{4}{9}$

11. $\dfrac{3}{7}$ 12. $\dfrac{1}{7}$ 13. $3\dfrac{1}{3}$ 14. $4\dfrac{2}{5}$ 15. $\dfrac{11}{4}$

16. $\dfrac{15}{8}$ 17. $\dfrac{3}{14}$ 18. $\dfrac{7}{16}$ 19. $5\dfrac{17}{21}$ 20. $7\dfrac{18}{25}$

21. $\dfrac{120}{11}$ 22. $\dfrac{235}{18}$ 23. $\dfrac{348}{103}$ 24. $\dfrac{795}{114}$

Some calculations may involve common fractions, decimals, and whole numbers. We follow the same rules for order of operations when evaluating such expressions.

Example 9 $(3.2)(0.5) + 24 \div 0.6 + 1.4 = 1.6 + 24 \div 0.6 + 1.4$
$= 1.6 + 40 + 1.4$
$= 43$

Example 10 $6.1 + \frac{1}{2}(0.5) - 1.35 \div \frac{3}{4} = 6.1 + .25 - 1.35 \div \frac{3}{4}$
$= 6.1 + .25 - 1.8$
$= 6.35 - 1.8$
$= 4.55$

▼ EXERCISE SET 3-7-3

Evaluate the following.

1. $8.12(0.5) - 6.9 \div 3$

2. $\frac{1}{2}(5.1) + 2.3 \div \frac{1}{4}$

3. $\frac{2}{3}(5.1) + 2.5(0.3)$

4. $\frac{1}{2}(0.5) + 1.6 \div \frac{1}{3}$

5. $3.4 + 0.18 \div 2 - 2.1$

6. $15.2 - 2.14 \div 0.2 + 2.18$

7. $4 + .012(5.1) \div 2$

8. $3.1 + 1.95 \div 1.3 - 2.8$

9. $11.2 \div 1.4 - \frac{1}{2}(3.5 + 2.1) + \frac{1}{3}(41.7)$

10. $7 + 0.2(.15) - \frac{3}{4}(6.12) + 0.153 \div .05$

ANSWERS

1. _____
2. _____
3. _____
4. _____
5. _____
6. _____
7. _____
8. _____
9. _____
10. _____

3-8 SCIENTIFIC NOTATION

OBJECTIVES

In this section you will learn to:
1. Write decimal numbers in expanded form using exponents.
2. Write numbers in scientific notation.

Very large or very small numbers are often written in **scientific notation.** Before defining scientific notation, we will review three definitions from the previous chapters and introduce one new definition.

Definition An **exponent** indicates the number of times a base is used as a factor.

Example 1 In 5^3, 5 is the base and 3 is the exponent. 5^3 means that 5 is used as a factor three times.

$$5^3 = 5 \times 5 \times 5 = 125$$

Definition Any number (except zero) that has an **exponent of zero** is equal to 1. In symbols, if a is any number except zero, then $a^0 = 1$.

Example 2 $5^0 = 1$; $2^0 = 1$; $10^0 = 1$

Definition The **reciprocal** of a nonzero number is 1 divided by that number. In other words, if a is a number (other than zero), then the reciprocal of a is $\dfrac{1}{a}$.

The new definition we wish to introduce is another "short hand" method that mathematicians use in writing certain numbers.

Definition An exponent with a minus sign preceding it indicates the **reciprocal.** In symbols, $a^{-n} = \dfrac{1}{a^n}$.

Example 3 $5^{-2} = \dfrac{1}{5^2}$; $2^{-5} = \dfrac{1}{2^5}$; $10^{-3} = \dfrac{1}{10^3}$

Recall that in chapter 1 we wrote whole numbers in expanded form using exponents.

Example 4 Write 235 in expanded form using exponents.

$$235 = 2 \times 10^2 + 3 \times 10^1 + 5 \times 10^0$$

Our new definition also allows us to write decimal numbers in expanded form using exponents.

Example 5 Write .527 in expanded form using exponents.

In section 3–1 we learned to write this as

$$.527 = 5 \times \frac{1}{10} + 2 \times \frac{1}{100} + 7 \times \frac{1}{1,000}.$$

Using the new definition, we now write it as

$$.527 = 5 \times 10^{-1} + 2 \times 10^{-2} + 7 \times 10^{-3}.$$

Example 6 Write 381.67 in expanded form using exponents.

$$381.67 = 3 \times 10^2 + 8 \times 10^1 + 1 \times 10^0 + 6 \times 10^{-1} + 7 \times 10^{-2}$$

Example 7 Write 5.6×10^2 as a decimal number.

$$5.6 \times 10^2 = 5.6 \times 100 = 560$$

Example 8 Write 2.37×10^3 as a decimal number.

$$2.37 \times 10^3 = 2.37 \times 1,000 = 2,370$$

Notice in examples 7 and 8 that we moved the decimal point to the right the same number of places as the value of the exponent of 10.

Example 9 $7.85 \times 10^4 = 7.85 \times 10,000 = 78,500$

Example 10 $7.85 \times 10^{-4} = 7.85 \times \frac{1}{10^4} = \frac{7.85}{10,000} = .000785$

Note that 10^{-4} indicates division, so the decimal is moved to the left.

EXERCISE SET 3-8-1

Write without exponents.

1. 7^0
2. 18^0
3. 2^{-2}
4. 2^{-3}

5. 5^{-3}
6. 6^{-2}
7. 10^{-5}
8. 10^{-6}

Write the following in expanded form using exponents.

9. .8
10. .27

11. .915
12. .605

13. 7.14
14. 60.3

15. 128.72
16. 5,124.17

Write each of the following as a decimal number.

17. $.4 \times 10$
18. $.73 \times 10$

19. 4.6×10^2
20. $.85 \times 10^2$

21. 3.16×10^3
22. 7.25×10^4

23. 61×10^{-1}
24. 34×10^{-2}

25. 605×10^{-4} **26.** 103×10^{-3}

27. 7.94×10^5 **28.** 5.21×10^7

29. 9.17×10^{-6} **30.** 4.32×10^{-9}

ANSWERS

25. _____
26. _____
27. _____
28. _____
29. _____
30. _____

Exponents will be used extensively in the study of algebra. We will now introduce one more idea that involves exponents.

> **Definition**   A number is expressed in **scientific notation** when it is written as a number equal to or greater than one and less than ten, multiplied by some power of ten.

Example 11 3.6×10^3; 7.25×10^{-3}; and 9.15×10^8 are all in scientific notation.

Notice that the definition contains two parts. First we must have a number equal to or greater than one and less than ten, and second we must multiply by a power of 10.

Example 12 26.5×10^3 is *not* in scientific notation since we do not have a number equal to or greater than one and less than ten.

Example 13 5.3×4^3 is *not* in scientific notation since we are not multiplying by a power of 10.

Example 14 Earth is approximately 93,000,000 miles from the sun. Express this distance in scientific notation.

> We know that scientific notation requires that we have 9.3 (a number equal to or greater than one and less than ten). Now we must determine the proper exponent for 10. The simplest way to determine this is to ask, "How many places, and in what direction, must we move the decimal point in 9.3 to get 93,000,000?" This way we arrive at 9.3×10^7.

Example 15 Write .0074 in scientific notation.

First we must have 7.4. Now, how many places, and in what direction, do we move the decimal point in 7.4 to get .0074? Thus $.0074 = 7.4 \times 10^{-3}$.

Remember, 10 to a *minus* power moves the decimal point to the left.

EXERCISE SET 3-8-2

ANSWERS

In problems 1–8 state whether or not the given number is in scientific notation.

1. 3.6×10^5
2. $.05 \times 10^8$
3. 2.78×10^{16}

4. 8.2×10^{-3}
5. 6.1×2^{10}
6. 25×10^{-3}

7. $.645 \times 10^4$
8. 5×10^8

In problems 9–16 write each number in scientific notation.

9. 5,000
10. 5,280
11. 728,000

12. 346,000,000
13. .000000235
14. .0000000052

15. 233,000,000,000,000
16. .0000000000739

In problems 17–26 write each number without using exponents.

17. 3.201×10^5
18. 7.28×10^{-6}
19. 1.07×10^{-9}

20. 6.23×10^{23}
21. 5.02×10^{10}
22. 3.58×10^{-1}

1. _____
2. _____
3. _____
4. _____
5. _____
6. _____
7. _____
8. _____
9. _____
10. _____
11. _____
12. _____
13. _____
14. _____
15. _____
16. _____
17. _____
18. _____
19. _____
20. _____
21. _____
22. _____

23. 4.07×10^{-5} **24.** 5.3762×10^2 **25.** 3.6×10^0

26. 9.9×10^1

27. The planet Mars is approximately 49,000,000 miles from Earth. Express this distance in scientific notation.

28. An enormous cloud of hydrogen gas eight million miles in diameter was discovered around the comet Bennet by NASA in 1970. Express this distance in scientific notation.

29. A light-year (the distance light travels in a year) is approximately 6.0×10^{12} miles. Express this distance without exponents.

30. The mass of the Earth is 1.3×10^{25} pounds. Express this without exponents.

31. An angstrom is a unit of length. One angstrom is approximately 1×10^{-8} cm. Express this length without exponents.

32. The diameter of the nucleus of an average atom is approximately 3.5×10^{-12} cm. Express this distance without exponents.

33. A red blood cell is approximately .001 cm in diameter. Express this length in scientific notation.

34. The Rubik's cube puzzle has over 43,000,000,000,000,000,000 color combinations. Express this number in scientific notation.

ANSWERS

23. _____

24. _____

25. _____

26. _____

27. _____

28. _____

29. _____

30. _____

31. _____

32. _____

33. _____

34. _____

3-9 SQUARE ROOTS

OBJECTIVES

In this section you will learn to:
1. Find a square root of a perfect square.
2. Find an approximate value of the square root of a number.
3. Simplify a radical.

Another symbol often used in mathematics is that to indicate a square root.

Definition If a is a number and $a^2 = x$, then a is said to be a **square root** of x.

Example 1 $5^2 = 25$, so 5 is a square root of 25.

Example 2 $4^2 = 16$, so 4 is a square root of 16.

The symbol $\sqrt{}$ (often referred to as a radical sign) is used to represent a square root of a number. For instance, in example 1 the statement "5 is a square root of 25" can be written as $5 = \sqrt{25}$. Also, in example 2 the statement "4 is a square root of 16" can be written as $4 = \sqrt{16}$.

Since we are only dealing with the numbers of arithmetic, our use of the square root symbol will be more limited than you will encounter in algebra. At this point, we wish to discuss only square roots involving the numbers of arithmetic. In future courses you will learn about other roots and other numbers.

Example 3 Find $\sqrt{49}$.

Since $7^2 = 49$, then $\sqrt{49} = 7$.

Definition The numbers of arithmetic that have whole numbers as square roots are called **perfect squares.**

Example 4 16 is a perfect square since the whole number 4 is a square root.

EXERCISE SET 3-9-1

ANSWERS

1. _____
2. _____
3. _____
4. _____
5. _____
6. _____

1. Starting with 0, list the first twenty-five perfect squares.

Find the following.

2. $\sqrt{4}$ 3. $\sqrt{9}$ 4. $\sqrt{0}$ 5. $\sqrt{81}$ 6. $\sqrt{100}$

7. $\sqrt{121}$ 8. $\sqrt{400}$ 9. $\sqrt{225}$ 10. $\sqrt{36}$ 11. $\sqrt{144}$

12. $\sqrt{1}$ 13. $\sqrt{64}$ 14. $\sqrt{169}$ 15. $\sqrt{256}$ 16. $\sqrt{289}$

17. $\sqrt{441}$ 18. $\sqrt{484}$ 19. $\sqrt{529}$ 20. $\sqrt{900}$ 21. $\sqrt{10{,}000}$

ANSWERS

7. ___
8. ___
9. ___
10. ___
11. ___
12. ___
13. ___
14. ___
15. ___
16. ___
17. ___
18. ___
19. ___
20. ___
21. ___

Not all numbers have square roots that are whole numbers. Square roots of numbers that are not perfect squares can be found correct to any number of decimal places by several methods. The easiest method is to use a calculator, and the next easiest method is to use a table of square roots, such as the one printed on the inside back cover of this book.

Our purposes here do not require these estimations, but you should have some concept as to the value of these numbers. As an exercise, you will be asked to guess or estimate the value of some square roots and then check your answers.

Example 5   Guess the value of $\sqrt{13}$.

Your thinking should proceed in this manner: "I know that 13 is between the perfect squares 9 and 16. Since $\sqrt{9} = 3$ and $\sqrt{16} = 4$, $\sqrt{13}$ must be between 3 and 4. Since 13 is a little more than halfway between 9 and 16, my guess is 3.6." If you look at the table, you see $\sqrt{13}$ is 3.6056 (to four decimal places), so your guess is pretty good.

EXERCISE SET 3-9-2

Guess the values of the following to one decimal place and then check the table on the inside back cover to see how close your guess is.

1. $\sqrt{2}$ 2. $\sqrt{5}$ 3. $\sqrt{30}$ 4. $\sqrt{50}$

5. $\sqrt{60}$ 6. $\sqrt{19}$ 7. $\sqrt{88}$ 8. $\sqrt{40}$

ANSWERS

1. ___
2. ___
3. ___
4. ___
5. ___
6. ___
7. ___
8. ___

ANSWERS

9. $\sqrt{15}$ 10. $\sqrt{22}$ 11. $\sqrt{75}$ 12. $\sqrt{110}$

9. _____
10. _____
11. _____
12. _____

Square roots of numbers that are not perfect squares can only be estimated. Whether you use a table of square roots or a calculator, the values obtained are only approximations. It can be shown, for instance, that $\sqrt{2}$ cannot be exactly expressed in decimal notation. So we either estimate the decimal approximation or simply leave such an answer in the square root or radical form. However, if we leave a number in radical form, we should leave it in simplified radical form. To do this we need the following theorem.

> **Theorem** If a and b are numbers of arithmetic, $\sqrt{a} \times \sqrt{b} = \sqrt{a \times b}$.

Example 6 $\sqrt{4} \times \sqrt{25} = \sqrt{4 \times 25} = \sqrt{100} = 10$

Note also that $\sqrt{4} \times \sqrt{25} = 2 \times 5 = 10$.

We now wish to use this theorem to simplify radicals.

> **Definition** If a is a number of arithmetic, $\sqrt{a}$ is in **simplified form** if no perfect square (other than 1) is a factor of a.

Example 7 Simplify $\sqrt{50}$.

We know that $\sqrt{50}$ is not simplified since 25 (a perfect square) is a factor of 50. We proceed as follows:

$$\sqrt{50} = \sqrt{25 \times 2} = \sqrt{25} \times \sqrt{2} = 5 \times \sqrt{2}.$$

We will agree that $5 \times \sqrt{2}$ will be written as $5\sqrt{2}$.

Example 8 Simplify $\sqrt{27}$.

$$\sqrt{27} = \sqrt{9 \times 3} = \sqrt{9} \times \sqrt{3} = 3\sqrt{3}$$

Example 9 Simplify $\sqrt{35}$.

$\sqrt{35}$ is already in simplified form since 35 is not divisible by a perfect square.

Example 10 Simplify $\sqrt{32}$.

We might notice that 32 is divisible by the perfect square 4 and write $\sqrt{32} = \sqrt{4 \times 8} = 2\sqrt{8}$. This answer is not completely simplified since 8 still has a factor that is a perfect square. We must be careful to choose the *largest* perfect square factor. So

$$\sqrt{32} = \sqrt{16 \times 2} = \sqrt{16} \times \sqrt{2} = 4\sqrt{2}.$$

Example 11 Simplify $\sqrt{288}$.

$$\sqrt{288} = \sqrt{144 \times 2} = \sqrt{144} \times \sqrt{2} = 12\sqrt{2}$$

▼ EXERCISE SET 3-9-3

Simplify the following.

1. $\sqrt{12}$
2. $\sqrt{20}$

3. $\sqrt{18}$
4. $\sqrt{14}$

5. $\sqrt{75}$
6. $\sqrt{48}$

7. $\sqrt{180}$
8. $\sqrt{80}$

9. $\sqrt{96}$
10. $\sqrt{72}$

11. $\sqrt{108}$
12. $\sqrt{102}$

13. $\sqrt{192}$
14. $\sqrt{176}$

15. $\sqrt{175}$
16. $\sqrt{396}$

ANSWERS

1. _____
2. _____
3. _____
4. _____
5. _____
6. _____
7. _____
8. _____
9. _____
10. _____
11. _____
12. _____
13. _____
14. _____
15. _____
16. _____

ANSWERS

17. _____
18. _____
19. _____
20. _____

17. $\sqrt{243}$

18. $\sqrt{294}$

19. $\sqrt{486}$

20. $\sqrt{1{,}000}$

CHAPTER 3 SUMMARY

Reading Decimal Numbers

- A decimal point is used in a number to indicate a fraction.
- Place values to the right of the decimal point are tenths, hundredths, thousandths, and so on.
- The decimal point is read as "and."

Rounding

- Rounding decimal numbers follows the same rule as for rounding whole numbers.

Adding and Subtracting Decimal Numbers

- When adding or subtracting decimal numbers, place them in column form with the decimal points in the same column.

Multiplying Decimal Numbers

- When multiplying decimal numbers, multiply as if they were whole numbers, and place the decimal point in the product so that the number of decimal places equals the sum of the decimal places in the two numbers being multiplied.

Dividing Decimal Numbers

- To divide a decimal number by a whole number use the long-division algorithm and place the decimal in the quotient directly above the decimal in the dividend.
- To divide a number by a decimal number move the decimal point in the divisor all the way to the right so that the divisor is a whole number. Then move the decimal point in the dividend the same number of places to the right.
- If a division is to be rounded to a certain number of decimal places, it must be carried to one more place to round back to the desired place.

Averaging

- An average is found by dividing the sum of a set of numbers by the number of numbers in the set.

Decimals–Fractions

- To change a decimal to a common fraction rewrite the decimal as a fraction with the proper denominator and reduce to simplest form.
- A fraction is changed to a decimal by dividing the numerator by the denominator.

Scientific Notation

- A number is expressed in scientific notation when it is a product of a number equal to or greater than one but less than ten, and a power of 10. For example, 4.2×10^{-3} is in scientific notation.

Square Roots

- If a is a number and $a^2 = x$, then a is said to be a square root of x.
- The symbol $\sqrt{}$ is used to indicate the square root of a number. For example, $\sqrt{25} = 5$.
- The numbers of arithmetic that have whole numbers as square roots are called perfect squares.
- If a and b are numbers of arithmetic, then $\sqrt{a} \times \sqrt{b} = \sqrt{ab}$.
- $\sqrt{a}$ is in simplified form if no factor (other than 1) of a is a perfect square.

CHAPTER 3 REVIEW

NAME: _____
CLASS / SECTION: _____   DATE: _____

1. Write .24 in expanded form.

2. Write .07 in expanded form.

3. Write 7.3074 in expanded form.

4. Write 12.1308 in expanded form.

5. Write .49 in words.

6. Write .513 in words.

7. Write 15.075 in words.

8. Write 8.079 in words.

9. Express "four hundred three thousandths" in decimal notation.

10. Express "two hundred eight ten thousandths" in decimal notation.

11. Round .8145 to the nearest hundredth.

12. Round 3.619 to the nearest whole number.

13. Round 6.3965 to the nearest thousandth.

14. Round 4.082 to the nearest tenth.

15. Round .9999 to the nearest hundredth.

16. Round .4495 to the nearest hundredth.

17. Round .548 to the nearest tenth.

18. Round .1496 to the nearest thousandth.

19. Round 9.523 to the nearest whole number.

20. Round 1.5555 to the nearest thousandth.

ANSWERS

1. _____
2. _____
3. _____
4. _____
5. _____
6. _____
7. _____
8. _____
9. _____
10. _____
11. _____
12. _____
13. _____
14. _____
15. _____
16. _____
17. _____
18. _____
19. _____
20. _____

ANSWERS

21. _____

22. _____

23. _____

24. _____

25. _____

26. _____

27. _____

28. _____

29. _____

30. _____

31. _____

32. _____

33. _____

34. _____

35. _____

36. _____

21. Change .83 to a common fraction.

22. Change .07 to a common fraction.

23. Change .288 to a common fraction.

24. Change .304 to a common fraction.

25. Change 17.48 to a mixed number.

26. Change 24.125 to a mixed number.

27. Change $\frac{23}{24}$ to a decimal rounded to three places.

28. Change $\frac{11}{34}$ to a decimal rounded to three places.

29. Change $5\frac{7}{9}$ to a decimal rounded to three places.

30. Change $11\frac{3}{8}$ to a decimal rounded to three places.

Perform the indicated operation. In division problems round the answer to three decimal places.

31. 3.14
$\underline{\times.02}$

32. 48.17
$\underline{+54.34}$

33. $4\overline{).72}$

34. $.3\overline{)29.1}$

35. 42.38
$\underline{+16.92}$

36. 5.23
$\underline{\times.04}$

37. .7)58.6 **38.** 6)1.44 **39.** 32.80
                                    − 5.14

40. 20.60   **41.** 7.34   **42.** 51.041
        − 8.34   +8.5   −36.28

43. 12)37.8 **44.** .04).005 **45.** 12.35
                                       × .18

46. 6.7   **47.** .008   **48.** 14)64.68
        +9.31   × .05

49. .09).306 **50.** 14.29   **51.** 40.029
                         × .14   −16.84

ANSWERS

37. _____

38. _____

39. _____

40. _____

41. _____

42. _____

43. _____

44. _____

45. _____

46. _____

47. _____

48. _____

49. _____

50. _____

51. _____

Answers

52. _____

53. _____

54. _____

55. _____

56. _____

57. _____

58. _____

59. _____

60. _____

61. _____

62. _____

63. _____

64. _____

65. _____

66. _____

52. .014
   × .07

53. 16.04
   − 9.6

54. 138.65
   + 92.77

55. 32.47
   × 100

56. 6.2$\overline{)\,.0186}$

57. 325.74
   +134.66

58. 24.01
   −13.09

59. 132.16
   − 94.39

60. 64.17
   × 100

61. 4.7$\overline{)\,.0376}$

62. 69.043
   +58.958

63. 17$\overline{)\,39.6}$

64. 540.38
   −359.49

65. 18.126
   × 4.38

66. 49.13
     135.624
   + 28.97

CHAPTER 3 REVIEW   217

67. 47.138
    +14.69

68. 23)78.4

69. 35.003
    −13.999

ANSWERS

67. _____

68. _____

69. _____

70. 28.019
    × 1.27

71. 2.14)62.4

72. 5.23)18.9

70. _____

71. _____

72. _____

73. 37.52
    468.054
    + 7.967

74. 68.004
    −48.995

75. 125.08
    × .156

73. _____

74. _____

75. _____

76. 319.26
    × .347

77. 15)397.12

78. 490.28
       6.847
    + 29.047

76. _____

77. _____

78. _____

79. 284.08
    −139.09

80. 405.16
    −285.77

81. 64.79
    153.8
    + 29.047

79. _____

80. _____

81. _____

82. 35)782.12

83. 47.16
    −29.362

84. 36)584.7

82. _____

83. _____

85. 5.284
   × 2.09

86. 7.4)81.6664

87. 42)129.4

84. _____

85. _____

88. 34.35
   −16.054

89. 8.7)26.7438

90. 4.635
   × 7.04

86. _____

87. _____

88. _____

91. A woman bought some party goods for her daughter's birthday. She paid $3.98 for plates, $2.65 for napkins, and $5.89 for a game. What was the total cost?

92. A man bought a tire for his car for $69.88. He also bought a battery for $74.99 and some oil for $10.98. What was the total cost?

89. _____

90. _____

91. _____

92. _____

93. A bank balance is $543.24. If a check is written for $295.65, what is the new balance?

94. If you bought three items for $6.25, $2.38, and $4.28, how much change would you receive from $20?

95. The balance in a checking account is $304.28. If three checks are written for $64.12, $18.39, and $112.18, what is the new balance?

96. Chuck steak is on sale for $3.19 per pound. What is the cost of a 5.34-pound steak? Round answer to the nearest cent.

97. What is the cost of a 21.15-pound turkey if the price is $.79 per pound? Round answer to the nearest cent.

98. What is the cost of four tires if they are priced at $69.98 each?

99. If the price of a certain material is $6.98 per yard, what is the cost of 5.5 yards?

100. A student has test scores of 71, 87, 56, 74, 83, 92, 95, 81, 78, 80, and 86. What is the average rounded to two decimal places?

ANSWERS

93. _____

94. _____

95. _____

96. _____

97. _____

98. _____

99. _____

100. _____

ANSWERS

101. _____

102. _____

103. _____

104. _____

105. _____

106. _____

107. _____

108. _____

109. _____

110. _____

111. _____

112. _____

101. During a storm there was 5.41 inches of rain in four hours. What was the average rainfall per hour rounded to the nearest hundredth?

102. During six football games, a running back rushed for 105, 86, 114, 72, 124, and 94 yards. What was the average number of yards per game rounded to the nearest tenth?

103. A package of ground beef weighing 3.16 pounds cost $6.60. What is the cost per pound rounded to the nearest cent?

104. Irena Szewinska of Poland ran the 400-meter dash in 49.29 seconds. What was her average number of meters per second rounded to two decimal places?

Write each of the following as a decimal number.

105. 42.6×10^3

106. 18.4×10^5

107. 139×10^{-4}

108. 42.71×10^{-6}

Write the following in scientific notation.

109. 5,180

110. 60,200

111. .8145

112. .0999

Determine what two consecutive whole numbers each of the following is between.

113. $\sqrt{38}$

114. $\sqrt{29}$

Simplify.

115. $\sqrt{28}$

116. $\sqrt{44}$

117. $\sqrt{56}$

118. $\sqrt{63}$

119. $\sqrt{136}$

120. $\sqrt{304}$

ANSWERS

113. _____

114. _____

115. _____

116. _____

117. _____

118. _____

119. _____

120. _____

SCORE: _____

CHAPTER 3 TEST

NAME: _____

CLASS / SECTION: _____   DATE: _____

ANSWERS

1. _____
2. _____
3. _____
4. _____
5. _____
6. _____
7. _____
8. _____
9. _____
10. _____
11. _____
12. _____
13. _____
14. _____
15. _____
16. _____
17. _____
18. _____
19. _____
20. _____
21. _____
22. _____
23. _____
24. _____

SCORE: _____

1. Write 52.649 in expanded form.

2. Write 298.37 in words.

3. Round 8.6531 to the nearest tenth.

4. Change .112 to a common fraction in simplest form.

5. Change $13\frac{5}{7}$ to a decimal rounded to three decimal places.

Perform the indicated operation. In division problems round the quotient to three decimal places.

6. 32.79
  + 9.63

7. 68.142
  − 9.374

8. $9\overline{)66.6}$

9. $.6\overline{)9.138}$

10. 6.27
  × .8

11. 7.465
  18.84
  + 9.356

12. 27.15
  − 9.758

13. $21\overline{)460.3}$

14. $.7\overline{)5.838}$

15. 240.15
  × 5.6

16. 28.64
  321.523
  + 36.048

17. 53.007
  −29.138

18. $18\overline{)127.44}$

19. 28.03
  × .049

20. $7.3\overline{).0657}$

21. Write 3.164×10^{-4} as a decimal number.

22. Write .00000214 in scientific notation.

23. Simplify: $\sqrt{12}$

24. Simplify: $\sqrt{112}$

CHAPTER 4

CHAPTER 4 PRETEST

ANSWERS

Answer as many of the following problems as you can before starting this chapter. When you finish the chapter, take the test at the end and compare your scores to see how much you have learned.

1. _____

2. _____

3. _____

4. _____

5. _____

6. _____

7. _____

8. _____

9. _____

10. _____

SCORE: _____

1. Write a literal expression for the product of the sum of a and b and the difference of x and y.

2. Express $a \div (a - b)$ in words.

3. Write an expression for and then find the value of the difference of 22 and 7, divided by the product of 3 and 5.

4. How many terms are there in the expression $5x^2y + 4(a - b) - 4x$?

5. Evaluate $5(2)^3 - 3^2 + 6$.

6. Evaluate $4a^2$ if $a = 3$.

7. Evaluate $3a^2b + 4ac - 6bc$ if $a = 5, b = 3, c = 1$.

8. A formula from physics is $V = 2\pi rf$. Find V when $\pi = 3.14, r = 2, f = 5$.

9. Simplify by combining like terms: $4x + 2y - 2x + 5y$.

10. Combine like terms: $16a^2b + 12ab - 9a^2b - 5ab + a^2b$.

CHAPTER 4

The Language of Algebra

A formula for changing Celsius temperature to Fahrenheit is $F = \frac{9}{5}C + 32$. Find F when $C = 30$.

Using letters to represent numbers is a fundamental characteristic of algebra. In this chapter letters will be used to represent the numbers of arithmetic. The skills you have learned in the first three chapters will be used along with some new definitions and symbols.

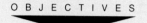 4-1 ALGEBRAIC EXPRESSIONS

OBJECTIVES

In this section you will learn to:
1. Write literal expressions involving operations of arithmetic.
2. Identify terms and factors of an expression.

All of the techniques used with the numbers of arithmetic are valid in algebra. The plus sign (+) is still used to indicate a sum. The sum of the numbers a and b is represented by $a + b$. The minus sign (−) still indicates a difference. The difference of a and b is represented by $a - b$. However, the operations of multiplication and division need some special attention.

The operation of multiplication (the product of two numbers) can be represented in four ways. You need to be familiar with each of these.

1. The times sign (×) used in arithmetic is still used in algebra. $a \times b$ (read as "a times b") represents the product of the two numbers a and b. This symbol is not usually used with numbers because it can be confused with the letter x.

2. $a \cdot b$ (read as "a times b") is also used to represent the product of two numbers a and b. This symbol is not usually used with numbers because it can be confused with the decimal point.

3. If two letters, or a number and a letter, are written together without an operation sign, multiplication is indicated.

   Example 1 ab means "a times b."

   Example 2 $5x$ means "5 times x."

   Example 3 $3y$ means "3 times y."

   This process of writing two letters or a number and a letter together without an operation sign is called **juxtaposition.** This method cannot be used with two numbers since 52 does not mean "5 times 2."

4. When two sets of parentheses have no operation sign between them, multiplication is indicated. Also, a letter or a number preceding a set of parentheses with no sign between them indicates multiplication. Of these four methods of indicating multiplication, the last two are most commonly used in algebra.

   Example 4 $(a)(b)$ means "a times b."

   Example 5 $a(b)$ means "a times b."

   Example 6 $4(a)$ means "4 times a."

Example 7 $3(x + y)$ means "3 times the sum of x and y."

Example 8 $(a + b)(x + y)$ means "the product of the sum of a and b and the sum of x and y."

The quotient of two numbers a and b can be written as $a \div b$ or as $\frac{a}{b}$. The fractional representation of division is most commonly used in algebra.

When letters are used to represent numbers, they are called **literal numbers.** If a number expression contains one or more literal numbers, it is called a **literal expression** or an **algebraic expression.**

Example 9   Write a literal expression for "the sum of x and 5, divided by 7."

$(x + 5)$ represents "the sum of x and 5" so we have $\frac{x + 5}{7}$ or $(x + 5) \div 7$.

Example 10   Write an algebraic expression for "the difference of a and b, multiplied by the sum of a and b."

The difference of a and b is $(a - b)$, and the sum of a and b is $(a + b)$, so we have $(a - b)(a + b)$.

Example 11   Express $c - (a + b)$ in words.

Since $(a + b)$ is in parentheses it is "the sum of a and b." So in words we have "the difference of c, and the sum of a and b."

▼ EXERCISE SET 4-1-1

Write literal expressions for each of the following (problems 1–12).

A N S W E R S

1. The sum of x and y

2. The difference of c and d

1. _____

2. _____

3. The product of a and p

4. The quotient of x and y

3. _____

4. _____

5. The product of a, and the sum of x and y

6. The quotient of a and the sum of x and y

5. _____

6. _____

7. The product of the sum of x and y, and the difference of x and y

8. The quotient of the product of a and b, and the sum of x and y

7. _____

8. _____

9. Five times *x*, subtracted from 3 times *y*

10. Seven times *a*, divided by eleven times *c*

11. *a* minus *b*, divided by *a* plus *b*

12. *a* divided by *b*, minus the sum of *a* and *b*

Express each of the following in words (problems 13–22).

13. $a + b$

14. xy

15. $x - y$

16. $a \div (b + c)$

17. $(a - b) \div c$

18. $a - (b \div c)$

19. $\dfrac{a + b}{a - b}$

20. $x(x + y)$

21. $(a + b)(a - b)$

22. $(x - y) \div (x + y)$

Write expressions for each of the following and then find its value (problems 23–30).

23. The sum of $\dfrac{3}{4}$ and $\dfrac{7}{8}$

24. The difference of 17 and 12

25. The sum of 19 and 23, divided by the product of 7 and 3

26. The quotient of $\dfrac{2}{3}$ and $\dfrac{1}{2}$

27. The product of $\frac{2}{3}$ and $\frac{3}{7}$, divided by $\frac{1}{2}$

28. One-half the sum of 9 and 19.

29. The product of $\frac{2}{3}$ and $\frac{5}{8}$, plus the quotient of $\frac{2}{3}$ and $\frac{5}{8}$

30. The sum of $\frac{1}{2}$ and $\frac{2}{3}$, subtracted from 10

27. _____

28. _____

29. _____

30. _____

Definition When an algebraic expression is composed of parts connected by addition or subtraction signs, these parts are called the **terms** of the expression.

Example 12 $a + b$ has two terms: a and b.

Example 13 $2x + 5y + 3$ has three terms: $2x$, $5y$, and 3.

Definition When an algebraic expression is composed of parts to be multiplied, these parts are called the **factors** of the expression.

Example 14 ab has two factors: a and b.

It is very important to be able to distinguish between terms and factors. Rules that apply to terms will not, in general, apply to factors. When naming terms or factors, it is necessary to regard the entire expression.

Example 15 The expression $3x - 2y - 7$ has three terms. Note that a term may contain factors (for example, the first term of this expression, $3x$, contains two factors), but the entire expression is made up of terms.

Example 16 $5xyz$ is one term made up of factors.

Example 17 $3(a + b)$ is one term having two factors. Note here that the factor $(a + b)$ has two terms, but the entire expression is made up of factors.

In section 1–2 we introduced the idea of exponent. We wrote 5^3 as $5 \times 5 \times 5$. Using parentheses to indicate multiplication, we write $5^3 = (5)(5)(5)$.

Example 18 x^5 means $(x)(x)(x)(x)(x)$.

Note the difference between $2x^3$ and $(2x)^3$. From the use of parentheses as grouping symbols we see that $2x^3$ means $2(x)(x)(x)$, whereas $(2x)^3$ means $(2x)(2x)(2x)$ or $8x^3$.

Definition In an expression such as $5x^4$

5 is called the **coefficient,**
x is called the **base,**
4 is called the **exponent.**

Note that only the base is affected by the exponent.

WARNING

Many students make the error of multiplying the base by the exponent. For example, they will say $3^4 = 12$ instead of the correct answer $3^4 = (3)(3)(3)(3) = 81$.

Example 19 In the expression ax^3

a is the coefficient,
x is the base,
3 is the exponent,
ax^3 means $(a)(x)(x)(x)$.

Example 20 In the expression $(ax)^3$

1 is the coefficient (understood),
ax is the base (because of the parentheses),
3 is the exponent,
$(ax)^3$ means $(ax)(ax)(ax)$.

When we write a literal number such as *x*, it will be understood that the coefficient is one and the exponent is one. This can be very important in many operations.

$$x \text{ means } 1x^1$$

If an expression has grouping symbols, the operations therein are performed first. If there are terms with exponents, these are evaluated next. Finally, the usual order of operations, multiplication and division, then addition and subtraction, is followed.

EXERCISE SET 4-1-2

State the number of terms in each expression for problems 1–20.

1. $x + y$
2. $2x + y$
3. $2x$

4. $8xy$
5. $x + 3$
6. $25 - a$

7. $3x^2$
8. $x^2 + 1$
9. $5a - b + 3c$

10. $x - y - 3$
11. $3x^2 - 2x + 4$
12. $4a^2$

13. $a + 2b - c + 1$
14. $4a^2 + ab - 2b$
15. $6x^3 - 4x^2 + x - 8$

16. $3a(a - 1) + 1$
17. $3(a^4 + b^2)$
18. $10x^2yz^3$

19. $x + (2x + y)$
20. $3(a + b)(a - b)$

ANSWERS

1. _____
2. _____
3. _____
4. _____
5. _____
6. _____
7. _____
8. _____
9. _____
10. _____
11. _____
12. _____
13. _____
14. _____
15. _____
16. _____
17. _____
18. _____
19. _____
20. _____

Evaluate each of the following.

21. 3^2
22. 5^2
23. 2^3
24. 7^3

25. 5^4
26. 2^4
27. 3^5
28. 4^5

29. $\left(\dfrac{1}{2}\right)^3$
30. $\left(\dfrac{1}{3}\right)^4$
31. $2(5)^2$
32. $[2(5)]^2$

33. $[2(3)]^5$
34. $2(3)^5$
35. $5(3)^4$
36. $[5(3)]^4$

37. $10 + 5^2$
38. $3^4 - 2^5$

39. $3(5)^2 + 2(5) - 8$
40. $35 + 3^2 - 5(2)^3$

4–2 EVALUATING LITERAL EXPRESSIONS

OBJECTIVES

In this section you will learn to:
1. Substitute numbers for letters in literal expressions.
2. Evaluate the expression once the substitution has been made.

The **principle of substitution** states that any quantity may be substituted for its equal in any process. This principle is used extensively in algebra and we will use it here to evaluate literal expressions. In this process we substitute numbers for letters and find a numerical value.

Example 1 Evaluate $x + 3$ if $x = 5$.

We substitute 5 for x in the expression obtaining

$$5 + 3 = 8.$$

Example 2 Evaluate $4a - 1$ if $a = 3$.

We substitute 3 for a in the expression obtaining

$$4(3) - 1 = 12 - 1$$
$$= 11.$$

Example 3 Evaluate $x^2 + 2x$ if $x = 3$.

We substitute 3 for x in the expression obtaining

$$(3)^2 + 2(3).$$

Evaluating this, we obtain

$$(3)^2 + 2(3) = 9 + 6$$
$$= 15.$$

Example 4 Evaluate $3a^3 - 2a^2 + a$ if $a = 2$.

We substitute 2 for a in the expression obtaining

$$3(2)^3 - 2(2)^2 + 2 = 24 - 8 + 2$$
$$= 18.$$

Example 5 Evaluate $3x + 7y$ if $x = 4$ and $y = 7$.

$$3x + 7y = 3(4) + 7(7)$$
$$= 12 + 49$$
$$= 61$$

Example 6 Evaluate $3x^2 - 5y$ if $x = 5$ and $y = 2$.

$$3x^2 - 5y = 3(5)^2 - 5(2)$$
$$= 3(25) - 5(2)$$
$$= 75 - 10$$
$$= 65$$

When substituting a number for a letter, it is a good practice to enclose the number in parentheses so that the proper operation will be performed.

If the same letter appears in more than one term of an expression, it must have the same value each time it occurs. For instance, in the expression $2x + 3xy$, the x could not have one value in the first term and another in the second term.

"Evaluate" will always mean that you are to obtain a number value.

Example 7 If $x = 3$, $y = 2$, and $z = 5$, evaluate $5x^2y - 3y - z^2$.

$$5x^2y - 3y - z^2 = 5(3)^2(2) - 3(2) - (5)^2$$
$$= 5(9)(2) - 3(2) - 25$$
$$= 90 - 6 - 25$$
$$= 59$$

One of the most common uses of evaluating literal expressions is in working with formulas.

Example 8 The perimeter (distance around) of a rectangle is found by using the formula $P = 2\ell + 2w$ where ℓ represents the length and w represents the width.

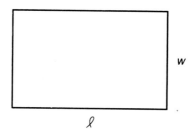

If the length of a rectangle is 10 and the width is 6, we may find the perimeter by substituting 10 for ℓ and 6 for w.

$$P = 2\ell + 2w$$
$$= 2(10) + 2(6)$$
$$= 20 + 12$$
$$= 32$$

▼ EXERCISE SET 4-2-1

Evaluate the following.

1. $x + 10$ if $x = 8$
2. $x - 5$ if $x = 9$

3. $5a$ if $a = 4$
4. $8b$ if $b = 2$

5. $x - 3$ if $x = 3$
6. $x + 4$ if $x = 1$

7. $2a + 1$ if $a = 4$
8. $3x + 5$ if $x = 2$

9. $5x - 10$ if $x = 3$
10. $5b - 7$ if $b = 2$

11. x^4 if $x = 2$
12. a^5 if $a = 3$

13. $3a^2$ if $a = 5$

14. $8x^3$ if $x = 4$

15. $x^2 + x$ if $x = 7$

16. $a^4 - 3a$ if $a = 2$

17. $2a^3 - 3a$ if $a = 6$

18. $4x^5 + 3x$ if $x = 3$

19. $3x^4 + x^3 - x^2$ if $x = 3$

20. $5x^5 - 2x^3 + x^2$ if $x = 2$

21. $x + y$ if $x = 3, y = 5$

22. $2a - b$ if $a = 9, b = 10$

23. $5x + 3y$ if $x = 6, y = 4$

24. $5x^2y$ if $x = 3, y = 2$

25. $4a^3b^2$ if $a = 2, b = 3$

26. $a^3 - a^2$ if $a = 5$

27. $3x^2 - x$ if $x = 4$

28. $x^3 - y^3$ if $x = 5, y = 4$

29. $3a^4 + b^2$ if $a = 2, b = 5$

30. $2x^2yz^3$ if $x = 3, y = 7, z = 2$

31. $3a^5b^2 - 4ab^3$ if $a = 2, b = 3$

32. $x^2 + 3x - 8$ if $x = 5$

33. $4a^2 - 2a + 5$ if $a = 3$

34. $4a^2bc - 5a + 4bc$ if $a = 5, b = 3, c = 6$

ANSWERS

13. _____
14. _____
15. _____
16. _____
17. _____
18. _____
19. _____
20. _____
21. _____
22. _____
23. _____
24. _____
25. _____
26. _____
27. _____
28. _____
29. _____
30. _____
31. _____
32. _____
33. _____
34. _____

35. $6x^3y + 2z - 5xy^2$ if $x = 5$, $y = 2$, $z = 1$

36. $10a^4b^8c^5d^9$ if $a = 4$, $b = 2$, $c = 0$, $d = 1$

37. $5x^3y + xy^2z - 10z$ if $x = 4$, $y = 5$, $z = 10$

38. $6a^2b^3 - 10ab + b^2$ if $a = 5$, $b = 3$

39. $4x^4y^5z + 11xz - 4y$ if $x = 3$, $y = 0$, $z = 2$

40. $5x^3y + 7x^2z - 3yz$ if $x = 2$, $y = 3$, $z = 4$

41. The perimeter of a rectangle is given by $P = 2\ell + 2w$. Find P when $\ell = 12$, $w = 7$.

42. The area of a rectangle is given by $A = bh$. Find A when $b = 9$, $h = 3$.

43. The perimeter of a square is given by $P = 4s$, where s represents the length of one side. Find P when $s = 8$.

44. The area of a square is given by $A = s^2$. Find A when $s = 3.5$.

45. The circumference of a circle is given by $C = \pi d$, where d represents the diameter of the circle and π is a constant number approximately equal to $\frac{22}{7}$. Find C when $d = 14$ and $\pi = \frac{22}{7}$.

46. The area of a circle is given by $A = \pi r^2$, where r represents the radius of the circle. Find A when $\pi = \frac{22}{7}$ and $r = 7$.

47. A formula for changing Fahrenheit temperature to Celsius is given by $C = \frac{5}{9}(F - 32)$. Find C when $F = 68$.

48. A formula for changing Celsius temperature to Fahrenheit is $F = \frac{9}{5}C + 32$. Find F when $C = 30$.

49. A formula from physics is $s = h + vt - 490t^2$. Find s when $h = 5{,}500$; $v = 100$; $t = 3$.

50. A formula from mathematics is $S = \frac{a}{1 - r}$. Find S when $a = 12$, $r = \frac{1}{4}$.

47. _____

48. _____

49. _____

50. _____

Definition **Like terms** are terms that have exactly the same literal factors.

OBJECTIVES

In this section you will learn to:
1. Identify like terms.
2. Combine like terms.

Example 1 $5x$ and $3x$ are like terms since they have the same literal factors. (x is the literal factor in each term.)

Example 2 $5x$ and $3y$ are *not* like terms since the literal factors are not the same. (x and y are the literal factors.)

Example 3 $3x^2y$ and x^2y are like terms since they have the same literal factors (x^2 and y).

Example 4 $3x^2y$ and $2xy$ are *not* like terms. They have different literal factors. (Note that x^2 and x are *not* the same.)

Rule Only like terms can be combined. When two terms are like terms, combine the numerical coefficients to obtain the coefficient of the like common factors.

Example 5 If we add the like terms $5x + 3x$, we combine the coefficients 5 and 3 obtaining

$$5x + 3x = (5 + 3)x$$
$$= 8x.$$

Example 6 $5x^2y - 2x^2y = (5 - 2)x^2y$
$= 3x^2y$

Example 7 $7A + 2A = 9A$

> **WARNING**
>
> Many students make the mistake of interchanging capital and lowercase letters in an expression. In algebra a capital "A" and a lowercase "a" are considered just as different as x and y. Therefore, terms such as 5A and 3a would *not* be considered like terms. Be careful. Be consistent with the letters you use.

Example 8 $7A + 3a - 2A + 5a = 5A + 8a$

Example 9 $3x^2y + 5xy - 2xy + 3x^2y = 6x^2y + 3xy$

Example 10 $7x + 2y + x + 5y - 3y = 8x + 4y$

Notice that the numerical coefficient of x is 1 and in this example must be added to the 7 of $7x$.

Example 11 $12ab + 4ba - 6ab = 10ab$

By the commutative law of multiplication $4ba = 4ab$.

EXERCISE SET 4-3-1

Simplify the following by combining like terms.

1. $4x + 7x$
2. $5a + 8a$
3. $9x - 5x$
4. $11a - 6a$
5. $3a + a$
6. $x + 5x$

7. $5x^2 + 3x^2$
8. $3x^2 + x^2$
9. $7a^3 - a^3$

10. $10a^3 - 5a^3$
11. $4xy + 12xy$
12. $12ab - 5ab$

13. $8xy - 7xy$
14. $3a^2b + 8a^2b$
15. $9a + 6b - 5a$

16. $5x - 4y + 6x$
17. $6ab^2 + 9a^2b + 5ab^2$

18. $13xy^2 - 7x^2y - 5xy^2$
19. $14abc + 3ab + 8abc - 2ab$

20. $4xy + 9x^2y + 5xy - 3x^2y$
21. $5ab + 11ac - 4ab - 10ac$

22. $6x^2y + 21xy^2 + 4x^2y - 17xy^2$

23. $13a^2 + 14a - 11a^2 - 6a + a^2$

24. $8xyz + 3xz - 5xyz - 2xz - 3xyz$

25. $14x^3y + 3x^3y^2 - 9x^3y + 5x^3y^2 - 5x^3y$

26. $5a^2b + 10ab - 4a^2b - 4ab + 2ab^2$

ANSWERS

7. _____
8. _____
9. _____
10. _____
11. _____
12. _____
13. _____
14. _____
15. _____
16. _____
17. _____
18. _____
19. _____
20. _____
21. _____
22. _____
23. _____
24. _____
25. _____
26. _____

27. $4x^2y - 5xy^2 + 15x^2y^2 + 3x^2y - 6x^2y^2$

28. $15ab^2 + 9a^2b - 5ab + 4a^2b - 6ab^2$

29. $20m^2n + 18n^2m - 17nm^2 - 12mn^2$

30. $7xyz - 2yxz - 3zxy - zyx - xzy$

31. $12M + 2m + 6m - 11M$

32. $8T + 9t - 7T - 8t$

Find the perimeter of each of the following figures. (The perimeter is the sum of the lengths of the sides of a figure.)

33. Rectangle with length ℓ and width w.

34.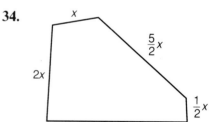

CHAPTER 4 SUMMARY

Operations on Literal Numbers

- The usual operation symbols are used to indicate the sum or difference of two literal numbers. For example, $a + b$ indicates the sum of a and b, and $x - y$ indicates the difference of x and y.
- The product of two literal numbers is usually indicated by placing them next to each other. For example, xy indicates the product of x and y, and $3a$ indicates the product of 3 and a. We sometimes use parentheses to indicate a product.

Terms

- Terms are those parts of an algebraic expression that are being added or subtracted.

Factors

- Factors are those parts of an algebraic expression that are being multiplied.

Combining Like Terms

- Like terms are terms having exactly the same literal factors.
- Like terms may be combined by combining the numerical coefficients to obtain the coefficient of the like common factors.

CHAPTER 4 REVIEW

ANSWERS

Write literal expressions for problems 1–6.

1. The sum of x and y.

2. The product of d and f.

3. The quotient of $3a$ and $4b$.

4. The product of the sum of a and b, and the difference of x and 3.

5. The quotient of the sum of x and y, and the product of a and c.

6. Eight times a number x, divided by the difference of 9 and a.

7. Find the value of the sum of $1\frac{2}{3}$ and $5\frac{1}{4}$.

8. Find the value of the product of 10 and $3\frac{5}{8}$.

9. Find the value of the product of $\frac{2}{3}$ and the quotient of $\frac{2}{5}$ and $\frac{4}{15}$.

10. Find the value of the sum of $\frac{3}{4}$ and $\frac{2}{7}$, divided by $\frac{1}{4}$.

11. How many terms are there in the expression $2a^2b - 4ab + 5$?

12. How many terms are there in the expression $4a(b + c)$?

13. Evaluate 6^3.

14. Evaluate $5(2)^4$.

15. Evaluate $3(5)^2 - 4^3 + 2(7)^2$.

16. Evaluate $4a^2b$ if $a = 3$, $b = 2$.

CHAPTER 4 REVIEW

17. Evaluate $2x^3 + 4y^2$ if $x = 2$, $y = 5$.

18. Evaluate $a^2 + 3ab - b^2$ if $a = 6, b = 4$.

19. Evaluate $5x^3y + 3yz - 10xy^4$ if $x = 0, y = 4, z = 5$.

20. Evaluate $2a^2bc + 4bc^2 - 5ac$ if $a = 1, b = 3, c = 10$.

21. The perimeter of a rectangle is given by the formula $P = 2\ell + 2w$. Find P when $\ell = 7$ and $w = 3\frac{1}{2}$.

22. The area of a triangle is given by the formula $A = \frac{1}{2}bh$. Find A when $b = \frac{3}{4}$ and $h = 8$.

23. A formula from physics is $s = vt - 5t^2$. Find s when $v = 18$ and $t = 3$.

24. A formula from mathematics is $S = \frac{n}{2}(a + b)$. Find S when $n = 15, a = 3$, and $b = 11$.

Combine like terms in problems 25–30.

25. $4a + 5a - 6b$

26. $5x^2 + 3x - 4y + 6x$

27. $2xy + 3x^2y + 5xy - x^2y$

28. $3a^2b + 9a^2 - 2a^2b - 5a^2$

29. $20xy^2 + 4xy - 15xy^2 - 3xy - 5xy^2$

30. $8x^2y^3 + 21x^2y + 5x^2y^3 - 14x^2y - 3x^2y - x^2y^2$

ANSWERS

17. _____
18. _____
19. _____
20. _____
21. _____
22. _____
23. _____
24. _____
25. _____
26. _____
27. _____
28. _____
29. _____
30. _____

SCORE: _____

CHAPTER 4 TEST

NAME: _____

CLASS / SECTION: _____ DATE: _____

ANSWERS

1. _____

2. _____

3. _____

4. _____

5. _____

6. _____

7. _____

8. _____

9. _____

10. _____

SCORE: _____

1. Write a literal expression for the product of x and y, divided by the difference of a and b.

2. Express $a(a + b)$ in words.

3. Write an expression for and then find the value of the difference of 24 and 9, divided by the sum of 2 and 3.

4. How many terms are there in the expression $5x^2(x - y)$?

5. Evaluate: $4(3)^2 - 2^3 + 4(5)^2$.

6. Evaluate xy^2 if $x = 3$, $y = 4$.

7. Evaluate $3x^2y + 5xz - 4yz^2$ if $x = 4$, $y = 2$, $z = 3$.

8. A formula from physics is $a = \dfrac{v^2}{r}$. Find a when $v = 9$, $r = 3$.

9. Simplify by combining like terms: $8a + 3b - 5a + b$.

10. Combine like terms: $17x^2y + 8xy - 15x^2y + 10xy - 18xy$.

CHAPTERS 1-4 CUMULATIVE TEST

NAME: _____
CLASS / SECTION: _____   DATE: _____

ANSWERS

1. Round 8,306 to the nearest ten.

2. Add: 1,906
          242
         3,054

3. Multiply: 2,516
              × 218

4. Find the greatest common factor of 70 and 84.

5. Find the least common multiple of 16 and 24.

6. Reduce $\dfrac{84}{90}$.

7. Find the LCD of $\dfrac{2}{3}, \dfrac{1}{12}, \dfrac{3}{8}$.

8. Divide: $\dfrac{3}{8} \div \dfrac{9}{10}$

9. Subtract: $\dfrac{3}{4} - \dfrac{5}{12}$

10. Multiply: $3\dfrac{1}{2} \times 4\dfrac{2}{7}$

11. Change .264 to a common fraction in simplest form.

12. Multiply: 25.3
              × .42

13. Divide: $21.4 \overline{)93.518}$

14. Write .0000631 in scientific notation.

15. Simplify $\sqrt{45}$.

16. Write a literal expression for the sum of *a* and *b*, divided by the product of *c* and *d*.

1. _____
2. _____
3. _____
4. _____
5. _____
6. _____
7. _____
8. _____
9. _____
10. _____
11. _____
12. _____
13. _____
14. _____
15. _____
16. _____

245

ANSWERS

17. _____

18. _____

19. _____

20. _____

SCORE: _____

17. Evaluate $4(3)^2$.

18. Evaluate $x^3 - 2x$ if $x = 5$.

19. Evaluate $3a^2 - 5b + 2ab$ if $a = 4$, $b = 1$.

20. Combine like terms: $25x^2y + 3y^2 - 5x^2y + y^2 - 11x^2y$

CHAPTER 5 PRETEST

NAME: _____

CLASS / SECTION: _____   DATE: _____

ANSWERS

Answer as many of the following problems as you can before starting this chapter. When you finish the chapter, take the test at the end and compare your scores to see how much you have learned.

1. _____

2. _____

3. _____

4. _____

5. _____

6. _____

7. _____

8. _____

9. _____

10. _____

SCORE: _____

1. Classify $5x = 4x + 1$ as a conditional equation or an identity.

2. Are the two equations $5x + 1 = 3x + 4$ and $x = 1\frac{1}{2}$ equivalent?

Solve for x in each of the equations in problems 3–9.

3. $x + 6 = 36$

4. $\frac{2}{3}x = \frac{4}{9}$

5. $3x - 5 = 20$

6. $4x = 64$

7. $5x - 2 = 2x + 10$

8. $\frac{2}{3}x + 1 = \frac{1}{5}x + 3$

9. $\frac{4}{5}x - \frac{1}{3} = \frac{1}{6}x + \frac{1}{2}$

10. The area (A) of a triangle is 48 square inches and the base (b) is $1\frac{1}{2}$ times as long as the height (h). Find the base and height. (Use the formula $A = \frac{1}{2}bh$.)

CHAPTER 5

Solving Equations

The selling price (S) of an automobile is $8,952.30. If the dealer has set the margin (M) to be one-twentieth of the cost (C), what was the cost of the vehicle? Use the formula $C + M = S$.

5-1 CONDITIONAL AND EQUIVALENT EQUATIONS

In this section you will learn to:
1. Classify an equation as conditional or an identity.
2. Solve simple equations mentally.
3. Determine if certain equations are equivalent.

Definition An **equation** is a statement in symbols that two number expressions are equal.

Equations can be classified in two main types:

   a. An **identity** is true for all values of the literal and arithmetical numbers in it.

Example 1 $5 \times 4 = 20$ is an identity.

Example 2 $2 + 3 = 5$ is an identity.

Example 3 $2x + 3x = 5x$ is an identity since any value substituted for x will yield an equality.

   b. A **conditional** equation is true for only certain values of the literal numbers in it.

Example 4 $x + 3 = 9$ is true only if the literal number $x = 6$.

Example 5 $3x - 4 = 11$ is true only if $x = 5$.

The literal numbers in an equation are sometimes referred to as **variables**.

Finding the values that make a conditional equation true is one of the main objectives of this text.

Definition A **solution** or **root** of an equation is the value of the variable or variables that make the equation a true statement.

The solution or root is said to satisfy the equation.

Solving an equation means finding the solution or root.

5-1 CONDITIONAL AND EQUIVALENT EQUATIONS

Many equations can be solved mentally. Ability to solve an equation mentally will depend on the ability to manipulate the numbers of arithmetic. The better you know the facts of multiplication and addition, the more adept you will be at mentally solving equations.

Example 6 Solve for x: $x + 3 = 7$

To have a true statement we need a value for x that, when added to 3, will yield 7. Our knowledge of arithmetic indicates that 4 is the needed value. Therefore, the solution to the equation is $x = 4$.

Example 7 Solve for x: $x - 5 = 3$

What number do we subtract 5 from to obtain 3? Again our experience with arithmetic tells us that $8 - 5 = 3$. Therefore, the solution is $x = 8$.

Example 8 Solve for x: $3x = 15$

What number must be multiplied by 3 to obtain 15? Our answer is $x = 5$.

Example 9 Solve for x: $2x - 1 = 5$

We would subtract 1 from 6 to obtain 5. Thus $2x = 6$. Then $x = 3$.

Regardless of how an equation is solved, the solution should always be checked for correctness.

> **WARNING**
>
> Many students think that when they have found the solution to an equation, the problem is finished. Not so! The final step should always be to check the solution.

Example 10 A student solved the equation $5x - 3 = 4x + 2$ and found an answer of $x = 6$. Was this right or wrong?

Does $x = 6$ satisfy the equation $5x - 3 = 4x + 2$? To check we substitute 6 for x in the equation to see if we obtain a true statement.

$$5x - 3 = 4x + 2$$
$$5(6) - 3 = 4(6) + 2$$
$$30 - 3 = 24 + 2$$
$$27 = 26$$

This is not a true statement, so the answer $x = 6$ is wrong.

Another student solved the same equation and found $x = 5$.

$$5x - 3 = 4x + 2$$
$$5(5) - 3 = 4(5) + 2$$
$$25 - 3 = 20 + 2$$
$$22 = 22$$

This is a true statement, so $x = 5$ is correct.

Definition Two equations are **equivalent** if they have the same solution or solutions.

Example 11 $3x = 6$ and $2x + 1 = 5$ are equivalent because in both cases $x = 2$ is a solution.

Techniques for solving equations will involve processes for changing an equation to an equivalent equation. If a complicated equation such as $2x - 4 + 3x = 7x + 2 - 4x$ can be changed to a simple equation $x = 3$, and the equation $x = 3$ is equivalent to the original equation, then we have solved the equation.

Two questions now become very important.

    a. Are two equations equivalent?

    b. How can we change an equation to another equation that is equivalent to it?

The answer to the first question is found by using the substitution principle.

Example 12 Are $5x + 2 = 6x - 1$ and $x = 3$ equivalent equations?

The question becomes, does $5(3) + 2 = 6(3) - 1$?
$$15 + 2 = 18 - 1$$
$$17 = 17$$

Answer: Yes, they are equivalent.

The answer to the second question involves the techniques for solving equations that will be discussed in the next few sections.

▼ EXERCISE SET 5-1-1

ANSWERS

In problems 1–10 classify each equation as conditional or an identity.

1. $3x + x = 4x$

2. $2x + 1 = 5$

3. $x - 2 = 11$

4. $6x = 4x + 2x$

5-1 CONDITIONAL AND EQUIVALENT EQUATIONS

5. $7x = 10x - 3x$

6. $6x + 9x = 15x$

7. $20x - 8x = 7x + 5x$

8. $x + 5 = 2x - 1$

9. $4x - 3 = 17$

10. $3x + 2x = 4x + 9$

Solve equations 11–30 mentally and check by substitution.

11. $x + 4 = 7$

12. $2x = 6$

13. $x - 3 = 8$

14. $x + 1 = 10$

15. $3x = 12$

16. $2x = 1$

17. $x - 7 = 3$

18. $10 - x = 6$

19. $5x = 30$

20. $x + 9 = 11$

21. $5x = 3$

22. $2x + 1 = 11$

23. $9x = 3$

24. $12 + x = 25$

25. $x + 5 = 5$

26. $3x - 5 = 22$

27. $2x - 1 = 13$

28. $x - 3 = 0$

ANSWERS

5. _____
6. _____
7. _____
8. _____
9. _____
10. _____
11. _____
12. _____
13. _____
14. _____
15. _____
16. _____
17. _____
18. _____
19. _____
20. _____
21. _____
22. _____
23. _____
24. _____
25. _____
26. _____
27. _____
28. _____

ANSWERS

29. $3x + 4 = 22$

30. $3x + 2 = 2x + 7$

29. _____
30. _____

Determine which of the following pairs of equations are equivalent.

31. $x + 7 = 2x + 6$
    and $x = 1$

32. $3x - 5 = 2x + 1$
    and $x = 5$

31. _____
32. _____

33. $2x + 1 = x$
    and $x = 1$

34. $4x - 3 = x + 6$
    and $x = 3$

33. _____
34. _____

35. $5x - 4 = 3x + 2$
    and $x = 2$

36. $11 - 2x = 7 - x$
    and $x = 4$

35. _____
36. _____

37. $2x - 2 + x = 6 + x$
    and $x = 5$

38. $3x + 1 = x - 5 + 4x$
    and $x = 6$

37. _____
38. _____

39. $3x + 1 = x + 5$
    and $x = 2$

40. $2x + 17 = 31 - 5x$
    and $x = 2$

39. _____
40. _____

5-2 THE DIVISION RULE

OBJECTIVES

In this section you will learn to:
1. Use the division rule to solve equations.
2. Solve some basic applied problems whose solutions involve using the division rule.

Rule If each term of an equation is divided by the same nonzero number, the resulting equation is equivalent to the original equation.

To prepare to use the division rule for solving equations we must make note of the following process:

Just as 18 inches ÷ 2 = 9 inches; $18x \div 2 = 9x$.

And as 18 inches ÷ 6 = 3 inches; $18x \div 6 = 3x$.

Also, as 18 inches ÷ 18 = 1 inch; $18x \div 18 = 1x = x$.

(We usually write $1x$ as x, with the coefficient 1 understood.)

5-2 THE DIVISION RULE

Example 1 Solve for x: $3x = 10$

Our goal is to obtain $x = $ some number. The division rule allows us to divide each term of $3x = 10$ by the same number, and our goal of finding a value of x would indicate that we divide by 3. This would give us a coefficient of 1 for x.

$$3x = 10$$
$$\frac{3x}{3} = \frac{10}{3}$$
$$x = \frac{10}{3} = 3\frac{1}{3}$$

Check: $3x = 10$ and $x = \frac{10}{3}$. Are these equivalent equations?

We substitute $\frac{10}{3}$ for x in the first equation obtaining

$$3x = 10$$
$$3\left(\frac{10}{3}\right) = 10$$
$$10 = 10.$$

The equations are equivalent, so the solution is correct.

Example 2 Solve for x: $5x = 20$

$$5x = 20$$
$$\frac{5x}{5} = \frac{20}{5}$$
$$x = 4$$

Check:
$$5x = 20$$
$$5(4) = 20$$
$$20 = 20$$

Example 3 Solve for x: $8x = 4$

$$8x = 4$$
$$\frac{8x}{8} = \frac{4}{8}$$
$$x = \frac{1}{2}$$

Check:
$$8x = 4$$
$$8\left(\frac{1}{2}\right) = 4$$
$$4 = 4$$

Example 4 Solve for x: $0.5x = 6$

$$0.5x = 6$$
$$\frac{0.5x}{0.5} = \frac{6}{0.5}$$
$$x = 12$$

Check:
$$0.5x = 6$$
$$0.5(12) = 6$$
$$6 = 6$$

Example 5 Solve for x: $\frac{3}{4}x = 9$

$$\frac{3}{4}x = 9$$
$$\frac{\frac{3}{4}x}{\frac{3}{4}} = \frac{9}{\frac{3}{4}}$$
$$x = 9\left(\frac{4}{3}\right) = 12$$

Check:
$$\frac{3}{4}x = 9$$
$$\frac{3}{4}(12) = 9$$
$$9 = 9$$

Example 6 The formula for finding the circumference (C) of a circle is $C = 2\pi r$, where r represents the radius of the circle and π is approximately 3.14. Find the radius of a circle if the circumference is measured to be 40.72 cm. Give the answer correct to two decimal places.

To solve a problem involving a formula we first use the substitution principle.

$$C = 2\pi r$$
$$40.72 = 2(3.14)r$$
$$40.72 = 6.28r$$
$$\frac{40.72}{6.28} = r$$
$$r = 6.48 \text{ cm}$$

Check:
$$C = 2\pi r$$
$$40.72 = 2(3.14)(6.48)$$
$$40.72 = 40.72$$

EXERCISE SET 5-2-1

Solve for x and check in problems 1–36.

1. $2x = 24$
2. $2x = 18$
3. $3x = 12$
4. $3x = 15$
5. $5x = 15$
6. $4x = 24$
7. $3x = 1$
8. $2x = 1$
9. $4x = 2$
10. $6x = 2$
11. $7x = 21$
12. $8x = 64$
13. $21x = 7$
14. $15x = 5$

ANSWERS

1. _____
2. _____
3. _____
4. _____
5. _____
6. _____
7. _____
8. _____
9. _____
10. _____
11. _____
12. _____
13. _____
14. _____

15. $15x = 15$

16. $9x = 9$

15. _____

17. $3x = 9$

18. $5x = 25$

16. _____

17. _____

19. $6x = 3$

20. $25x = 5$

18. _____

19. _____

21. $12x = 180$

22. $6x = 132$

20. _____

21. _____

23. $10x = 4$

24. $12x = 10$

22. _____

23. _____

25. $52x = 13$

26. $70x = 14$

24. _____

25. _____

27. $0.2x = 5$

28. $0.3x = 9$

26. _____

27. _____

28. _____

29. $1.5x = 4.8$

30. $2.4x = 15.6$

31. $\frac{2}{3}x = 10$

32. $\frac{3}{5}x = 18$

33. $\frac{1}{2}x = 4$

34. $\frac{1}{3}x = 9$

35. $\frac{3}{5}x = \frac{9}{20}$

36. $\frac{2}{3}x = \frac{8}{15}$

37. The formula for the area of a rectangle is $A = \ell w$, where A represents the area, ℓ the length, and w the width. If the area of a rectangle is 391 square meters and the length is 23 meters, find the width.

38. The formula for the perimeter (P) of a square is given by $P = 4s$, where s represents the length of one side. Find the length of a side of a square whose perimeter is 52 inches.

39. The formula for the area of a triangle is $A = \frac{1}{2}bh$, where A represents the area, b the base, and h the altitude. If the area of the triangle is 279 square inches and the altitude is 18 inches, find the length of the base.

40. The formula $d = rt$ gives the distance d traveled at a constant rate r in the time t. If a person drives 347 kilometers in five hours, what was the average rate?

ANSWERS

29. _____
30. _____
31. _____
32. _____
33. _____
34. _____
35. _____
36. _____
37. _____
38. _____
39. _____
40. _____

 5-3 THE SUBTRACTION RULE

OBJECTIVES

In this section you will learn to use the subtraction rule to solve equations.

Rule If the same quantity is subtracted from both sides of an equation, the resulting equation will be equivalent to the original equation.

Example 1 Solve for x if $x + 7 = 12$.

Even though this equation can easily be solved mentally, we wish to illustrate the subtraction rule. We should think in this manner: "I wish to solve for x so I need x by itself on one side of the equation. But I have $x + 7$. So if I subtract 7 from $x + 7$, I will have x alone on the left side." (Remember that a quantity subtracted from itself gives zero.) But if we subtract 7 from one side of the equation, the rule requires us to subtract 7 from the other side as well. So we proceed as follows.

$$x + 7 = 12$$
$$x + 7 - 7 = 12 - 7$$
$$x + 0 = 5$$
$$x = 5$$

Check:
$$x + 7 = 12$$
$$5 + 7 = 12$$
$$12 = 12$$

Example 2 Solve for x: $5x = 4x + 3$

Here our thinking should proceed in this manner. "I wish to obtain all unknown quantities on one side of the equation and all numbers of arithmetic on the other so I have an equation of the form $x = $ some number. I thus need to subtract $4x$ from both sides."

$$5x = 4x + 3$$
$$5x - 4x = 4x - 4x + 3$$
$$5x - 4x = 0 + 3$$
$$x = 3$$

Check:
$$5x = 4x + 3$$
$$5(3) = 4(3) + 3$$
$$15 = 12 + 3$$
$$15 = 15$$

Example 3 Solve for x: $3x + 6 = 2x + 11$

Here we have a more involved task. First subtract 6 from both sides.

$$3x + 6 = 2x + 11$$
$$3x + 6 - 6 = 2x + 11 - 6$$
$$3x = 2x + 5$$

Now we must eliminate $2x$ on the right side by subtracting $2x$ from both sides.

$$3x - 2x = 2x + 5 - 2x$$
$$x = 0 + 5$$
$$x = 5$$

Check:
$$3x + 6 = 2x + 11$$
$$3(5) + 6 = 2(5) + 11$$
$$15 + 6 = 10 + 11$$
$$21 = 21$$

We now look at a solution that requires the use of both the subtraction rule and the division rule.

Example 4 Solve for x: $3x + 2 = 17$

We first use the subtraction rule to subtract 2 from both sides obtaining

$$3x + 2 = 17$$
$$3x + 2 - 2 = 17 - 2$$
$$3x = 15.$$

Then we use the division rule to obtain

$$\frac{3x}{3} = \frac{15}{3}$$
$$x = 5.$$

Check:
$$3x + 2 = 17$$
$$3(5) + 2 = 17$$
$$15 + 2 = 17$$
$$17 = 17$$

Example 5 Solve for x: $7x + 1 = 5x + 9$

We first use the subtraction rule.

$$7x + 1 = 5x + 9$$
$$7x + 1 - 1 = 5x + 9 - 1$$
$$7x = 5x + 8$$
$$7x - 5x = 5x + 8 - 5x$$
$$2x = 8$$

Then the division rule gives us

$$\frac{2x}{2} = \frac{8}{2}$$
$$x = 4.$$

Check: $7x + 1 = 5x + 9$
$7(4) + 1 = 5(4) + 9$
$28 + 1 = 20 + 9$
$29 = 29$

Example 6 The perimeter (P) of a rectangle is found by using the formula $P = 2\ell + 2w$, where ℓ stands for the length and w stands for the width. If the perimeter of a rectangle is 54 cm and the length is 15 cm, what is the width?

Using the formula $P = 2\ell + 2w$, we have

$$54 = 2(15) + 2w$$
$$54 = 30 + 2w$$
$$54 - 30 = 30 + 2w - 30$$
$$24 = 2w$$
$$\frac{24}{2} = \frac{2w}{2}$$
$$12 = w$$
$$w = 12.$$

Check: $54 = 2(15) + 2(12)$
$54 = 30 + 24$
$54 = 54$

EXERCISE SET 5-3-1

Solve for x and check.

1. $x + 3 = 5$

2. $x + 4 = 11$

3. $x + 1 = 9$

4. $x + 5 = 13$

5. $x + 3.4 = 8.1$

6. $x + 2.8 = 3.5$

7. $x + \dfrac{3}{4} = 4\dfrac{2}{5}$

8. $x + 1\dfrac{1}{2} = 3\dfrac{1}{3}$

9. $x + 15 = 52$

10. $x + 12 = 21$

11. $x + 5 = 5$

12. $x + 8 = 8$

13. $2x = x + 1$

14. $3x = 2x + 7$

15. $7x = 6x + 1$

16. $8 + x = 2x$

17. $5x + 2 = 4x + 10$

18. $7x + 5 = 6x + 14$

19. $2x + 3 = 15$

20. $3x + 1 = 13$

ANSWERS

7. _____

8. _____

9. _____

10. _____

11. _____

12. _____

13. _____

14. _____

15. _____

16. _____

17. _____

18. _____

19. _____

20. _____

ANSWERS

21. _____

22. _____

23. _____

24. _____

25. _____

26. _____

27. _____

28. _____

29. _____

30. _____

31. _____

32. _____

33. _____

34. _____

21. $5x + 1 = 21$

22. $4x + 3 = 27$

23. $8x + 3 = 19$

24. $7x + 4 = 25$

25. $3x + 10 = 13$

26. $6x + 5 = 11$

27. $7x = 2x + 15$

28. $12x = 5x + 14$

29. $11x = 7x + 20$

30. $8x = 3x + 5$

31. $4x + 3 = x + 18$

32. $5x + 4 = 2x + 25$

33. $8x + 5 = 2x + 5$

34. $17x + 15 = 20x + 3$

35. $x + 9 = 5x + 1$

36. Two sides of a triangle are 18.1 m and 13.5 m. If the perimeter (P) of the triangle is 47 m, find the length of the third side. Use the formula $P = a + b + c$.

37. If the perimeter (P) of a rectangle is 39.8 cm and the width is 7.5 cm, what is the length? Use $P = 2\ell + 2w$.

38. What is the cost (C) of an item whose selling price (S) is $12.95 if the margin ($M$) is $3.50? Use the formula $S = C + M$.

39. The total earnings (E) of a waitress are equal to the sum of the tips (t), the product of the hourly rate (r), and the number of hours (h) worked ($E = rh + t$). If Joan worked eight hours at $1.50 per hour and her total earnings were $50.75, what did she make in tips?

40. $5F = 9C + 160$ is a relationship between Fahrenheit (F) and Celsius (C) temperatures. Find the Celsius temperature if the Fahrenheit temperature is 86 degrees.

ANSWERS

35. _____

36. _____

37. _____

38. _____

39. _____

40. _____

5-4 THE ADDITION RULE

Rule If the same quantity is added to both sides of an equation, the resulting equation will be equivalent to the original equation.

OBJECTIVES

In this section you will learn to use the addition rule to solve equations.

Example 1 Solve for x if $x - 7 = 2$.

As always, in solving an equation we wish to arrive at the form of "$x = $ some number." We observe that 7 has been subtracted from x, so to obtain x alone on the left side of the equation, we add 7 to both sides.

$$x - 7 = 2$$
$$x - 7 + 7 = 2 + 7$$
$$x = 9$$

Check: $$\begin{aligned} x - 7 &= 2 \\ 9 - 7 &= 2 \\ 2 &= 2 \end{aligned}$$

Example 2 Solve for x: $2x - 3 = 6$

Keeping in mind our goal of obtaining x alone, we observe that since 3 has been subtracted from $2x$, we add 3 to both sides of the equation.

$$\begin{aligned} 2x - 3 &= 6 \\ 2x - 3 + 3 &= 6 + 3 \\ 2x &= 9 \end{aligned}$$

Now we must use the division rule.

$$\begin{aligned} \frac{2x}{2} &= \frac{9}{2} \\ x &= 4\frac{1}{2} \end{aligned}$$

Check: $$\begin{aligned} 2x - 3 &= 6 \\ 2\left(4\frac{1}{2}\right) - 3 &= 6 \\ 9 - 3 &= 6 \\ 6 &= 6 \end{aligned}$$

Example 3 Solve for x: $3x - 4 = 11$

We first use the addition rule.

$$\begin{aligned} 3x - 4 &= 11 \\ 3x - 4 + 4 &= 11 + 4 \\ 3x &= 15 \end{aligned}$$

Then using the division rule, we obtain

$$\begin{aligned} \frac{3x}{3} &= \frac{15}{3} \\ x &= 5. \end{aligned}$$

Check: $$\begin{aligned} 3x - 4 &= 11 \\ 3(5) - 4 &= 11 \\ 15 - 4 &= 11 \\ 11 &= 11 \end{aligned}$$

Example 4 Solve for x: $5x = 14 - 2x$

Here our goal of obtaining x alone on one side would suggest we eliminate the $2x$ on the right, so we add $2x$ to both sides of the equation.

$$\begin{aligned} 5x &= 14 - 2x \\ 5x + 2x &= 14 - 2x + 2x \\ 7x &= 14 \end{aligned}$$

We next apply the division rule.

$$\frac{7x}{7} = \frac{14}{7}$$
$$x = 2$$

Check:
$$5x = 14 - 2x$$
$$5(2) = 14 - 2(2)$$
$$10 = 14 - 4$$
$$10 = 10$$

Example 5 Solve for x: $3x - 2 = 8 - 2x$

Here our task is more involved. We must think of eliminating the number 2 from the left side of the equation and also the $2x$ from the right side to obtain x alone on one side. We may do either of these first. If we choose to first add $2x$ to both sides, we obtain

$$3x - 2 = 8 - 2x$$
$$3x - 2 + 2x = 8 - 2x + 2x$$
$$5x - 2 = 8.$$

We now add 2 to both sides.

$$5x - 2 + 2 = 8 + 2$$
$$5x = 10$$

Finally the division rule gives

$$\frac{5x}{5} = \frac{10}{5}$$
$$x = 2.$$

Check:
$$3x - 2 = 8 - 2x$$
$$3(2) - 2 = 8 - 2(2)$$
$$6 - 2 = 8 - 4$$
$$4 = 4$$

▼ EXERCISE SET 5-4-1

Solve for x.

1. $x - 4 = 6$
2. $x - 7 = 2$
3. $x - 5 = 1$

4. $x - 8 = 8$
5. $x - \dfrac{1}{2} = 4$
6. $x - \dfrac{2}{3} = \dfrac{3}{5}$

ANSWERS

1. _____
2. _____
3. _____
4. _____
5. _____
6. _____

7. _____

8. _____

9. _____

10. _____

11. _____

12. _____

13. _____

14. _____

15. _____

16. _____

17. _____

18. _____

19. _____

20. _____

21. _____

22. _____

23. _____

24. _____

7. $x - 3.6 = 2.7$ **8.** $x - 11.5 = 4.6$ **9.** $2x - 3 = 5$

10. $2x - 7 = 3$ **11.** $5x - 9 = 31$ **12.** $3x - 2 = 19$

13. $3x - 1 = 1$ **14.** $5x - 2 = 1$ **15.** $14x - 5 = 9$

16. $8x - 3 = 1$ **17.** $3x = 10 - 2x$ **18.** $6x = 21 - x$

19. $4x = 5 - x$ **20.** $3x = 12 - x$ **21.** $2x = 5 - 4x$

22. $5x = 2 - 3x$ **23.** $3x - 4 = 8 - x$ **24.** $2x - 3 = 12 - 3x$

25. $x + 2 = 10 - 3x$

26. $3x - 5 = 1 - 5x$

27. Using the formula $M = S - C$, determine the selling price (S) of an article that cost (C) $38.75 and sold with a margin (M) of $5.65.

28. Find the margin (M) on the sale of an item if the profit (P) was $10.35 and the overhead ($O$) was $3.54. Use $P = M - O$.

29. A relationship between Fahrenheit (F) and Celsius (C) temperatures is given by $5F - 160 = 9C$. Find the Fahrenheit temperature if the Celsius temperature is 32 degrees.

30. The distance (s) in meters above the ground at a given time (t) in seconds of an object dropped from a height (h) is given by the formula $s = h - 4.9t^2$. Find the height that an object was dropped from if it strikes the ground in 3 seconds.

ANSWERS

25. _____

26. _____

27. _____

28. _____

29. _____

30. _____

5-5 THE MULTIPLICATION RULE

Rule If each term of an equation is multiplied by the same nonzero number, the resulting equation is equivalent to the original equation.

OBJECTIVES

In this section you will learn to use the multiplication rule to solve equations.

In elementary arithmetic some of the most difficult operations are those involving fractions. The multiplication rule allows us to avoid these operations when solving an equation involving fractions by finding an equivalent equation that contains only whole numbers.

Remember that when we multiply a whole number by a fraction, we use the rule $a\left(\dfrac{b}{c}\right) = \dfrac{ab}{c}$.

Example 1 $3\left(\dfrac{2}{7}\right) = \dfrac{6}{7}$

Example 2 $6\left(\dfrac{2}{3}\right) = \dfrac{12}{3} = 4$

Example 3 $20\left(\dfrac{3}{5}\right) = \dfrac{60}{5} = 12$

We are now ready to solve an equation involving fractions.

Example 4   Solve for x: $\dfrac{2}{3}x = 18$

Keep in mind that we wish to obtain x alone on one side of the equation. We also would like to obtain an equation in whole numbers that is equivalent to the given equation. To eliminate the fraction in the equation we need to multiply by a number that is divisible by the denominator 3. We thus use the multiplication rule and multiply each term of the equation by 3.

$$(3)\left(\dfrac{2}{3}x\right) = (3)(18)$$
$$2x = 54$$

We now have an equivalent equation that contains only whole numbers. Using the division rule, we obtain

$$\dfrac{2x}{2} = \dfrac{54}{2}$$
$$x = 27.$$

Check: $\dfrac{2}{3}x = 18$
$\dfrac{2}{3}(27) = 18$
$18 = 18$

Example 5   Solve for x: $\dfrac{5}{8}x = 12$

$$\dfrac{5}{8}x = 12$$
$$(8)\left(\dfrac{5}{8}x\right) = (8)(12)$$
$$5x = 96$$
$$x = 19\dfrac{1}{5}$$

Check:
$$\frac{5}{8}x = 12$$
$$\frac{5}{8}\left(19\frac{1}{5}\right) = 12$$
$$\frac{5}{8}\left(\frac{96}{5}\right) = 12$$
$$12 = 12$$

Example 6 Solve for x: $\frac{2}{3}x = \frac{3}{5}$

Here our task is the same but a little more complex. We have two fractions to eliminate. We must multiply each term of the equation by a number that is divisible by both 3 and 5. It is best to use the least of such numbers, which you will recall is the **least common multiple**. We will therefore multiply by 15.

$$\frac{2}{3}x = \frac{3}{5}$$
$$(15)\left(\frac{2}{3}x\right) = (15)\left(\frac{3}{5}\right)$$
$$10x = 9$$
$$x = \frac{9}{10}$$

Check:
$$\frac{2}{3}x = \frac{3}{5}$$
$$\left(\frac{2}{3}\right)\left(\frac{9}{10}\right) = \frac{3}{5}$$
$$\frac{3}{5} = \frac{3}{5}$$

Example 7 Solve for x: $\frac{5}{8}x + 3 = 5\frac{1}{2}$

The least common multiple for 8 and 2 is 8, so we multiply each term of the equation by 8.

$$\frac{5}{8}x + 3 = 5\frac{1}{2}$$
$$(8)\left(\frac{5}{8}x\right) + (8)(3) = (8)\left(\frac{11}{2}\right)$$

> **WARNING**
>
> Be careful to note that *each* term of the equation must be multiplied by the same number.

$$5x + 24 = 44$$

We now use the subtraction rule.

$$5x + 24 - 24 = 44 - 24$$
$$5x = 20$$

Finally the division rule gives us

$$x = \frac{20}{5}$$
$$x = 4.$$

Check:
$$\frac{5}{8}x + 3 = \frac{11}{2}$$
$$\frac{5}{8}(4) + 3 = \frac{11}{2}$$
$$\frac{5}{2} + 3 = \frac{11}{2}$$
$$\frac{11}{2} = \frac{11}{2}$$

EXERCISE SET 5-5-1

Solve for x.

1. $\frac{1}{2}x = 6$
2. $\frac{1}{3}x = 12$
3. $\frac{2}{3}x = 10$

4. $\frac{3}{5}x = 9$
5. $\frac{4}{5}x = 4$
6. $\frac{3}{8}x = 6$

7. $\frac{3}{4}x = 8$
8. $\frac{2}{3}x = 5$
9. $\frac{7}{8}x = 42$

10. $\frac{5}{6}x = 20$
11. $\frac{3}{7}x = 14$
12. $\frac{1}{2}x = \frac{3}{4}$

13. $\frac{1}{3}x = \frac{6}{7}$
14. $\frac{2}{3}x = \frac{5}{8}$
15. $\frac{3}{5}x = \frac{5}{6}$

16. $\frac{5}{8}x = \frac{1}{4}$ 17. $\frac{5}{6}x = \frac{2}{3}$ 18. $\frac{2}{5}x = \frac{4}{15}$

19. $\frac{2}{3}x + 2 = 2\frac{5}{6}$ 20. $\frac{4}{5}x - 5 = \frac{1}{2}$ 21. $\frac{1}{4}x - \frac{2}{3} = \frac{1}{2}$

22. $\frac{7}{8}x + 4 = \frac{21}{4}$ 23. $3x - \frac{1}{2} = x + \frac{3}{4}$ 24. $2x - \frac{3}{5} = \frac{3}{10}x + 8$

25. $\frac{4}{9}x - 2 = \frac{1}{3}x + \frac{1}{2}$ 26. $\frac{5}{6}x + \frac{2}{3} = \frac{1}{2}x + \frac{3}{4}$

27. Find the altitude (h) of a triangle if the area (A) is 12 in.² and the base (b) is 3 inches. Use $A = \frac{1}{2}bh$.

28. The distance (s) in meters that a falling object travels in t seconds is given by $s = \frac{1}{2}gt^2$, where g is the acceleration due to gravity. Find g if an object was measured to fall $61\frac{1}{4}$ meters in 5 seconds.

29. A relationship between Fahrenheit (F) and Celsius (C) temperatures is given by $F = \frac{9}{5}C + 32$. Find C if $F = 212°$.

30. The volume (V) of a rectangular solid is given by the formula $V = \ell wh$, where ℓ is the length, w is the width, and h is the height. If a room is $16\frac{1}{2}$ feet long and 15 feet wide, how high must it be to have a volume of 2,970 cubic feet?

5-6 COMBINING RULES FOR SOLVING EQUATIONS

OBJECTIVES

In this section you will learn to:
1. Use combinations of the various rules to solve more complex equations.
2. Apply the orderly steps established in this section to systematically solve equations.

Many of the exercises in previous sections have required the use of more than one rule in the solution process. In fact, it is possible that a single problem could involve all the rules.

There is no mandatory process for solving equations involving more than one rule, but experience has shown that the following order gives a smoother, more mistake-free procedure.

First Eliminate fractions, if any, by multiplying each term of the equation by the least common multiple of all denominators of fractions in the equation.

Second Simplify by combining like terms on each side of the equation.

Third Add or subtract the necessary quantities to obtain the unknown quantity on one side and the numbers of arithmetic on the other side.

Fourth Divide by the coefficient of the unknown quantity.

Fifth Check your answer.

Example 1 Solve for x: $2x + 5 - \frac{2}{3}x = x + 7$

$(3)(2x) + (3)(5) - (3)\left(\frac{2}{3}x\right) = (3)(x) + (3)(7)$ (Multiply by 3 to eliminate the fraction.)

$6x + 15 - 2x = 3x + 21$
$4x + 15 = 3x + 21$ (Combine like terms.)
$x + 15 = 21$ (Subtract $3x$ from both sides.)
$x = 6$ (Subtract 15 from both sides.)

Check: $2x + 5 - \frac{2}{3}x = x + 7$

$2(6) + 5 - \frac{2}{3}(6) = 6 + 7$

$12 + 5 - 4 = 6 + 7$

$13 = 13$

Example 2 Solve for x: $\frac{3}{5}x + \frac{2}{3} = \frac{1}{3}x + 1$

Multiplying each term by 15 yields

$9x + 10 = 5x + 15$
$4x + 10 = 15$
$4x = 5$
$x = \frac{5}{4} = 1\frac{1}{4}.$

You may want to leave your answer as an improper fraction instead of a mixed number. Either form is correct, but the improper fraction form will be more useful in checking your solution.

Check:
$$\frac{3}{5}x + \frac{2}{3} = \frac{1}{3}x + 1$$
$$\frac{3}{5}\left(\frac{5}{4}\right) + \frac{2}{3} = \frac{1}{3}\left(\frac{5}{4}\right) + 1$$
$$\frac{3}{4} + \frac{2}{3} = \frac{5}{12} + 1$$
$$\frac{17}{12} = \frac{17}{12}$$

Example 3 The selling price (S) of a certain article was $30.00. If the margin ($M$) was one-fifth of the cost (C), find the cost of the article. Use the formula $C + M = S$.

Since the margin was one-fifth of the cost, we may write

$$M = \frac{1}{5}C.$$

Then
$$C + M = S$$
becomes
$$C + \frac{1}{5}C = S.$$

Since $S = 30$, we have

$$C + \frac{1}{5}C = 30.$$

Multiplying each term by 5 gives

$$5C + C = 150$$
$$6C = 150$$
$$C = 25.$$

Check: $C + M = S$
$25 + 5 = 30$
$30 = 30$

▼ EXERCISE SET 5-6-1

Solve for x in each of the following.

1. $3x = \frac{1}{2}x + 5$

2. $2x = \frac{2}{3}x + 4$

3. $\frac{1}{2}x + \frac{3}{4} = 5$

4. $x = \frac{1}{2} - \frac{2}{3}x$

ANSWERS

1. _____

2. _____

3. _____

4. _____

ANSWERS

5. _____

6. _____

7. _____

8. _____

9. _____

10. _____

11. _____

12. _____

13. _____

14. _____

15. _____

16. _____

17. _____

18. _____

19. _____

20. _____

21. _____

22. _____

5. $\dfrac{2}{3}x + \dfrac{1}{2} = \dfrac{5}{6}$

6. $2x - \dfrac{3}{5} = \dfrac{2}{3}x$

7. $2x + \dfrac{2}{3} = 3 - \dfrac{1}{3}x$

8. $\dfrac{1}{4}x + 5 = \dfrac{1}{2}x + 4$

9. $2x - \dfrac{2}{3} = \dfrac{3}{5}x + 2$

10. $4x - \dfrac{3}{8} = \dfrac{3}{4}x + 3$

11. $\dfrac{3}{4}x + \dfrac{2}{3} = \dfrac{5}{6} + \dfrac{1}{2}x$

12. $3x - \dfrac{5}{7} = \dfrac{3}{4}x + 4$

13. $\dfrac{2}{3}x - 5 = \dfrac{2}{9}x + 1$

14. $5x + \dfrac{3}{4} = 5 - \dfrac{1}{3}x$

15. $\dfrac{3}{8} + 5x = \dfrac{3}{4}x + 2$

16. $\dfrac{5}{8}x - \dfrac{1}{2} = \dfrac{3}{5}x + 3$

17. $\dfrac{4}{5}x + 6 - \dfrac{1}{3} = \dfrac{2}{5}x + 9$

18. $x + \dfrac{6}{7} + \dfrac{2}{3}x = \dfrac{3}{5}x + 4$

19. $\dfrac{2}{3}x = 11 + \dfrac{1}{2}x - \dfrac{3}{5}$

20. $\dfrac{4}{5}x - 10 = \dfrac{2}{3}x + 2 - \dfrac{1}{5}x$

21. $6x - \dfrac{2}{3} = \dfrac{5}{8}x + 4 + \dfrac{2}{3}x$

22. $\dfrac{4}{7}x = 28 - \dfrac{2}{3}x$

23. $2x + \dfrac{1}{3} = \dfrac{5}{8}x + 1$

24. $13 - \dfrac{2}{5}x = 25 - \dfrac{3}{4}x$

25. $\dfrac{2}{3}x - 1 = \dfrac{3}{5}x + \dfrac{1}{6}$

26. $x - \dfrac{5}{9} = \dfrac{5}{12}x + \dfrac{3}{4}$

27. The width (w) of a rectangle is two-thirds the length (ℓ). If the perimeter (P) of the rectangle is 250 inches, find the length. Use the formula $P = 2\ell + 2w$.

28. An article cost (C) a dealer $63.00. If the margin ($M$) is to be one-fourth of the selling price (S), what should the article sell for? Use the formula $C + M = S$.

29. The selling price (S) of an automobile is $8,952.30. If the dealer has set the margin (M) to be one-twentieth of the cost (C), what was the cost of the vehicle? Use the formula $C + M = S$.

30. The second side (b) of a triangle is one-half the first side (a) and the third side (c) is two-thirds the first side. If the perimeter (P) of the triangle is 39 inches, find the three sides. Use the formula $P = a + b + c$.

ANSWERS

23. _____
24. _____
25. _____
26. _____
27. _____
28. _____
29. _____
30. _____

CHAPTER 5 SUMMARY

Equations

- An equation is a statement in symbols that two number expressions are equal.
- An identity is true for all values of the literal and arithmetical numbers in it.
- A conditional equation is true for only certain values of the literal numbers in it.
- A solution or root of an equation is the value of the variable(s) that makes the equation a true statement.
- Two equations are equivalent if they have the same solution set.

Steps for Solving an Equation

1. Eliminate fractions by multiplying each term by the least common multiple of all denominators in the equation.
2. Combine like terms on each side of the equation.
3. Add or subtract terms to obtain the unknown quantity on one side and the numbers or arithmetic on the other.
4. Divide each term by the coefficient of the unknown quantity.
5. Check your answer.

CHAPTER 5 REVIEW

Classify the equations in problems 1–6 as conditional or as an identity.

1. $2x = x + 3$
2. $5x - x = 4x$
3. $6x - 2x = 3x + x$
4. $7x - 3 = 4x$
5. $6x = 3x + 3$
6. $2x + 3 - x = x + 3$

In problems 7–12 determine which pairs of equations are equivalent.

7. $3x - 4 = 2x$ and $x = 4$
8. $5x + 3 = 18$ and $x = 3$
9. $2x - 1 = 11$ and $x = 5$
10. $4x + 3 = 2x + 15$ and $x = 9$
11. $3x + 1 = x + 2$ and $x = \frac{1}{2}$
12. $6x + 1 - 2x = x + 1$ and $x = 0$

Solve for x.

13. $7x = 56$
14. $x + 6 = 25$
15. $x - 12 = 36$
16. $2x = 104$
17. $x + 13 = 52$
18. $x - 7 = 68$
19. $3x = 25$
20. $3x = x + 8$
21. $\frac{2}{3}x = 4$

22. $5x = 41$

23. $7x = 5x + 9$

24. $\dfrac{1}{5}x = 20$

25. $\dfrac{1}{3} + 2x = \dfrac{1}{2}x + 4$

26. $6x = 78$

27. $4x + 5 = 2x + 17$

28. $\dfrac{2}{5}x = \dfrac{4}{7}$

29. $x - \dfrac{2}{5} = \dfrac{1}{7}$

30. $4x + \dfrac{3}{4} = \dfrac{1}{2}x + 3$

31. $\dfrac{5}{6}x - 3 = \dfrac{1}{2}$

32. $14x = 168$

33. $\dfrac{2}{3}x - 18 = \dfrac{1}{2}x + 5$

34. $5x + 2 = x + 19$

35. $3x - 4 = 19$

36. $\dfrac{3}{8}x = \dfrac{5}{6}$

37. $\dfrac{4}{7}x - 5 = \dfrac{1}{2}x$

38. $7x - \dfrac{1}{2} = \dfrac{4}{5}x + 4$

39. $11x = 166$

40. $4x - 1 = 3x + 5$

ANSWERS

22. _____
23. _____
24. _____
25. _____
26. _____
27. _____
28. _____
29. _____
30. _____
31. _____
32. _____
33. _____
34. _____
35. _____
36. _____
37. _____
38. _____
39. _____
40. _____

41. $\dfrac{2}{3} + 5x = \dfrac{3}{2} + 3x$

42. $8x - 5 = x + 37$

43. $3x + 5 = 2x + 5$

44. $\dfrac{5}{6}x - \dfrac{2}{3} = \dfrac{1}{2}x + 1$

45. $3\dfrac{1}{2}x = 4\dfrac{1}{3}$

46. $x - 6 = 14 - 3x$

47. $\dfrac{3}{2}x - 2 = \dfrac{1}{5} + \dfrac{1}{3}x$

48. $\dfrac{1}{2}x - \dfrac{1}{2} = \dfrac{2}{3} - \dfrac{1}{2}x$

49. $\dfrac{3}{4} + \dfrac{2}{3}x = \dfrac{1}{2}x + 2$

50. $x - \dfrac{7}{8} = \dfrac{3}{5}x + \dfrac{3}{4}$

51. The area (A) of a rectangle is 225 square feet. Find the length (ℓ) if the width (w) is 12.5 feet. Use the formula $A = \ell w$.

52. The volume (V) of a rectangular solid is 60 cubic meters. Find the length (ℓ) of the solid if the width (w) is 5 meters and the height (h) is $\dfrac{2}{3}$ meter. Use the formula $V = \ell w h$.

53. The wages (*w*) made by a worker are equal to the product of the hourly rate (*r*) and the number of hours worked (*t*). The formula is given by $w = rt$. Find the hourly rate of a worker earning $217.50 in $37\frac{1}{2}$ hours.

54. A relationship between Fahrenheit (*F*) and Celsius (*C*) temperatures is given by $9C = 5F - 160$. Find the Fahrenheit temperature if the Celsius temperature is 23.5°.

55. Find the width (*w*) of a rectangle if the perimeter (*P*) is 400 cm and the length (ℓ) is 115.8 cm. Use the formula $P = 2\ell + 2w$.

56. The perimeter (*P*) of a certain rectangle is $20\frac{5}{6}$ meters. Find the width (*w*) of the rectangle if the width is two-thirds the length (ℓ). Use the formula $P = 2\ell + 2w$.

53. _____

54. _____

55. _____

56. _____

SCORE: _____

CHAPTER 5 TEST

NAME: _____

CLASS / SECTION: _____   DATE: _____

ANSWERS

1. Classify $5x - 2 = 3x - 2 + 2x$ as a conditional equation or an identity.

2. Are the two equations $4x - 3 = x + 1$ and $x = 4$ equivalent?

1. _____

2. _____

Solve for x in each of the equations in problems 3–9.

3. $x + 3 = 12$

4. $4x - 9 = 27$

3. _____

5. $6x = 78$

6. $\dfrac{3}{4}x = \dfrac{1}{8}$

4. _____

5. _____

7. $3x + 4 = 10 - x$

8. $\dfrac{1}{2}x - 1 = \dfrac{1}{3}x + 4$

6. _____

7. _____

9. $\dfrac{7}{8}x + \dfrac{1}{2} = \dfrac{2}{3}x + \dfrac{3}{4}$

10. The length (ℓ) of a rectangle is $2\dfrac{1}{2}$ times the width (w). If the perimeter (P) of the rectangle is 35 inches, find the length and width. Use the formula $P = 2\ell + 2w$.

8. _____

9. _____

10. _____

SCORE: _____

CHAPTER 6 PRETEST

NAME: _____

CLASS / SECTION: _____ DATE: _____

ANSWERS

Answer as many of the following problems as you can before starting this chapter. When you finish the chapter, take the test at the end and compare your scores to see how much you have learned.

1. _____

2. _____

3. _____

4. _____

5. _____

6. _____

7. _____

8. _____

9. _____

10. _____

11. _____

12. _____

13a. _____

13b. _____

14. _____

15. _____

1. Write the ratio 24:84 as a fraction in reduced form.

2. Are the ratios 16:36 and 24:56 equal?

3. Is $\dfrac{6}{16} = \dfrac{15}{40}$ true or false?

4. Is $\dfrac{16}{24} = \dfrac{12}{18}$ true or false?

5. Solve for x: $\dfrac{x}{16} = \dfrac{3}{12}$

6. Solve for x: $\dfrac{6}{x} = \dfrac{8}{12}$

7. Solve for x: $\dfrac{3}{4} = \dfrac{x}{14}$

8. If the ratio of passing grades to failing grades is 7 to 2, how many students failed if 28 passed?

9. If an automobile uses $8\dfrac{1}{4}$ gallons of gasoline to travel 231 miles, how many miles can it travel on 12 gallons?

10. If 2 teaspoons of an instant coffee are used for three cups of coffee, how many are needed for 18 cups?

11. The speed limit on some interstate highways is 65 miles per hour. What is the speed limit in kilometers per hour? Round answer to the nearest tenth. (1 mile = 1.609 kilometers)

12. Mr. Smith's lawn mower has a 7-liter gasoline tank. How many gallons will the tank hold? Round answer to nearest hundredth. (1 gallon = 3.785 liters)

13. a. Change $\dfrac{3}{8}$ to percent.

   b. Change .002 to percent.

14. The school service club has 40 members. A proposal to change the meeting time was defeated by a vote of 23 to 17. What percent of the members voted for the change?

15. Mr. Jones must pay .5% interest each month on the balance of his home loan. If the balance is now $17,850, how much interest will he have to pay this month?

SCORE: _____

CHAPTER 6

Ratio, Proportion, and Percent

The ratio of teeth on gear A to those on gear B is 5 to 7. If gear B has 35 teeth, how many teeth does gear A have?

CHAPTER 6 RATIO, PROPORTION, AND PERCENT

Ratios and proportions arise continually in our daily living and we may have used them without knowing them by name. In this chapter we will define ratios and show how they are used in proportions. The techniques established in solving equations will be utilized to find solutions to proportions. We will also examine various problems that use ratios and proportions in their solutions. Percent will be introduced as a ratio and problems involving percent will be solved using equations.

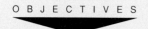

6-1 RATIOS

OBJECTIVES

In this section you will learn to:
1. Express a ratio as a fraction.
2. Determine if two ratios are equal.

Definition A **ratio** is the quotient of two numbers, usually written as a fraction.

The ratio of a to b can be written as $a:b$ or as $\dfrac{a}{b}$, but the fraction form is more common. Note the value of b cannot be zero as division by zero is not allowed.

Example 1   Write the ratio of 8 to 12 as a fraction.

The answer is $\dfrac{8}{12}$.

Example 2   Write the ratio of 2 to 3 as a fraction.

The answer is $\dfrac{2}{3}$.

Are the ratios in examples 1 and 2 equal? The answer is yes since $\dfrac{8}{12} = \dfrac{2}{3}$.

Definition Two ratios are **equal** if they are the same when expressed as fractions in reduced form.

Example 3   Is the ratio of 15 to 18 the same as the ratio of 10 to 12?

$$\dfrac{15}{18} = \dfrac{5}{6} \text{ in reduced form.}$$

$$\dfrac{10}{12} = \dfrac{5}{6} \text{ in reduced form.}$$

Therefore 15 to 18 is the same as 10 to 12.

Example 4 Write the ratio of $\frac{1}{2}$ to $\frac{3}{8}$ as a common fraction.

$$\frac{\frac{1}{2}}{\frac{3}{8}} = \frac{1}{2} \cdot \frac{8}{3} = \frac{4}{3}$$

Example 5 Write the ratio of .04 to .12 as a common fraction.

$$\frac{.04}{.12} = \frac{4}{12} = \frac{1}{3}$$

> **Definition** When a ratio expresses the amount of one thing to the units of another, we refer to it as a **rate.**

An automobile traveling at a *rate* of 55 miles per hour expresses the ratio of miles (distance) to one hour (a unit of time).

Example 6 If Jane rode her bicycle in a race and traveled 60 miles in 3 hours, express the ratio of miles to hours.

$$60 \text{ to } 3 \text{ or } \frac{60}{3}$$

Notice that this ratio is $\frac{\text{miles}}{\text{hours}}$ and can be read as miles per hour when the denominator is 1. Since $\frac{60}{3} = \frac{20}{1} = 20$, we say that Jane rode at the average speed or rate of 20 miles per hour (abbreviated mph).

Example 7 Batting averages in baseball are computed as the ratio of the number of hits to the number of times at bat. This ratio is then expressed as a decimal to the nearest thousandth. If Johnny B. had 14 hits in 38 times at bat, express his batting average as a fraction and also as a decimal to the nearest thousandth.

$$\frac{\text{hits}}{\text{at bat}} = \frac{14}{38} = \frac{7}{19}, \text{ as a fraction.}$$

$7 \div 19 = .368$ to the nearest thousandth.

EXERCISE SET 6-1-1

ANSWERS

Write the following ratios as common fractions in reduced form.

1. 5 to 7
2. 2 to 5
3. 8 to 9

4. 1 to 4
5. 6 to 15
6. 10 to 55

7. 3:17
8. 12:30
9. 16:48

10. 72:90
11. $\frac{3}{4}$ to $\frac{5}{8}$
12. $\frac{1}{3}:\frac{3}{5}$

13. .03 to .27
14. .12:3.6
15. $2\frac{1}{2}$ to $3\frac{1}{8}$

16. $.5:3\frac{1}{4}$
17. $1.4:\frac{3}{5}$

Determine if the following pairs of ratios are equal.

18. $\frac{1}{2}, \frac{5}{10}$
19. $\frac{2}{3}, \frac{8}{12}$
20. $\frac{18}{30}, \frac{3}{5}$
21. $\frac{18}{24}, \frac{3}{4}$

22. $\frac{7}{21}, \frac{2}{3}$
23. $\frac{3}{7}, \frac{36}{84}$
24. $\frac{32}{64}, \frac{7}{14}$
25. $\frac{12}{16}, \frac{16}{24}$

26. $\frac{42}{98}, \frac{48}{84}$
27. $\frac{24}{32}, \frac{36}{48}$

28. It rained 2 inches in 3 hours. Express the ratio of inches to hours.

29. A car traveled 120 miles in 3 hours. Express the ratio of miles to hours in reduced form.

30. An 8-ounce can of vegetables costs 77 cents. Express the ratio of cents to ounces and find the quotient to the nearest hundredth of a cent.

31. Baseball player Hank Aaron had 3,771 hits from 12,364 times at bat. Find his batting average by expressing the ratio of hits to the number of times at bat and give the quotient as a decimal rounded to three decimal places.

32. Edith McGuire, of the United States, ran the 200-meter dash in 23 seconds. Express the ratio of meters to seconds and give the quotient as a decimal rounded to one decimal place.

ANSWERS

28. _____

29. _____

30. _____

31. _____

32. _____

6–2 PROPORTIONS

Definition A **proportion** is a statement that two ratios are equal.

OBJECTIVES

In this section you will learn to:
1. Write a proportion.
2. Determine if a proportion is true.

Example 1 $\frac{4}{6} = \frac{2}{3}$ is a proportion.

Example 2 $\frac{3}{4} = \frac{75}{100}$ is a proportion.

Example 3 $\frac{5}{10} = \frac{10}{20}$ is a proportion.

In section 6–1 we found that two ratios are equal if they are the same when expressed as fractions in reduced form. Another method, called **cross multiplication,** is derived as follows.

Given the proportion $\frac{a}{b} = \frac{c}{d}$, we use the method of solving fractional equations and multiply both sides by the LCD, which is (bd).

$$(bd)\left(\frac{a}{b}\right) = (bd)\left(\frac{c}{d}\right)$$

This gives $ad = bc$.

This useful fact is stated as: if $\frac{a}{b} = \frac{c}{d}$, then $ad = bc$.

WARNING

Note that this "cross product" applies *only* when a *single* fraction occurs on each side of the equation.

Example 4 Is $\frac{8}{12} = \frac{6}{9}$ a proportion?

We find the cross products $(8)(9) = 72$ and $(12)(6) = 72$. Since the cross products are equal, the statement is a proportion.

Example 5 Is $\frac{7}{15} = \frac{9}{19}$ true or false?

$(7)(19) = 133$ and $(15)(9) = 135$. Since the cross products are not equal, the statement is false.

Example 6 If Kathy cycled 2 miles in 8 minutes and Bobby cycled 5 miles in 20 minutes, were they traveling at the same rate?

We need to determine if $\frac{2}{8} = \frac{5}{20}$ is true or false.

WARNING

Make sure that *both* ratios are miles to minutes.

Since $(2)(20) = 40$ and $(8)(5) = 40$, we see the statement is true. Therefore, they were traveling at the same rate.

EXERCISE SET 6-2-1

Determine if each of the following is true or false.

1. $\dfrac{1}{2} = \dfrac{3}{6}$
2. $\dfrac{2}{3} = \dfrac{6}{9}$
3. $\dfrac{3}{5} = \dfrac{6}{10}$
4. $\dfrac{3}{4} = \dfrac{12}{16}$

5. $\dfrac{12}{15} = \dfrac{4}{5}$
6. $\dfrac{6}{14} = \dfrac{3}{7}$
7. $\dfrac{2}{5} = \dfrac{6}{10}$
8. $\dfrac{8}{9} = \dfrac{4}{3}$

9. $\dfrac{6}{12} = \dfrac{5}{10}$
10. $\dfrac{3}{6} = \dfrac{5}{8}$
11. $\dfrac{4}{12} = \dfrac{8}{16}$
12. $\dfrac{14}{16} = \dfrac{21}{24}$

13. $\dfrac{12}{28} = \dfrac{6}{14}$
14. $\dfrac{5}{9} = \dfrac{35}{63}$
15. $\dfrac{25}{49} = \dfrac{5}{7}$
16. $\dfrac{14}{21} = \dfrac{12}{18}$

17. $\dfrac{15}{20} = \dfrac{16}{24}$
18. $\dfrac{36}{39} = \dfrac{24}{26}$
19. $\dfrac{36}{42} = \dfrac{24}{28}$
20. $\dfrac{24}{42} = \dfrac{47}{74}$

21. Do the ratios 3:8 and 12:24 form a proportion?

22. Do the ratios 5:9 and 40:72 form a proportion?

23. Do the ratios 18:21 and 30:35 form a proportion?

24. Do the ratios 42:72 and 63:99 form a proportion?

25. In a certain class there were 24 women and 30 men. In a second class there were 20 women and 25 men. Do the ratios of women to men in the two classes form a proportion?

26. In a certain sample 21 people favored a sales tax increase while 28 were opposed to it. In another sample 36 were in favor and 48 were opposed. Do the ratios in the two samples form a proportion?

ANSWERS

1. _____
2. _____
3. _____
4. _____
5. _____
6. _____
7. _____
8. _____
9. _____
10. _____
11. _____
12. _____
13. _____
14. _____
15. _____
16. _____
17. _____
18. _____
19. _____
20. _____
21. _____
22. _____
23. _____
24. _____
25. _____
26. _____

27. Jim cycled 3 miles in 15 minutes and George cycled 7 miles in 35 minutes. Did they travel at the same rate?

28. One car traveled 132 miles in 3 hours. A second car traveled 88 miles in two hours. Did they travel at the same rate?

29. A 6-ounce jar of mustard sells for 90 cents and a 4-ounce jar sells for 60 cents. Are they priced at the same rate per ounce?

30. A 12-ounce can of beans sells for 98 cents and an 8-ounce can sells for 67 cents. Are they priced at the same rate per ounce?

6-3 PROBLEMS INVOLVING PROPORTIONS

In this section you will learn to:
1. Solve a proportion.
2. Use proportions to solve various applied problems.

We can find a missing number in a proportion by treating the proportion as a fractional equation.

Example 1 Find x so that $\dfrac{x}{8} = \dfrac{5}{4}$.

Multiplying both sides of the equation by the LCD gives

$$(8)\left(\frac{x}{8}\right) = (8)\left(\frac{5}{4}\right)$$
$$x = 10.$$

Example 2 Find x so that $\dfrac{2}{6} = \dfrac{x}{15}$.

The LCD is 30.

$$(30)\left(\frac{2}{6}\right) = (30)\left(\frac{x}{15}\right)$$
$$10 = 2x$$
$$x = 5$$

We could use the method of cross multiplication, discussed in the previous section, to solve a proportion.

Example 3 Find x such that $\dfrac{5}{x} = \dfrac{2}{30}$.

Cross multiplying, we obtain

$$2x = (5)(30)$$
$$2x = 150$$
$$x = 75.$$

Example 4 If Sally earned $37.00 in 5 hours, how much would she earn (at the same hourly rate) in 7 hours?

Using the ratio of $\dfrac{\text{earnings}}{\text{hours}}$, we set up the proportion $\dfrac{37}{5} = \dfrac{x}{7}$.

Cross multiplying gives $5x = 259$
$x = 51.80.$

So Sally would earn $51.80 in 7 hours.

Example 5 Map makers use ratios so that maps will be accurate. If a certain road map states that 1 inch = 30 miles, what is the distance (in miles) between two cities that measure $2\dfrac{1}{4}$ inches on the map?

$$\dfrac{1}{30} = \dfrac{2\frac{1}{4}}{x}$$
$$(1)(x) = (30)\left(\dfrac{9}{4}\right)$$
$$x = \dfrac{135}{2} = 67\dfrac{1}{2} \text{ miles}$$

▼ EXERCISE 6-3-1

Solve for the missing number x in each of the following.

1. $\dfrac{1}{2} = \dfrac{x}{8}$
2. $\dfrac{x}{4} = \dfrac{9}{12}$
3. $\dfrac{1}{5} = \dfrac{4}{x}$
4. $\dfrac{2}{3} = \dfrac{x}{12}$

5. $\dfrac{5}{2} = \dfrac{x}{8}$
6. $\dfrac{4}{7} = \dfrac{x}{28}$
7. $\dfrac{x}{72} = \dfrac{5}{12}$
8. $\dfrac{3}{4} = \dfrac{27}{x}$

9. $\dfrac{3}{x} = \dfrac{21}{28}$
10. $\dfrac{4}{5} = \dfrac{24}{x}$
11. $\dfrac{x}{15} = \dfrac{3}{5}$
12. $\dfrac{16}{72} = \dfrac{x}{9}$

13. $\dfrac{4}{x} = \dfrac{3}{15}$
14. $\dfrac{x}{12} = \dfrac{12}{18}$
15. $\dfrac{x}{10} = \dfrac{6}{4}$
16. $\dfrac{x}{2} = \dfrac{3}{5}$

ANSWERS

1. _____
2. _____
3. _____
4. _____
5. _____
6. _____
7. _____
8. _____
9. _____
10. _____
11. _____
12. _____
13. _____
14. _____
15. _____
16. _____

17. $\dfrac{6}{x} = \dfrac{5}{9}$ **18.** $\dfrac{35}{28} = \dfrac{15}{x}$

19. What number has the same ratio to 15 as 2 has to 3?

20. What number has the same ratio to 49 as 2 has to 7?

Use proportions to solve the following.

21. If Ted earns $35 in 5 hours, how much would he earn in 8 hours?

22. If Brenda earns $360 in 8 days, how much does she earn in 5 days?

23. If 3 cans of spray paint are needed to paint 2 lawn chairs, how many cans are needed to paint 12 chairs?

24. A 6-foot man has a 5-foot shadow. A tree has a 20-foot shadow. How tall is the tree?

25. The distance between two cities on a road map is 5 inches. If 1 inch represents 50 miles, what is the distance between the two cities?

26. If 2 pounds of grass seed cover 300 square feet, how many pounds are needed for 1,875 square feet?

27. If 3 pounds of hamburger cost $3.87, what is the cost of 5 pounds?

28. The ratio of teeth on gear A to those on gear B is 5 to 7. If gear B has 35 teeth, how many teeth does gear A have?

29. If two cities on a map are 5 inches apart and the actual distance between them is 275 miles, what is the actual distance between two cities that are 8 inches apart?

30. The sales tax on an item costing $75 is $5.25. What is the sales tax on an item costing $114?

31. If the property tax on a home valued at $74,000 is $814, what is the tax on a home valued at $55,000?

32. On a road map two cities are 4 inches apart and the actual distance between them is 280 miles. What is the actual distance between two cities that are $2\frac{3}{4}$ inches apart on the map?

33. A company that manufactures light bulbs found that from a sample of 1,250 light bulbs 3 were defective. How many could they expect to be defective out of 25,000 light bulbs?

34. An investment of $12,500 earned $1,687.50 interest in a year. How much interest would an investment of $40,000 have earned?

35. If an automobile uses $5\frac{1}{2}$ gallons of gasoline to travel 176 miles, how many gallons are needed to travel 254 miles? Round answer to the nearest tenth.

36. A woman's investments of stocks to bonds were in the ratio of 5 to 7. If she had $18,000 invested in stocks, how much did she have invested in bonds?

37. Hockey player Phil Esposito played 1,241 games in 17 seasons. How many games did he average playing in 3 seasons?

38. If the legal restrictions for room capacity require 3 cubic meters of air space per person, how many people can legally occupy a room containing 144 cubic meters?

39. A wildlife management team, conducting a study on a particular lake, caught, tagged, and released 75 black bass. Several weeks later they caught 125 black bass and observed that 3 of them were tagged. What is the total number of black bass they could expect to be present in the lake?

40. The "golden ratio" is approximately 1 to 1.618. This ratio is believed by some to be the width-to-length ratio of a rectangle most pleasing to the eye. Assuming this to be true, what width would a book cover be if the length is 10.5 inches and it is to be most pleasing to the eye? Round answer to one decimal place.

ANSWERS

31. _____

32. _____

33. _____

34. _____

35. _____

36. _____

37. _____

38. _____

39. _____

40. _____

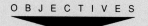
6-4 METRIC MEASUREMENT

OBJECTIVES

In this section you will learn to convert between the American and metric systems of measurement using proportions.

Metric measurement is becoming more common in the United States. Converting measurements from American to metric or metric to American makes use of ratio and proportion. The following tables give the equivalent values of various metric measurements.

Metric Length

1 kilometer (km) = 1,000 meters (m)
1 hectometer (hm) = 100 meters
1 dekameter (dam) = 10 meters
1 decimeter (dm) = .1 meter
1 centimeter (cm) = .01 meter
1 millimeter (mm) = .001 meter

The most commonly used units of metric length are the kilometer, meter, centimeter, and millimeter.

Metric Weight

1 kilogram (kg) = 1,000 grams (g)
1 milligram (mg) = .001 gram

Metric Volume

1 kiloliter (kl) = 1,000 liters (l)
1 milliliter (ml) = .001 liter

The most common relationships between the American and metric systems of measurement are listed here.

1 inch = 2.54 centimeters	1 centimeter = .3937 inch
1 yard = .9144 meter	1 meter = 1.094 yards
1 mile = 1.609 kilometers	1 kilometer = .6214 mile
1 ounce = 28.35 grams	1 gram = .0353 ounce
1 pound = .4536 kilogram	1 kilogram = 2.205 pounds
1 quart = .9464 liter	1 liter = 1.057 quarts
1 gallon = 3.785 liters	

The measurements given have been rounded and are therefore approximate.

With these conversion factors we can use proportions to convert from one system to the other.

Example 1 An article weighs 30 kilograms. How many pounds does it weigh? Round answer to two decimal places.

We let x = the number of pounds. From the table we see that 1 pound = .4536 kilogram. Using this information, we set up a proportion.

Notice that we must be careful in setting up the proportion to have like ratios on both sides of the equal sign. In this case we will have

$$\frac{\text{pounds}}{\text{kilograms}} = \frac{\text{pounds}}{\text{kilograms}}$$

$$\frac{x}{30} = \frac{1}{.4536}$$

$$x = \frac{30}{.4536}$$

$$x = 66.14 \text{ pounds.}$$

Example 2 A sign at Carnstown, Florida, states "Miami 89 miles." How many kilometers is it from Carnstown to Miami? Round answer to one decimal place.

Since 1 mile = 1.609 kilometers, we have

$$\frac{1}{1.609} = \frac{89}{x}$$

$$x = (89)(1.609)$$

$$x = 143.2 \text{ kilometers.}$$

Example 3 It took 43 liters to fill a car's gas tank. How many gallons was this? Round answer to two decimal places.

$$\frac{1}{3.785} = \frac{x}{43}$$

$$\frac{43}{3.785} = x$$

$$x = 11.36 \text{ gallons.}$$

▼ EXERCISE SET 6-4-1

Convert each of the following. Round answer to one decimal place.

1. 30 centimeters to inches
2. 36 inches to centimeters
3. 40 yards to meters
4. 10 meters to yards

ANSWERS

1. _____

2. _____

3. _____

4. _____

ANSWERS

1. _____
2. _____
3. _____
4. _____
5. _____
6. _____
7. _____
8. _____
9. _____
10. _____
11. _____
12. _____
13. _____
14. _____
15. _____
16. _____
17. _____
18. _____

5. 350 centimeters to inches

6. 2 pounds to kilograms

7. 2 quarts to liters

8. 500 centimeters to inches

9. 5 yards to meters

10. 75 meters to yards

11. 30 miles to kilometers

12. 75 centimeters to inches

13. 310 grams to ounces

14. 4 ounces to grams

15. 2,500 kilometers to miles

16. 5 pounds to kilograms

17. 100 meters to yards

18. 100 yards to meters

19. 1.75 liters to quarts

20. 125 kilograms to pounds

ANSWERS

19. _____

21. 950 kilometers to miles

22. 16 ounces to grams

20. _____

21. _____

22. _____

23. The cooling system of a certain automobile contains 13.5 liters. How many quarts does it contain?

24. The speed limit in Canada is 100 kilometers per hour. What is the speed limit in miles per hour?

23. _____

24. _____

25. Find your weight in kilograms.

26. Find your height in centimeters.

25. _____

26. _____

27. A station sells gasoline by the liter. If you bought 45 liters, how many gallons did you buy?

28. A fuel tank in a Buick Century has a capacity of 15.5 gallons. How many liters will it hold?

27. _____

28. _____

6–5 PERCENT

The symbol %, read as "percent," is encountered in many areas of our daily life. We hear statements such as: "The prime rate rose 1%," "14% of the class made A's," "57% of the people surveyed disagreed with the proposal," and so on.

Percent means "per-hundred" and indicates a ratio of a number to 100.

Example 1   75% means $\dfrac{75}{100}$.

Example 2   150% means $\dfrac{150}{100}$.

The meaning of percent therefore allows us to change any percent to a common fraction form.

OBJECTIVES

In this section you will learn to:
1. Change a percent to a decimal number.
2. Change a decimal number to a percent.
3. Change a fraction to a percent.

Example 3 Change 24% to a common fraction in reduced form.

$$24\% = \frac{24}{100} = \frac{6}{25}$$

Example 4 Change $33\frac{1}{3}\%$ to a common fraction in reduced form.

$$33\frac{1}{3}\% = \frac{33\frac{1}{3}}{100} = \frac{100}{3} \cdot \frac{1}{100} = \frac{1}{3}$$

Example 5 Change 12.5% to a common fraction in reduced form.

$$12.5\% = \frac{12.5}{100} = \frac{125}{1,000} = \frac{1}{8}$$

From our study of decimals we know that $\frac{75}{100}$ can be written as .75.

In other words, we note that 75% = .75. When working problems containing percent, we usually change the percent form to decimal form. Hence the following rule.

> **Rule** To change a percent to a decimal remove the % sign and move the decimal point two places to the left. Conversely, when changing a decimal number to a percent, move the decimal point two places to the right and add the % sign.

Example 6 Change 37% to a decimal number.

$$37\% = .37 \text{ (two places to the left)}$$

Example 7 Change .61 to a percent.

$$.61 = 61\% \text{ (two places to the right)}$$

> **WARNING**
>
> The rule states *two* places. Don't make the mistake of moving the decimal more or less than two places.

Example 8 Change 273% to a decimal number.

$$273\% = 2.73 \text{ (two places to the left)}$$

Example 9 Change .5% to a decimal number.

$$.5\% = .005 \text{ (two places to the left)}$$

Example 10 Change 1.6 to a percent.

$$1.6 = 160\% \text{ (two places to the right)}$$

Example 11 Change .003 to a percent.

$$.003 = .3\% \text{ (two places to the right)}$$

We sometimes need to change a fraction to a percent. Since we have already learned to change a fraction to a decimal number, we have the following rule.

> **Rule** To change a fraction to a percent first change the fraction to a decimal number and then change the decimal number to a percent.

Example 12 Change $\frac{7}{8}$ to a percent correct to one decimal place.

Dividing 8 into 7, we obtain

$$\frac{7}{8} = .875 = 87.5\%.$$

Example 13 Change $\frac{1}{3}$ to a percent correct to one decimal place.

$$\frac{1}{3} = .333 = 33.3\%$$

Note that this answer is approximate.

EXERCISE SET 6-5-1

Change each of the following percents to a common fraction in reduced form.

1. 50% 2. 20% 3. 65% 4. 150% 5. $24\frac{1}{2}\%$

1. _____
2. _____
3. _____
4. _____
5. _____

ANSWERS

6. $31\frac{1}{5}\%$ 7. $83\frac{1}{3}\%$ 8. 1.75% 9. 87.5% 10. 112.8%

Change each of the following percents to a decimal number.

11. 15% 12. 30% 13. 3% 14. 1.8% 15. 15.6%

16. .19% 17. 238% 18. 100% 19. .09% 20. 18.64%

Change each of the following decimal numbers to a percent.

21. .92 22. .5 23. 3.2 24. .06 25. .403

26. 8 27. .072 28. 2.03 29. .004 30. 1.018

Change each of the following fractions to a percent. If not exact, round the answer to one decimal place.

31. $\frac{1}{2}$ 32. $\frac{1}{5}$ 33. $\frac{3}{4}$ 34. $\frac{9}{10}$ 35. $\frac{7}{100}$

36. $\frac{1}{6}$ 37. $\frac{3}{8}$ 38. $\frac{4}{7}$ 39. $\frac{7}{4}$ 40. $\frac{10}{3}$

6–6 PROBLEMS INVOLVING PERCENTS

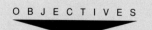

OBJECTIVES

In this section you will learn to solve various problems involving percents.

Many types of problems involving percent are solved by setting up a simple equation based on the fact that "the part equals the percent of the whole" (part = % of whole). The word "of" in problems usually means multiply. Hence we have the equation.

$$\text{part} = \% \times \text{whole}.$$

6-6 PROBLEMS INVOLVING PERCENTS

Example 1 Find 37% of 243.

If we write 37% as the decimal .37 and let x represent the part, then we have the equation

$$x = .37(243)$$
$$= 89.91.$$

Example 2 53 is 20% of what number?

$$53 = (.20)x$$
$$x = \frac{53}{.20}$$
$$x = 265$$

Example 3 15 is what percent of 12?

$$15 = x(12)$$
$$x = \frac{15}{12}$$
$$x = 1.25$$

Changing this decimal to a percent gives

$$x = 125\%.$$

Example 4 In a math class of 35 students 7 students earned a grade of A. What percent of the class obtained an A?

The problem here is, "7 is what percent of 35?"

$$7 = x(35)$$
$$x = \frac{7}{35}$$
$$x = .2$$
$$x = 20\%$$

Example 5 Bill has saved $1,285 toward buying a new outboard motor for his boat. He claims this is 80% of his goal. What is his goal?

The problem is, "1,285 is 80% of what number?"

$$1{,}285 = (.80)x$$
$$x = \frac{1{,}285}{.80}$$
$$x = \$1{,}606.25$$

Example 6 The population of a certain town increased from 36,400 in 1970 to 42,800 in 1980. What was the percent of increase rounded to the nearest tenth of a percent?

> **WARNING**
>
> Percent of increase or decrease is always based on the *original* number, which in this case is 36,400.

$$\text{Increase} = 42{,}800 - 36{,}400 = 6{,}400.$$

The problem now is, "6,400 is what percent of 36,400?"

$$6{,}400 = x(36{,}400)$$
$$x = \frac{6{,}400}{36{,}400}$$
$$x = .176$$
$$x = 17.6\% \text{ (rounded to the nearest tenth)}$$

EXERCISE SET 6-6-1

1. Find 20% of 80.

2. Find 35% of 70.

3. What is 16% of 60?

4. Find 28% of 150.

5. 66% of 230 is what number?

6. Find 83% of 420.

7. 6 is 60% of what number?

8. 8 is 20% of what number?

9. 93 is 75% of what number?

10. 18 is 4% of what number?

11. 36 is 16% of what number?

12. 162 is 9% of what number?

13. 3 is what percent of 4?

14. 7 is what percent of 15?

15. What percent of 24 is 5?

16. What percent of 25 is 8?

17. 25 is what percent of 28?

18. What percent of 55 is 42?

19. Frank borrowed $5,000 at 14% annual interest and was to pay interest only the first year. How much should he pay the first year?

20. A person makes $320 during a certain week. Twenty percent of the wages is withheld for federal taxes. How much is withheld?

21. There were 45 members present at a club meeting. If that number represents 60% of the total membership, what is the total membership?

22. A community group has collected donations amounting to $120,000. If this represents 80% of its goal, what is its goal?

23. Ray has money invested at 14% annual interest. If his interest for the year was $3,360, how much did he have invested?

24. A credit card company charges interest at the rate of 1.5% per month on the balance. What was the amount of the balance during a month if the interest charge was $5.40?

25. An exploration party had traveled $78\frac{3}{10}$ miles up the Amazon River. They were told by their guide that this represents only 2% of the total length of the river. How long is the Amazon River?

26. During a snowstorm Buffalo, New York, received 7.09 inches of snow. If this represents 8% of the total annual average for the city, what is the total yearly snowfall? Round answer to the nearest tenth of an inch.

ANSWERS

15. _____

16. _____

17. _____

18. _____

19. _____

20. _____

21. _____

22. _____

23. _____

24. _____

25. _____

26. _____

27. A calculator regularly priced at $39.00 was offered with a $5.85 rebate from the manufacturer. What percent was the rebate of the regular price?

28. A discount of $10.50 was given on a pair of shoes regularly priced at $59.50. What percent of the regular price was the discount? Round answer to the nearest tenth of a percent.

29. In a math class consisting of 55 students 24 were females. What percent of the class was female? Round answer to the nearest tenth of a percent.

30. In one season the New York Yankees played a total of 162 games and won 103 of them. What percent of the games did they win? Round answer to the nearest tenth of a percent.

31. A certain model of an automobile that cost $9,800 last year is priced at $10,350 this year. What is the percent of increase in price rounded to the nearest tenth of a percent?

32. By installing insulation a homeowner was able to reduce the normal $120 per month utility bill to $96. What is the percent of decrease of the utility bill?

CHAPTER 6 SUMMARY

Ratio
- A ratio is the indicated quotient of two numbers.

Proportion
- A proportion is a statement that two ratios are equal.
- In a proportion the cross products are equal.

Metric Measurement
- To change from American measurement to metric measurement or from metric to American express as a proportion and solve the proportion.

Percent
- Percent means per-hundred or a ratio to 100.
- Ratios to 100 are usually written as decimal numbers.
- To change from a decimal number to percent move the decimal point two units to the right and add a percent sign.
- To change from a percent to a decimal number move the decimal point two units to the left and drop the percent sign.

CHAPTER 6 REVIEW

Write the following ratios as fractions in reduced form.

1. 3 to 8
2. 4 to 6

3. 12 : 27
4. 18 : 42

Determine if the following pairs of ratios are equal.

5. $\dfrac{3}{5}, \dfrac{15}{25}$
6. $\dfrac{12}{18}, \dfrac{18}{27}$

7. $\dfrac{12}{36}, \dfrac{21}{54}$
8. $\dfrac{36}{78}, \dfrac{24}{52}$

Determine if the following proportions are true or false.

9. $\dfrac{1}{2} = \dfrac{5}{10}$
10. $\dfrac{3}{8} = \dfrac{12}{34}$

11. $\dfrac{6}{9} = \dfrac{4}{6}$
12. $\dfrac{10}{14} = \dfrac{30}{42}$

13. $\dfrac{27}{72} = \dfrac{15}{45}$
14. $\dfrac{10}{24} = \dfrac{30}{72}$

15. $\dfrac{21}{35} = \dfrac{27}{45}$
16. $\dfrac{32}{52} = \dfrac{48}{78}$

ANSWERS

1. _____
2. _____
3. _____
4. _____
5. _____
6. _____
7. _____
8. _____
9. _____
10. _____
11. _____
12. _____
13. _____
14. _____
15. _____
16. _____

Solve for x in each of the following.

17. $\dfrac{1}{4} = \dfrac{5}{x}$

18. $\dfrac{8}{x} = \dfrac{2}{5}$

19. $\dfrac{7}{3} = \dfrac{x}{6}$

20. $\dfrac{4}{x} = \dfrac{6}{18}$

21. $\dfrac{12}{x} = \dfrac{15}{4}$

22. $\dfrac{15}{4} = \dfrac{x}{6}$

23. $\dfrac{x}{3} = \dfrac{6}{9}$

24. $\dfrac{x}{27} = \dfrac{54}{36}$

Convert each of the following. Round answers to one decimal place.

25. 7.5 yd to meters

26. 18 qt to liters

27. 215 cm to inches

28. 45 g to ounces

29. 20 km to miles

30. 135 lb to kilograms

31. 8 oz to grams

32. 2.5 gal to liters

Change each of the following percents to decimals.

33. 37%

34. 9%

35. 13.8%

36. 206% **37.** .7% **38.** 12.04%

Change each of the following decimals to percent.

39. .55 **40.** .3 **41.** .06

42. 5.2 **43.** .016 **44.** 4.198

Change each of the following fractions to a percent. If not exact, round the answer to the nearest tenth of a percent.

45. $\frac{2}{5}$ **46.** $\frac{3}{16}$ **47.** $\frac{7}{10}$

48. $\frac{21}{40}$ **49.** $\frac{5}{2}$ **50.** $\frac{12}{5}$

51. Find 40% of 90. **52.** Find 35% of 60.

53. Find 56% of 240. **54.** Find 120% of 40.

Find each of the following to the nearest tenth of a percent.

55. 3 is what percent of 15? **56.** What percent of 66 is 12?

57. 56 is what percent of 320? **58.** What percent of 900 is 150?

ANSWERS

36. _____
37. _____
38. _____
39. _____
40. _____
41. _____
42. _____
43. _____
44. _____
45. _____
46. _____
47. _____
48. _____
49. _____
50. _____
51. _____
52. _____
53. _____
54. _____
55. _____
56. _____
57. _____
58. _____

Find the following numbers. Round to two decimal places.

59. 3 is 60% of what number?

60. 14 is 83% of what number?

61. 128 is 11% of what number?

62. 17.4 is 55% of what number?

63. What number has the same ratio to 24 as 5 has to 8?

64. A car traveled 228 miles in 6 hours. Express the ratio of miles to hours in reduced form.

65. One car traveled 114 miles in 3 hours and a second car traveled 190 miles in 5 hours. Did they travel at the same average rate?

66. The distance between two cities on a road map is 3 inches. If the actual distance is 135 miles, what is the actual distance between two cities that are 5 inches apart?

67. Gear A has 30 teeth and gear B has 35. If this same ratio is to be maintained and gear A is replaced by a gear with 12 teeth, how many teeth must be on the gear replacing gear B?

68. A car travels 187.6 miles on 6.7 gallons of gasoline. How many gallons are needed to travel 266 miles?

69. How many quarts are contained in a 1.75-liter bottle? Round answer to two decimal places.

70. A swimming event in the Olympic Games is the 1,500-meter freestyle. What is the distance in yards? Round answer to one decimal place.

71. In a survey of 345 people 28% were under age 20. How many were under age 20? Round answer to the nearest whole number.

72. Find the interest paid on $28,000 for one year if the annual rate is 13.5%.

73. A dress regularly priced at $88.50 was marked down by $19.00. What percent was the discount of the regular price? Round answer to the nearest tenth of a percent.

74. During the 1985–86 season the Boston Celtics played 82 games and won 67 of them. What percent of the games did they win? Round answer to the nearest tenth of a percent.

75. Baseball player Ty Cobb had 4,191 hits in his career, giving him a batting average of .367, which means he got a hit 36.7% of his times at bat. Find the number of times he was at bat. Round answer to the nearest whole number.

76. A woman invested some money at an annual interest rate of 12.5%. If she received $1,687.50 interest during the year, find how much she had invested.

ANSWERS

73. _____

74. _____

75. _____

76. _____

SCORE: _____

CHAPTER 6 TEST

ANSWERS

1. Write the ratio 75:90 as a fraction in reduced form.

2. Are the ratios 12:28 and 15:35 equal?

3. Is $\dfrac{6}{15} = \dfrac{14}{35}$ true or false?

4. Is $\dfrac{12}{36} = \dfrac{6}{9}$ true or false?

5. Solve for x: $\dfrac{x}{6} = \dfrac{15}{9}$

6. Solve for x: $\dfrac{10}{x} = \dfrac{4}{14}$

7. Solve for x: $\dfrac{6}{5} = \dfrac{x}{4}$

8. The ratio of boys to girls in grade four of a certain school is 6 to 5. If there are 15 girls in the class, how many boys are in the class?

9. If an automobile uses 3 gallons of gasoline to travel 69 miles, how many gallons are necessary to travel 161 miles?

10. In a certain population 3 out of 5 people have brown eyes. If 75 people are selected at random from this population, how many could be expected to have brown eyes?

11. If the speed limit on a certain road is 55 miles per hour, what is the speed limit in kilometers per hour? Round answer to the nearest tenth.
(1 kilometer = .6214 mile)

12. It took 14 gallons of gasoline to fill a tank. How many liters would have filled the tank? Round answer to the nearest whole number.
(1 gallon = 3.785 liters)

13. a. Change $\dfrac{5}{8}$ to percent.

b. Change 1.3 to percent.

14. In a survey 560 people were asked to give a "yes" or "no" opinion on a question. 231 answered "yes." What percent answered "yes"?

15. On a loan of $800 Bill paid $76 interest for one year. What was the yearly rate of interest?

SCORE:

CHAPTERS 1-6 CUMULATIVE TEST

1. Subtract: 3,021
   −   534

2. Divide: $142\overline{)35{,}926}$

3. Evaluate:
   $2[5(6 - 2) - 7]$
   $- 4[23 - 5(3 + 1)]$

4. Reduce: $\dfrac{60}{84}$

5. Multiply: $\dfrac{8}{15} \times \dfrac{5}{6}$

6. Subtract: $3\dfrac{1}{8} - 1\dfrac{5}{6}$

7. Round 25.135 to the nearest hundredth.

8. Multiply: 12.8
   × .035

9. Subtract: 53.6
   − 8.634

10. Write 7.05×10^{-4} as a decimal number.

11. Write a literal expression for the difference of a and b divided by the sum of x and y.

12. Evaluate $5x^3y + 3xz - 4xz^2$, if $x = 2, y = 3, z = 4$.

13. Combine like terms:
    $6a^3b + 3ab^2 - 2a^3b - ab^2 + 5a^3b$

14. Solve for x: $7x = 98$

15. Solve for x: $\dfrac{5}{6}x = \dfrac{3}{4}$

16. Solve for x:
    $3x - \dfrac{1}{2} = \dfrac{3}{8}x + 1$

ANSWERS

1. _____
2. _____
3. _____
4. _____
5. _____
6. _____
7. _____
8. _____
9. _____
10. _____
11. _____
12. _____
13. _____
14. _____
15. _____
16. _____

ANSWERS

17. _____

18. _____

19. _____

20. _____

SCORE: _____

17. A relationship between Fahrenheit (F) and Celsius (C) temperatures is given by $9C = 5F - 160$. Find the Fahrenheit temperature if the Celsius temperature is zero degrees.

19. Change $\frac{3}{8}$ to percent.

18. A man 6 feet tall casts a 15-foot shadow. How tall is a tree that casts a 50-foot shadow?

20. The sales tax in a certain city is 7%. What is the sales tax on an automobile that costs $15,600?

CHAPTER 7 **PRETEST**

NAME:

CLASS / SECTION: DATE:

A N S W E R S

Answer as many of the following problems as you can before starting this chapter. When you finish the chapter, take the test at the end and compare your scores to see how much you have learned.

1a. _____

1b. _____

2. _____

3. _____

4a. _____

4b. _____

5a. _____

5b. _____

6. _____

7. _____

8. _____

9. _____

10a. _____

10b. _____

11. _____

12a. _____

12b. _____

13a. _____

13b. _____

14. _____

15. _____

SCORE: _____

1. Find the negative of: a. -5.9 b. $\dfrac{3}{7}$

2. Simplify: $-(-13)$

3. Use a signed number to represent the result of a deposit of $18 followed by a withdrawal of $12.

4. Add: a. $(+15) + (-28)$ b. $(-6) + (-16)$

5. Subtract: a. $(-9) - (+20)$ b. $(-5) - (-19)$

6. Simplify:
$-8 - 2 + 13 - (-5) + 1$

7. Remove parentheses and simplify: $16 - (12 - 31) - 6$

8. Combine like terms:
$9x^2y + 4x^2y - 6x^2y$

9. Subtract the difference of 11 and 7 from 12.

10. Find the product:
a. $(+5)(-14)$ b. $(-6)(-8)$

11. Find the product:
$(-2)(-3)(6)(-1)$

12. Find the quotient: a. $\dfrac{-36}{-9}$ b. $\dfrac{2}{3} \div \left(-\dfrac{4}{9}\right)$

13. Evaluate: a. $\dfrac{(-14)(-6)}{-21}$ b. $\dfrac{-6-15}{-7}$

14. Remove grouping symbols and simplify:
$13x - 2\{3x - 5[x - 2(3x - 5)]\}$

15. Evaluate $2a^3 - 4a^2b - b^2$ when $a = -5$, $b = -2$.

316

CHAPTER 7

Signed Numbers

The stock market gained ten points in the morning, then lost six points in the afternoon. What was the net gain or loss for the day?

In previous chapters we dealt with the operations on a set of numbers called the numbers of arithmetic. In this chapter we wish to introduce the set of **signed numbers** or **directed numbers** and establish the four basic operations on this set.

7-1 THE MEANING OF SIGNED NUMBERS

In this section you will learn to:
1. Locate signed numbers on a number line.
2. Find the negative of a given number.
3. Remove grouping symbols preceded by a minus sign.

We can represent the numbers of arithmetic on a **number line.** We start by placing zero at a point on the line. We then agree that we will place numbers at other points on the line such that a larger number is to the right and that all units (0 to 1, 4 to 5, 10 to 11, and so on) will be of equal length. This line is illustrated below.

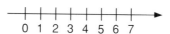

We now wish to extend our thinking to include numbers to the left of zero. Such numbers are called **negative numbers** and are written with a *minus* sign preceding them. Numbers such as -6, -1, and -10 are read as "negative six," "negative one," and "negative ten."

The whole numbers that we have studied up to this point and the negative whole numbers together make up the set of numbers called **integers.**

The set of numbers that can be expressed as the ratio of two integers is called the set of **rational numbers.** Fractions, such as $\frac{1}{2}$, $-\frac{2}{3}$, $\frac{8}{3}$, $-\frac{5}{2}$, will fall proportionally on the number line. The number line containing the rational numbers is called the **rational number line.**

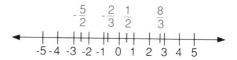

The numbers to the right of zero are called **positive numbers** and are written with a *plus* sign. If a nonzero number is written without a sign, it is assumed to be a positive number. Hence $+7$ or simply 7 would represent positive seven. Also note that "to the right" is a positive direction and "to the left" is a negative direction. Zero is neither positive nor negative.

These positive and negative numbers are referred to as **signed numbers** or **directed numbers.**

Most likely you have encountered signed numbers many times in your newspaper or on TV. For instance, if the temperature is ten degrees below zero, it is usually written as $-10°$. If the stock market falls four points, it is usually written as -4. You can probably think of other examples you have encountered.

Example 1 If a football team gained three yards on a play, how would you represent this using a signed number? How would you represent a loss of five yards?

$+3$ would represent a gain of three yards.
-5 would represent a loss of five yards.

Example 2 Jack had a bank balance of $23 and wrote a check for $30. Represent his balance with a signed number.

Since he spent seven dollars more than he had, his balance is $\$-7$ and the banker is probably not too happy.

To be successful in using operations on signed numbers a very important distinction must be made between **a negative number** and **the negative of a number.** We have already noted that a negative number is a number preceded by a minus sign. On the number line we noted that negative numbers are to the left of zero.

Definition The **negative of a number** is the opposite of that number or that number with the opposite sign.

Example 3 The negative of $+5$ is -5.

Example 4 The negative of -5 is $+5$.

Example 5 The negative of $+x$ is $-x$.

Example 6 The negative of $-x$ is $+x$.

Example 7 If $+7$ represents 7 steps north, then -7 represents 7 steps south.

Example 8 If $-3°$ represents three degrees below zero, then $+3°$ represents three degrees above zero.

Example 9 What is the negative of (-10)?

Since "the negative of" means "opposite of," the negative of -10 is $+10$.

The minus sign is often used as a symbol for "the negative of" or "opposite of." Therefore, $-(-10)$ means "the opposite of" -10.

Example 10 $-(+7) = -7$

Example 11 $-(-2) = +2$

EXERCISE SET 7-1-1

ANSWERS

Find the negative of each of the following.

1. $+3$
2. -4
3. $+\dfrac{1}{2}$
4. -5.3

5. $+y$
6. $-b$
7. 7
8. $-4\dfrac{1}{3}$

9. 0
10. 100
11. $5\dfrac{3}{4}$
12. -7.38

Simplify each of the following.

13. $-(+4)$
14. $-(+2.05)$
15. $-\left(\dfrac{1}{2}\right)$
16. $-\left(-\dfrac{3}{4}\right)$

Identify the signed numbers represented by the following points.

17.

Represent the signed numbers on the number line for each of the following.

18. -3

19. $-(-1)$

20. -4

21. $-(+2)$

22. If $+20$ represents a temperature of twenty degrees above zero, use a signed number to represent a temperature of twenty degrees below zero.

23. If -10 represents a drop of ten points on the stock market, use a signed number to represent a gain of ten points.

22. _____

24. A football team loses five yards on a play. Use a signed number to indicate this loss.

25. A news reporter stated that the Dow-Jones Average dropped thirteen points. Indicate this drop using a signed number.

23. _____

24. _____

26. A person's weight was recorded two weeks ago at 165 pounds. This week the person's weight is 159 pounds. Indicate the change from last week using a signed number.

25. _____

26. _____

In chapter 1 you learned that parentheses are used as grouping symbols. If we enclose a number expression in parentheses and place a minus sign before it, we are indicating the "negative" of the entire expression within the parentheses.

Example 12 $-(a + b)$ means "the negative of the sum of a and b."

Example 13 $-(x + 3 - b)$ means "the negative of, x plus 3 minus b."

> **Rule** To find the negative of an expression enclosed in grouping symbols find the negative of each term within the symbols.

Example 14 $-(a + b) = -a - b$

Example 15 $-(x + 3 - b) = -x - 3 + b$

Example 16 $-(a - b) = -a + b$

> **WARNING**
>
> A very common error in finding the negative of an expression is not changing *every* sign. Be sure to express the opposite of every term within the grouping symbols.

Example 17 $-(a - 3 + 4 - x) = -a + 3 - 4 + x$

Let us now restate the rule in more general terms.

> **Rule** To remove grouping symbols preceded by a minus sign change the sign of every term that was within the symbols. To remove grouping symbols preceded by a plus sign do not change the sign of any term that was within the symbols.

Example 18 $-(a + b - 3) = -a - b + 3$

Example 19 $+(a + b - 3) = a + b - 3$

Example 20 $(x + y - a + b) = x + y - a + b$

EXERCISE SET 7-1-2

Remove the grouping symbols in each of the following.

1. $-(x + y)$
2. $-(a + 4)$
3. $+(x - 5)$
4. $+(a - b)$
5. $-(a - b - c)$
6. $-(x + y - 3)$
7. $(a - b + 8)$
8. $+(a - c - 4)$
9. $-(a + b - c)$
10. $-(x - y - 15)$
11. $-(x + 4 - z)$
12. $-(a - b + c - 5)$
13. $(a - b)$
14. $-(b - a)$
15. $(a - 2b + c)$

16. $(-a + b - 4c)$ 17. $-(-x + y)$ 18. $-(-2x - y)$

19. $-(-a - 6b + 4)$ 20. $-(-3a + b - 16)$

21. $-(5x - y + 4z - 3)$ 22. $(x - 8y - 2z - 10)$

23. $-(a + 2b - 5c + 6)$ 24. $-(-2a + 3b - 4c - 9)$

ANSWERS

16. _____
17. _____
18. _____
19. _____
20. _____
21. _____
22. _____
23. _____
24. _____

7-2 COMBINING SIGNED NUMBERS

Example 1   If you start at a point and move six steps north then move three more steps north, where will you be in relation to your starting point?

The answer to this problem is nine steps north. If "steps north" is represented as positive, then six steps north followed by three steps north equals nine steps north could be written as $+6 + 3 = +9$.

Example 2   The temperature at 6:00 P.M. is zero degrees. From 6:00 P.M. until midnight the temperature falls three degrees, and from midnight until 4:00 A.M. it falls seven more degrees. What is the temperature at 4:00 A.M.?

If a fall in temperature is represented by a minus sign, then a fall of 3° followed by a fall of 7° can be written as

$$-3° - 7° = -10°.$$

Example 3   At 9:00 A.M. the temperature is $+20°$ and from 9:00 A.M. until 1:00 P.M. it rises 14°. Then from 1:00 P.M. until 5:00 P.M. it falls 18°. What is the temperature at 5:00 P.M.?

Starting at $+20°$, a rise of 14° takes the temperature to $+34°$. A fall of 18° then takes it to $+16°$. If a rise in temperature is positive and a fall in temperature is negative, then we can write

$$+20° + 14° - 18° = +16°.$$

OBJECTIVES

In this section you will learn to:
1. Determine the result of combining two or more signed numbers.
2. Use the number line to combine signed numbers.

EXERCISE SET 7-2-1

ANSWERS

Find the result of each of the following.

1. If we consider north as positive and south as negative, then ten steps north followed by three steps south equals

2. Five steps south followed by two steps north equals

3. Four steps north followed by six steps south equals

4. Three steps south followed by six steps south equals

5. Eight steps south followed by ten steps north equals

6. A profit of $10 combined with a profit of $12 equals

7. A profit of $16 combined with a loss of $9 equals

8. A loss of $20 combined with a profit of $12 equals

9. A loss of $6 combined with a loss of $18 equals

10. A profit of $30 combined with a loss of $36 equals

11. A gain of two yards in a football game followed by a loss of six yards equals

12. A loss of two yards followed by a loss of five yards equals

13. A loss of eight yards followed by a gain of twelve yards equals

14. A gain of twelve yards followed by a loss of fifteen yards equals

15. A fall of six degrees in temperature followed by a fall of two degrees equals

16. A rise of eight degrees in temperature followed by a fall of two degrees equals

1. _____
2. _____
3. _____
4. _____
5. _____
6. _____
7. _____
8. _____
9. _____
10. _____
11. _____
12. _____
13. _____
14. _____
15. _____
16. _____

17. If at 10:00 P.M. the temperature was 65° and from 10:00 P.M. until 4:00 A.M. it fell twelve degrees, what was the temperature at 4:00 A.M.?

18. If the temperature was 50° at 6:00 A.M. and from 6:00 A.M. to 2:00 P.M. it rose fifteen degrees, then from 2:00 P.M. to 7:00 P.M. it fell nine degrees, what was the temperature at 7:00 P.M.?

17. _____

18. _____

19. Ten steps north, followed by six steps south, followed by nine steps north, followed by twelve steps south equals

20. A temperature starts at 10°. It falls fifteen degrees, then rises eight degrees, then falls three degrees. What is the final temperature?

19. _____

20. _____

The number line is a useful tool in combining signed numbers. If a movement "to the right" is considered + and a movement "to the left" is considered −, then consider the following examples.

Example 4 $+3 + 4 = +7$ indicates "start at zero, move three units to the right, then move four units to the right." The result is a movement of seven units to the right.

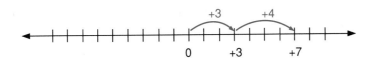

Example 5 $-2 - 4 = -6$ indicates "start at zero, move two units to the left, then move four units to the left." The result is a movement of six units to the left.

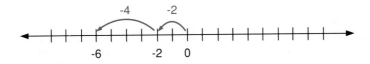

> **WARNING**
>
> Unless the problem states otherwise, always start at zero for the *first* move.

Example 6 $-6 + 8 = +2$ indicates "start at zero, move six units to the left, then move eight units to the right." The result is a movement of two units to the right.

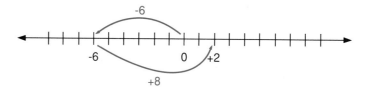

Example 7 Combine: $3 - 6 + 4 - 5 =$

Using the number line and starting at zero, first move three units to the right.

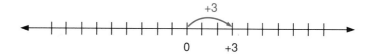

Then move six units to the left.

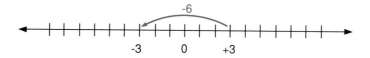

Then move four units to the right.

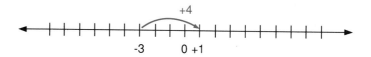

Finally move five units to the left.

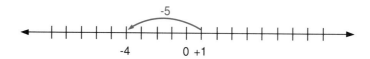

The final location is at -4.
Therefore, $3 - 6 + 4 - 5 = -4$.

Example 8 After losing nine yards on the first down, the Browns completed a pass for a twelve-yard gain on the second down. What was the total gain or loss?

First write the number statement $-9 + 12 =$

Using the number line, we obtain

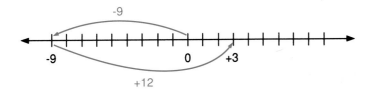

Thus $-9 + 12 = +3$ (a three-yard gain).

▼ EXERCISE SET 7-2-2

Use a number line to combine each of the following.

ANSWERS ▼

1. $+3 + 5 =$

1. _____

2. $-3 - 4 =$

2. _____

3. $-4 + 10 =$

3. _____

4. $9 - 6 =$

4. _____

5. $-2 + 7 =$

5. _____

6. $-11 + 5 =$

6. _____

7. $+8 - 8 =$

7. _____

8. $+13 - 20 =$

8. _____

ANSWERS

9. _____

10. _____

11. _____

12. _____

13. _____

14. _____

15. _____

16. _____

17. _____

18. _____

19. _____

20. _____

9. $9 - 15 =$

10. $-11 + 18 =$

11. $8 - 17 =$

12. $4 - 2 + 7 =$

13. $-6 + 13 - 7 =$

14. $-4 + 13 - 9 =$

15. $-5 + 8 - 4 + 1 =$

16. $3 - 8 + 5 - 4 =$

17. $3 - 9 + 6 - 14 =$

18. $-14 + 3 - 9 + 6 =$

19. $-8 + 6 - 5 + 7 =$

20. $-5 + 6 - 8 + 7 =$

21. The stock market gained ten points in the morning, then lost six points in the afternoon. What was the net gain or loss for the day?

◄─┼┼┼┼┼┼┼┼┼┼┼┼┼┼┼┼┼┼┼┼┼┼┼┼─►

A N S W E R S

22. The temperature rose nine degrees in the morning and dropped twelve degrees in the afternoon. What was the net gain or loss of temperature?

◄─┼┼┼┼┼┼┼┼┼┼┼┼┼┼┼┼┼┼┼┼┼┼┼┼─►

21. _____

23. A worker earned $25 in the morning and spent $18 in the afternoon. What was the net gain or loss for the day?

◄─┼┼┼┼┼┼┼┼┼┼┼┼┼┼┼┼┼┼┼┼┼┼┼┼─►

22. _____

24. A weight-watcher lost five pounds the first week, gained two pounds the second week, lost three pounds the third week, and lost two pounds the fourth week. What was the net gain or loss for the four weeks?

◄─┼┼┼┼┼┼┼┼┼┼┼┼┼┼┼┼┼┼┼┼┼┼┼┼─►

23. _____

25. The temperature in Denver was 28° C at 1:00 P.M. During the next hour the temperature rose four degrees. During the second hour the temperature dropped three degrees and during the third hour it dropped another six degrees. What was the temperature at 4:00 P.M.?

◄─┼┼┼┼┼┼┼┼┼┼┼┼┼┼┼┼┼┼┼┼┼┼┼┼─►

24. _____

25. _____

▭ 7-3 RULES FOR COMBINING SIGNED NUMBERS

In the preceding section we used the meaning of a signed number and the number line to work problems combining signed numbers. We are now ready to establish rules for performing this operation.

First we make the following observations from the discussion and examples of the preceding section.

1. A sign (plus or minus) affects only the number expression to its right.

Example 1 In $+6 - 7$ the $+$ sign affects only the 6 and the $-$ sign affects only the 7.

2. A sign preceding parentheses affects all terms inside the parentheses.

O B J E C T I V E S

In this section you will learn to use the rules for combining signed numbers.

Example 2 In $-(+6 - 8)$ the $-$ sign before the parentheses affects both the $+6$ and -8.

    3. We combine numbers only *two* at a time.

Example 3 Combine $+6 + 4 + 3$.

We first combine $+6 + 4$ to obtain $+10$. Then we combine $+10 + 3$ to obtain $+13$.

Observation 3 makes it clear that we only need rules for combining two signed numbers.

Rule To combine two numbers with the same sign add the numbers and attach the common sign.

Example 4 $+6 + 8 = +14$. We add 6 and 8 to obtain 14 and then attach the $+$ sign that is common to both 6 and 8.

Example 5 $-6 - 8 = -14$. We add the 6 and 8 to obtain 14 and then attach the $-$ sign that is common to both 6 and 8.

▼ EXERCISE SET 7-3-1

Combine the following.

1. $+5 + 7 =$
2. $+13 + 28 =$
3. $8 + 9 =$

4. $-2 - 6 =$
5. $-8 - 15 =$
6. $-3 - 18 =$

7. $12 + 49 =$
8. $-18 - 14 =$
9. $-15 - 16 =$

10. $17 + 18 =$
11. $+8 + 13 + 4 =$
12. $5 + 7 + 11 =$

13. $-4 - 8 - 16 =$
14. $-6 - 15 - 9 =$
15. $2 + 14 + 1 =$

16. −18 − 1 − 5 =   17. −9 − 15 − 3 =   18. 26 + 35 + 13 =

19. 8 + 4 + 15 + 2 =   20. −6 − 27 − 4 − 1 =

ANSWERS

16. _____
17. _____
18. _____
19. _____
20. _____

Rule To combine two numbers with unlike signs subtract the smaller number (without regard to sign) from the larger number (without regard to sign) and attach the sign of the larger number.

Example 6 −7 + 11 = +4. We subtract 7 from 11 and use the + sign because 11 is larger than 7.

Example 7 +7 − 11 = −4. We subtract 7 from 11 and use the − sign because 11 is the larger of the two numbers.

Example 8 −14 + 8 = −6. We subtract 8 from 14 and use the − sign of 14.

▼ **EXERCISE SET 7–3–2**

Combine each of the following pairs of numbers.

1. +8 − 3 =   2. 11 − 7 =   3. −6 + 15 =

4. +14 − 20 =   5. 6 − 8 =   6. 14 − 28 =

7. +18 − 18 =   8. +21 − 20 =   9. +1 − 1 =

10. −2 + 5 =   11. +11 − 19 =   12. 18 − 20 =

13. 10 − 11 =   14. −19 + 10 =   15. −35 + 49 =

ANSWERS

1. _____
2. _____
3. _____
4. _____
5. _____
6. _____
7. _____
8. _____
9. _____
10. _____
11. _____
12. _____
13. _____
14. _____
15. _____

ANSWERS

16. _____
17. _____
18. _____
19. _____
20. _____

16. $68 - 101 =$

17. $+17 - 17 =$

18. $-7 + 32 =$

19. $-23 + 22 =$

20. $-9 + 35 =$

Note again that rules for combining signed numbers apply to only two numbers at a time. If an expression contains several numbers, we must apply the rules more than once.

Example 9 Combine: $-6 + 4 + 8 - 3 =$

First combine -6 and $+4$ to obtain -2.

$$-6 + 4 + 8 - 3 = -2 + 8 - 3$$

Next combine -2 and $+8$ to obtain $+6$.

$$-2 + 8 - 3 = +6 - 3$$

Finally combine $+6$ and -3 to obtain the final result.

$$+6 - 3 = +3$$

Since the order of combining numbers will not change the answer (commutative property), we can, in example 9, proceed in a different way.

Example 10 Combine: $-6 + 4 + 8 - 3 =$

Using the commutative property, we can write

$$-6 + 4 + 8 - 3 = -6 - 3 + 4 + 8.$$

We then combine -6 and -3 to obtain -9.

$$-6 - 3 + 4 + 8 = -9 + 4 + 8$$

Next combine $+4$ and $+8$ to obtain $+12$.

$$-9 + 4 + 8 = -9 + 12$$

Finally combine -9 and $+12$ to obtain the final result.

$$-9 + 12 = +3$$

Both of these approaches should be practiced so that you can choose whichever method is easier for a particular problem.

Example 11 Combine: $+6 - 3 - 8 - 7 + 4 =$

$$+6 - 3 - 8 - 7 + 4 = +10 - 3 - 8 - 7$$
$$= +10 - 10 - 8$$
$$= -8$$

Note in this last example that this order gives easier combinations than combining left to right. The choice of order is left to the student, since the answer will be the same either way.

EXERCISE SET 7-3-3

Combine the following.

1. $3 - 6 + 9 =$
2. $2 - 7 + 5 =$
3. $-5 + 8 - 2 =$
4. $-11 + 3 - 2 =$
5. $5 - 6 + 4 =$
6. $6 - 8 - 4 =$
7. $-3 + 9 - 5 =$
8. $3 - 9 + 7 =$
9. $5 + 9 - 5 =$
10. $21 - 10 - 12 =$
11. $13 - 17 + 5 - 4 =$
12. $-9 + 10 - 6 + 3 =$
13. $-7 + 5 + 7 + 9 =$
14. $9 + 20 - 9 + 5 =$
15. $27 - 59 - 27 + 50 =$
16. $12 - 7 + 4 - 5 =$
17. $-20 + 17 - 5 + 25 =$
18. $8 - 7 + 2 - 4 + 9 =$

19. $34 - 16 + 9 + 16 - 3 =$

20. $6 - 4 + 8 + 16 - 9 - 8 =$

In section 1 we learned rules for removing grouping symbols. When combining numbers, we must first remove grouping symbols to be sure that each term involved has the proper sign.

> **Rule** Remove all grouping symbols before combining terms in an expression.

Example 12 $12 - (-3) = 12 + 3$
$= 15$

Example 13 $9 + (-6 - 4) = 9 - 6 - 4$
$= -1$

Example 14 $15 - (4 - 9) = 15 - 4 + 9$
$= 20$

Example 15 $-8 - (-10 - 7) = -8 + 10 + 7$
$= 9$

Recall that when we simplify an expression containing grouping symbols within grouping symbols, we remove the innermost set of symbols first.

Example 16 $10 - [-3 - (5 - 21)] = 10 - [-3 - 5 + 21]$
$= 10 + 3 + 5 - 21$
$= -3$

EXERCISE SET 7-3-4

Remove the grouping symbols and simplify.

1. $8 + (-5) =$

2. $7 + (-8) =$

3. $10 - (+4) =$

4. $6 - (+11) =$

5. $3 - (-8) =$

6. $5 - (-3) =$

7. $6 + (8 - 2) =$

8. $9 + (6 - 3) =$

9. $5 + (9 - 12) =$

10. $7 + (3 - 8) =$

11. $10 - (2 + 5) =$

12. $18 - (6 + 4) =$

13. $13 - (2 - 1) =$

14. $11 - (8 - 10) =$

15. $21 - (4 - 10) =$

16. $16 - (7 - 11) =$

17. $8 + (-5 - 4) =$

18. $4 + (-6 - 2) =$

19. $4 - (-5 + 9) =$

20. $9 - (-10 + 12) =$

21. $4 + [13 - (6 + 4)] =$

22. $8 + [10 - (3 + 4)] =$

23. $11 + [-6 - (3 + 8)] =$

24. $14 + [-8 - (4 + 7)] =$

25. $8 - [4 - (3 - 2)] =$

26. $9 - [5 - (4 - 7)] =$

27. $-3 - [7 + (4 - 8)] =$

28. $-6 - [4 - (9 - 4)] =$

ANSWERS

5. _____
6. _____
7. _____
8. _____
9. _____
10. _____
11. _____
12. _____
13. _____
14. _____
15. _____
16. _____
17. _____
18. _____
19. _____
20. _____
21. _____
22. _____
23. _____
24. _____
25. _____
26. _____
27. _____
28. _____

29. $10 - \{8 + [5 - (9 - 2)]\} =$ 30. $12 - \{10 - [4 - (5 - 9)]\} =$

7-4 COMBINING LIKE TERMS

OBJECTIVES

In this section you will learn to:
1. Combine like terms involving signed numbers.
2. Remove grouping symbols and combine like terms.

In chapter 4 we defined like terms as terms that have exactly the same literal factors. We also noted that only like terms can be combined. We will now apply those same rules to terms involving signed numbers.

Example 1 $-3x - 7x = -10x$

Example 2 $-5x + 3x - 4x = -6x$

Example 3 $3x - (2x + 6x) = 3x - 2x - 6x$
$ = -5x$

Example 4 $12x^2 + 3y - 4x^2 - 4y = 8x^2 - y$

Example 5 $5x - [3x - (2x + 3)] = 5x - [3x - 2x - 3]$
$ = 5x - 3x + 2x + 3$
$ = 4x + 3$

EXERCISE SET 7-4-1

In each of the following remove any grouping symbols, then combine like terms.

1. $8x + 5x =$

2. $5a - 3a =$

3. $6x - 9x =$

4. $11x^2 + 7x^2 =$

5. $8xy - 12xy =$

6. $11a - 9b =$

7. $17x^3 + 3x^2 =$

8. $9abc - 3abc =$

9. $-7ab^2c - 5abc =$

10. $3x^2y^2 - 2x^2y^2 =$

11. $4x + 6x - 10x =$

12. $8y - 10y - 6y =$

13. $3ab - 9ab + 4ab =$

14. $3x - (7x - 5x) =$

15. $16a^2b - 31a^2b + 5ab^2 =$

16. $2ab + 5bc - 6ac =$

17. $7ab - (6ab - 3ac) =$

18. $4xy + 13a - (7xy + 5a) =$

19. $8xy - (5x^2y - 2xy) - 10xy + 5x^2y =$

20. $21xyz + 15xy - (17xyz - 7xy^2) =$

21. $2x + [3x - (2x + 1)] =$

22. $4x - [3x - (x + 1)] =$

23. $2x - [5x - (3x + 1)] =$

24. $2x - [1 - (3x + 4)] =$

25. $5 - [x + (8 - 3x)] =$

26. $[3a - (4 - 2a)] - 3a =$

27. $8x + [3x - 4 - (x + 5)] =$

28. $5a - [-4 - (2a - 3)] =$

29. $2x - 3 + [4x - (x - 7)] =$

30. $7a + 4 - [3a - (5 - 2a)] =$

ANSWERS

11. _____
12. _____
13. _____
14. _____
15. _____
16. _____
17. _____
18. _____
19. _____
20. _____
21. _____
22. _____
23. _____
24. _____
25. _____
26. _____
27. _____
28. _____
29. _____
30. _____

ANSWERS

31. $6a + 3b + [a + (3a - 4b)] =$

31. _____

32. $10x - y - [3x - (2x - 5y)] =$

32. _____

33. $a + b - [a - b - (a + b)] =$

33. _____

34. $6x + \{3x - [4x - 3 - (x - 1)]\} =$

34. _____

35. $2x^2 - 3x + \{x - 3x^2 - [4x - (2x^2 - x)]\} =$

35. _____

 7-5 A WORD ABOUT SUBTRACTION

OBJECTIVES

In this section you will learn to:
1. Define subtraction in terms of addition.
2. Subtract signed numbers.

The symbol $-$ is used to indicate both **subtraction** and **the negative of.** The expression $5x - (-4x)$ can be thought of as "subtract the negative of $4x$ from $5x$" or as "add the negative of negative $4x$ to $5x$." We have combined these like terms by thinking of this in the second way. We changed $5x - (-4x)$ to $5x + 4x$ and obtained $9x$.

The following definition should make it clear that both statements are really the same.

> **Definition** **Subtraction** is adding the negative. In symbols
> $$a - b = a + (-b).$$
> (b subtracted from a is the same as the negative of b added to a.)

Example 1 $\quad 8 - (+6) = 8 + (-6)$
$\qquad\qquad\qquad\quad = 2$

> **WARNING**
>
> Don't forget, you can never write two signs together without using parentheses.

Example 2 $3 - (+7) = 3 + (-7)$
$= -4$

Example 3 $8 - 5 = 8 + (-5)$
$= 3$

Example 4 $6 - (-2) = 6 + 2$
$= 8$

Example 5 Subtract -8 from 5.

$$5 - (-8) = 5 + 8$$
$$= 13$$

Example 6 Subtract the sum of 5 and -2 from 10.

$$10 - [5 + (-2)] = 10 - [5 - 2]$$
$$= 10 - 5 + 2$$
$$= 7$$

Example 7 Subtract $(5x - 3y + 4)$ from $(2x - y + 6)$.

$$(2x - y + 6) - (5x - 3y + 4) = 2x - y + 6 - 5x + 3y - 4$$
$$= -3x + 2y + 2$$

▼ EXERCISE SET 7-5-1

Rewrite each of the following subtraction problems as an addition problem and determine the answer.

A N S W E R S

1. $6 - (+4) =$

2. $9 - (+12) =$

1. _____

2. _____

3. $13 - (-2) =$

4. $-16 - (-9) =$

3. _____

4. _____

5. $22 - (-6) =$

6. $3 - (+5) =$

5. _____

6. _____

ANSWERS

7. $-11 - (-1) =$

8. $24 - (+24) =$

7. _____

9. $-25 - (-3) =$

10. $18 - (+18) =$

8. _____

9. _____

11. $8 - 3 =$

12. $8 - 21 =$

10. _____

11. _____

Simplify the following.

12. _____

13. $4 - (2 + 3) =$

14. $15 - (3 - 9) =$

13. _____

14. _____

15. $5a - 3 - (4a - 7) =$

16. $2x^2 - 3x + 1 - (3x^2 - 5x + 4) =$

15. _____

16. _____

17. Subtract 12 from 8.

18. Subtract -4 from 9.

17. _____

18. _____

19. Subtract 23 from -18.

20. Subtract -14 from -8.

19. _____

20. _____

21. Subtract the sum of 5 and -11 from 16.

22. Subtract the sum of -6 and -2 from -18.

21. _____

22. _____

23. Subtract the difference of 7 and 3 from 10.

24. Subtract the difference of 9 and -4 from 12.

23. _____

24. _____

25. Subtract $(4x - 3y + 1)$ from $(8x - y - 4)$.

26. Subtract $(5x^2y + 3xy - y^2)$ from $(2x^2y - xy + 5y^2)$.

25. _____

26. _____

7-6 MULTIPLICATION AND DIVISION OF SIGNED NUMBERS

Multiplication can be thought of as "short-cut addition." For instance, if we wish to multiply $(7)(5)$, we can think of this as "seven fives" or $5 + 5 + 5 + 5 + 5 + 5 + 5$ and obtain the result of 35. Or we could think of it as "five sevens" or $7 + 7 + 7 + 7 + 7$ and also obtain the same result of 35.

Applying this same technique to signed numbers will lead to one of the rules for their multiplication.

Think of $(7)(-5)$ as seven (-5)s or

$$(-5) + (-5) + (-5) + (-5) + (-5) + (-5) + (-5) = -35.$$

Thus $(7)(-5) = -35$.

> **OBJECTIVES**
>
> In this section you will learn to:
> 1. Apply the rule for multiplying signed numbers.
> 2. Apply the distributive property of multiplication over addition.
> 3. Divide signed numbers.

> **Rule** The product of a positive number and a negative number yields a negative number.

Example 1 $(-2)(5) = -10$

Example 2 $(3)(-12) = -36$

Example 3 $\left(\dfrac{2}{3}\right)\left(-\dfrac{3}{4}\right) = -\dfrac{1}{2}$

From arithmetic we know the following rule.

> **Rule** The product of two positive numbers is positive.

In each of these rules be careful to note that the word "two" is very important. Since we can only multiply two numbers at a time, we only need rules for two numbers.

Example 4 Find the product $(3)(-2)(7)$.

There are several ways this could be done. We could first multiply $(3)(-2)$ obtaining

$$(3)(-2)(7) = (-6)(7)$$
$$= -42.$$

We could also first multiply $(3)(7)$ obtaining

$$(3)(-2)(7) = (21)(-2)$$
$$= -42.$$

Still another approach would be to first multiply $(-2)(7)$ obtaining

$$(3)(-2)(7) = (3)(-14)$$
$$= -42.$$

It should be clear that the order in which the numbers are multiplied does not matter. The result will be the same.

EXERCISE SET 7-6-1

Find the following products.

1. $(+8)(-5) =$
2. $(-5)(+8) =$
3. $(7)(-3) =$

4. $(-4)(6) =$
5. $14(-13) =$
6. $\left(-\dfrac{1}{2}\right)(8) =$

7. $\left(\dfrac{2}{3}\right)\left(-\dfrac{9}{10}\right) =$
8. $(-3)(2.5) =$
9. $(5.1)(-2.3) =$

10. $\left(\dfrac{2}{3}\right)\left(-\dfrac{3}{8}\right) =$
11. $\left(-\dfrac{1}{3}\right)(6.3) =$
12. $(5)(-2)(4) =$

13. $(-3)(2)(8) =$
14. $(5)(3)(-4) =$
15. $(-10)\left(\dfrac{1}{2}\right)(7) =$

16. $(-16)(121)(0) =$
17. $(4)\left(-\dfrac{1}{2}\right)(3) =$
18. $\left(-\dfrac{2}{7}\right)(14)\left(2\dfrac{3}{8}\right) =$

19. $(18)(2.3)(-4) =$
20. $(5.4)\left(-3\dfrac{4}{5}\right)(0) =$

To establish a rule for the product or quotient of two negative numbers we will need to use a property of the operations on numbers called the **distributive property of multiplication over addition**. In symbols the distributive property of multiplication over addition is

$$a(b + c) = ab + ac.$$

We may express this property in words by saying "to multiply terms enclosed in parentheses by a number we must multiply each term in the parentheses by that number."

Also, since $b - c = b + (-c)$, then

$$\begin{aligned} a(b - c) &= a[b + (-c)] \\ &= ab + a(-c) \\ &= ab - ac. \end{aligned}$$

Example 5 Evaluate $5(4 + 3)$.

Of course, we could just add $4 + 3$ and obtain

$$\begin{aligned} 5(4 + 3) &= 5(7) \\ &= 35. \end{aligned}$$

If we use the distributive property, we have

$$\begin{aligned} 5(4 + 3) &= 5(4) + 5(3) \\ &= 20 + 15 \\ &= 35. \end{aligned}$$

Example 6 Remove parentheses and combine like terms.

$$-5(x + 2y) + 3x + 8y$$

$$\begin{aligned} -5(x + 2y) + 3x + 8y &= -5(x) - 5(2y) + 3x + 8y \\ &= -5x - 10y + 3x + 8y \\ &= -2x - 2y \end{aligned}$$

▼ EXERCISE SET 7-6-2

Remove parentheses and combine like terms.

1. $5(x + 4)$
2. $2(x - 9)$
3. $2(3x + 4)$
4. $-3(2x + 5)$
5. $14(3x - 2)$
6. $-8(2x + 5)$

ANSWERS

1. _____
2. _____
3. _____
4. _____
5. _____
6. _____

7. _____

8. _____

9. _____

10. _____

11. _____

12. _____

13. _____

14. _____

15. _____

16. _____

17. _____

18. _____

19. _____

20. _____

7. $3(2x^2 + 4x + 2)$

8. $5(2x^2 - x + 3)$

9. $-2(x^2 + 3x + 1)$

10. $-4(3a + 2b + c)$

11. $2x + 3(x + 5)$

12. $7x - 2(x + 3)$

13. $9 + (12x - 7)$

14. $-4(2x + 3) + 15$

15. $-3(4a + 1) + 12a$

16. $6(4a - 3) - 10a + 18$

17. $3(2x - y) + x - 2y$

18. $-4(3x + 2y) + 9x + 10y$

19. $3(5x - 4) - 2(x + 3)$

20. $3(x - 4) - 2(x^2 + 2x + 1)$

We now wish to establish a rule for multiplying two negative numbers.

By using our rules for adding signed numbers we have worked problems such as $(-15) + (+15) = -15 + 15 = 0$.

We also know that multiplying by zero gives zero as a result. For instance, $(-3)(0) = 0$.

Now consider this problem. Find the result of $(-3)[(5) + (-5)]$. If we note that $(5) + (-5) = 0$, then we have

$$(-3)[(5) + (-5)] = (-3)(0)$$
$$= 0.$$

However, if we first use the distributive property, we obtain

$$(-3)[(5) + (-5)] = (-3)(5) + (-3)(-5)$$
$$= -15 + (-3)(-5).$$

We know this result must be zero. That is, $-15 + (-3)(-5) = 0$. Therefore $(-3)(-5) = +15$ since $+15$ is the only number that can be added to -15 to obtain a result of zero.

The choice of numbers in this discussion would not change the conclusion. Hence we have the following rule.

Rule The product of two negative numbers is positive.

WARNING

Do not confuse the rule for *adding* two negative numbers with the rule for *multiplying* them.

Example 7 $(-2)(-4) = 8$

Example 8 $(-4)(-6) = 24$

Example 9 $\left(-\dfrac{2}{3}\right)\left(-\dfrac{3}{4}\right) = \dfrac{1}{2}$

Once more we must realize the importance of the word "two." This rule, as all others, must be applied to only *two* numbers at a time.

Example 10 Find the product of $(-3)(-2)(-5)$.

If we first multiply $(-3)(-2)$, we obtain $(+6)$. Then multiplying $(+6)(-5)$, we obtain the final result.

$$(-3)(-2)(-5) = (+6)(-5)$$
$$= -30$$

Example 11 Find the product of $(-3)(4)(-5)$.

$$(-3)(4)(-5) = (-12)(-5)$$
$$= 60$$

Example 12 Find the product of $(-2)(5)(-3)(-1)(4)$.

$$(-2)(5)(-3)(-1)(4) = (-10)(-3)(-1)(4)$$
$$= (+30)(-1)(4)$$
$$= (-30)(4)$$
$$= -120$$

EXERCISE SET 7-6-3

Find the following products.

1. $(-2)(-3) =$

2. $(-1)(-8) =$

3. $(-3)(-7) =$

4. $(-5)(-9) =$

5. $\left(-\dfrac{1}{2}\right)(-30) =$

6. $\left(-\dfrac{1}{3}\right)\left(-\dfrac{3}{5}\right) =$

7. $(-5)(+8) =$

8. $7(-3) =$

9. $(-1.2)(-5.0) =$

10. $(6)(-2)(-5) =$

11. $(-1)(-3)(-4) =$

12. $(-5)(6)(-2) =$

13. $(-1)(-5)\left(-\dfrac{1}{5}\right) =$

14. $(-3)(5)(0) =$

15. $(3)(-1)(-5)(10) =$

16. $(-1)(-5)(-2)(-3) =$

17. $(-2)(-7)(3)(4) =$

18. $(-10)(10)(-1)(-3) =$

19. $(-2)(-1)(-10)(4)(1) =$

20. $(27)\left(-\dfrac{1}{9}\right)\left(-\dfrac{1}{3}\right)(-37) =$

7-6 MULTIPLICATION AND DIVISION OF SIGNED NUMBERS

Division is defined as "multiplying by the inverse." The **multiplicative inverse** of a number is sometimes referred to as the **reciprocal** of the number.

In a problem such as $12 \div 6$ we are dividing by 6. Using the definition, the problem is the same as multiplying by the multiplicative inverse (or reciprocal) of 6, which is $\frac{1}{6}$.

Thus: $$12 \div 6 = 12\left(\frac{1}{6}\right).$$

This relationship ties division and multiplication together so that the rules for division are the same as the rules for multiplication. For convenience we will restate the rules as a single rule.

> **Rule** In multiplying or dividing signed numbers the product or quotient of two numbers with like signs is positive and the product or quotient of two numbers with unlike signs is negative.

Example 13 $-12 \div (-6) = +2$

Example 14 $-12 \div (+6) = -2$

Example 15 $+12 \div (-6) = -2$

Example 16 $\dfrac{-8}{-2} = +4$

Example 17 $\dfrac{-8}{+2} = -4$

Example 18 $\left(-\dfrac{2}{3}\right) \div \left(-\dfrac{3}{4}\right) = \left(-\dfrac{2}{3}\right)\left(-\dfrac{4}{3}\right)$
$$= +\dfrac{8}{9}$$

Example 19 $\dfrac{(-6)(-7)}{(-3)} = -14$

EXERCISE SET 7-6-4

Find the following quotients.

1. $\dfrac{-6}{-2} =$

2. $\dfrac{-6}{+2} =$

3. $\dfrac{-15}{-3} =$

4. $\dfrac{+15}{-3} =$

5. $\dfrac{-20}{+5} =$

6. $\dfrac{-42}{-6} =$

7. $(-20) \div (-5) =$

8. $(+81) \div (-3) =$

9. $(-28) \div (+4) =$

10. $\left(-\dfrac{1}{2}\right) \div \left(-\dfrac{1}{8}\right) =$

11. $\left(-\dfrac{2}{5}\right) \div \left(\dfrac{3}{7}\right) =$

12. $\left(\dfrac{3}{8}\right) \div \left(-\dfrac{3}{16}\right) =$

13. $(-3) \div \left(-\dfrac{6}{7}\right) =$

14. $\left(-\dfrac{5}{7}\right) \div 15$

15. $\dfrac{(-5)(-3)}{(-6)} =$

16. $\dfrac{(-3)(-6)}{-9} =$

17. $\dfrac{(+4)(-6)}{-2} =$

18. $\dfrac{(-6)(+4)}{12} =$

19. $\dfrac{3-15}{-6} =$

20. $\dfrac{9-27}{6} =$

7-7 EXERCISES USING SIGNED NUMBERS

We can use the distributive property, discussed in the last section, to simplify expressions in which grouping symbols are used within grouping symbols.

OBJECTIVES

In this section you will learn to:
1. Simplify expressions involving signed numbers.
2. Evaluate expressions involving signed numbers.

Example 1 Simplify: $3x - 2[5(x - y) + 8y]$

$$\begin{aligned}3x - 2[5(x - y) + 8y] &= 3x - 2[5x - 5y + 8y] \\ &= 3x - 10x + 10y - 16y \\ &= -7x - 6y\end{aligned}$$

Example 2 Simplify: $3\{2x - 4[3(4 - 5x)]\}$

$$\begin{aligned}3\{2x - 4[3(4 - 5x)]\} &= 3\{2x - 4[12 - 15x]\} \\ &= 3\{2x - 48 + 60x\} \\ &= 6x - 144 + 180x \\ &= 186x - 144\end{aligned}$$

In chapter 4 we evaluated literal expressions using only the numbers of arithmetic. We now are able to evaluate expressions that involve signed numbers.

Example 3 Evaluate $x^2 - 4x + 2$ when $x = -3$.

Replacing x with -3, we obtain

$$\begin{aligned}(-3)^2 - 4(-3) + 6 &= 9 + 12 + 6 \\ &= 27.\end{aligned}$$

Example 4 Evaluate $3x^2 - 2y + x$ when $x = 1, y = -3$.

We replace x with 1 and y with -3 obtaining

$$\begin{aligned}3(1)^2 - 2(-3) + 1 &= 3 + 6 + 1 \\ &= 10.\end{aligned}$$

EXERCISE SET 7-7-1

Simplify each of the following.

1. $4x - 3[(x + y) - 3y]$

2. $a - [b - 5(2a - 3b)]$

3. $2x + 5[y - 4(2x - 3y)]$

4. $3a + 4[4b - 2(2b - a)]$

5. $x - 3[2x - (3 + 3x)]$

6. $6a - 3[4 - 5(a - 2)]$

ANSWERS

1. _____
2. _____
3. _____
4. _____
5. _____
6. _____

7. $4[5x - 3(x - 7)] + 21$

8. $-2[3(2a - 3) - 5(a - 4)]$

9. $3(x + 2y) - [2(2x + y) - 6y]$

10. $19a + 2\{5b - [4(2a - 3b) + 6b]\}$

11. $2[x - (3y - 5)] - 7[3x + 2(y - 8)]$

12. $8x - 3[3x + (2a + 4x) - 6a]$

13. $2[4x - (x + 1)] - 3\{2[3x + (x + 1)] - 5x\}$

14. $5\{3x + 4[x - 3(2x - 1)] - 4\}$

15. $2\{2(x + y) - 9[3x - (x + y)]\}$

Evaluate the following.

16. $3x + 5$ when $x = -7$

17. $16 - 2x$ when $x = -3$

18. x^2 when $x = -5$

19. $-x^2$ when $x = -5$

20. $(-x)^2$ when $x = -5$

21. $3x^3$ when $x = -2$

ANSWERS

7.
8.
9.
10.
11.
12.
13.
14.
15.
16.
17.
18.
19.
20.
21.

22. $(3x)^3$ when $x = -2$

23. $4x^2 - 5x + 2$ when $x = -2$

24. $2x - 3y$ when $x = 4, y = -2$

25. $3x^2 + xy - 6$ when $x = -2, y = -1$

26. $2x^3 + 3xyz - 2z^2$ when $x = 2, y = 5, z = -1$

27. $x^3 - 2xz - 3y$ when $x = -2, y = -3, z = 4$

28. $a^3 - b^3$ when $a = -2, b = -3$

29. $C = \dfrac{5}{9}(F - 32)$. Find C when $F = 5$.

30. $F = \dfrac{9}{5}C + 32$. Find F when $C = -10$.

ANSWERS

22. _____
23. _____
24. _____
25. _____
26. _____
27. _____
28. _____
29. _____
30. _____

SUMMARY

Signed Numbers
- Signed numbers are preceded by a plus or minus sign. Numbers preceded by a plus sign are **positive numbers** and those preceded by a minus sign are **negative numbers.**
- A **number line** is useful for visualizing the relative positions of numbers.
- Negative means opposite. To find the negative of a number change its sign. To remove parentheses preceded by a minus sign find the negative of each term within the parentheses.

Operations on Signed Numbers
- To combine two numbers with like signs add the numbers and use the common sign.
- To combine two numbers with unlike signs subtract the smaller from the larger (without regard to sign) and attach the sign of the larger.
- To multiply or divide signed numbers two like signs give a positive product or quotient and two unlike signs give a negative product or quotient.

CHAPTER 7 REVIEW

NAME: _____

CLASS / SECTION: _____ DATE: _____

ANSWERS

1. _____
2. _____
3. _____
4. _____
5. _____
6. _____
7. _____
8. _____
9. _____
10. _____
11. _____
12. _____
13. _____
14. _____
15. _____
16. _____
17. _____
18. _____
19. _____
20. _____
21. _____

Find the negative of each of the following.

1. $+10$
2. $+\dfrac{2}{3}$
3. $-\dfrac{3}{8}$
4. -5.1

Combine.

5. $(+7)+(-9)$
6. $(-17)+(-5)$
7. $\left(+\dfrac{5}{9}\right)+\left(-\dfrac{2}{9}\right)$
8. $\left(-\dfrac{1}{2}\right)+\left(+\dfrac{2}{3}\right)$
9. $(-5)+(+8)+(-10)$
10. $(+6)+(-15)+(+4)$
11. $5-8+2$
12. $4-(-6)-1$
13. $-3+(-4)+9$
14. $-13+5-(-16)-2$
15. $-4+9+2-8-7+6$
16. $4-(9-2)$
17. $5-(-6-9)+2$
18. $7-(5-16)-4$

Evaluate.

19. $(-4)(+12)$
20. $(-6)\left(+\dfrac{1}{3}\right)$
21. $\left(-\dfrac{3}{4}\right)\left(-\dfrac{2}{7}\right)$

22. $(-2.3)(-3.5)$ **23.** $(8)(-5)(-7)$ **24.** $(-4)(3)(10)$

25. $\dfrac{-36}{-3}$ **26.** $\dfrac{-32}{+8}$ **27.** $\left(\dfrac{4}{9}\right) \div \left(-\dfrac{2}{3}\right)$

28. $(-15) \div \left(-\dfrac{3}{5}\right)$ **29.** $\dfrac{(-6)(5)}{-10}$ **30.** $\dfrac{(-8)(-6)}{+12}$

31. $(-3)(-4)(-5)$ **32.** $(-3)(-1)(5)(-10)\left(-\dfrac{2}{5}\right)(3)$

33. $\dfrac{(-4)(-9)}{-6}$ **34.** $\dfrac{-18 + 3}{-5}$

Remove grouping symbols and combine like terms.

35. $3x - (2x + 5) - 7$ **36.** $10a - (9 - 4a) + 16$

37. $8 - (3x + 5) + 12x$ **38.** $15 + (5x - 7) - (9x - 2)$

39. $14 - 3(2x - 1)$ **40.** $5(3a - 2) - 4(3a + 1)$

41. $3(a - 2) - 5(2a - 4)$ **42.** $8(3x - 4) + 3x - 2(6x + 5)$

43. $13x - 5[x - (3x + 4)]$ **44.** $5a - [-3 - 2(3a - 2)]$

ANSWERS

22. _____
23. _____
24. _____
25. _____
26. _____
27. _____
28. _____
29. _____
30. _____
31. _____
32. _____
33. _____
34. _____
35. _____
36. _____
37. _____
38. _____
39. _____
40. _____
41. _____
42. _____
43. _____
44. _____

ANSWERS

45. _____

46. _____

47. _____

48. _____

49. _____

50. _____

51. _____

52. _____

53. _____

54. _____

SCORE: _____

45. $2[3a - 2(a + 3) + a]$

46. $10x - 3[2x + 5(x - 3)]$

47. $a + 4\{3a - [5(a + 1) - a]\}$

48. $15x - 2\{2x - 4[3 - 4(x - 5)]\}$

Evaluate.

49. $3x^3$ when $x = -5$

50. $-x^3$ when $x = -2$

51. $2x^2 - 3x + 1$ when $x = -4$

52. $x^2 - y^2$ when $x = -3$, $y = -4$

53. $5x^3 - 2xy^2 - 3y$ when $x = -3$, $y = 4$

54. $S = \dfrac{a}{1 - r}$. Find S when $a = -12$, $r = \dfrac{1}{4}$.

CHAPTER 7 TEST

NAME: _____
CLASS / SECTION: _____
DATE: _____

ANSWERS

1. Find the negative of: a. $-\dfrac{5}{8}$ b. 6.9

2. Simplify: $-(-28)$

3. Use a signed number to represent the result of a credit of $100 followed by a debit of $60.

4. Add: a. $(-32) + (+18)$ b. $(-13) + (-26)$

5. Subtract: a. $(-6) - (+14)$ b. $(-19) - (-23)$

6. Simplify: $4 - 11 - 8 - (-3) + 5$

7. Remove parentheses and simplify: $11 - (-2 - 8) + 15$

8. Combine like terms: $16ab + 9a^2 - 12ab - 6a^2$

9. Subtract the sum of 13 and 6 from 40.

10. Find the product:
    a. $(-11)(-5)$ b. $(-8)(+12)$

11. Find the product: $(-3)(8)(-4)(-2)$

12. Find the quotient: a. $\dfrac{-26}{-2}$ b. $\left(-\dfrac{4}{5}\right) \div 8$

1a. _____
1b. _____
2. _____
3. _____
4a. _____
4b. _____
5a. _____
5b. _____
6. _____
7. _____
8. _____
9. _____
10a. _____
10b. _____
11. _____
12a. _____
12b. _____

13. Evaluate: a. $\dfrac{(-8)(-9)}{-6}$ b. $\dfrac{-21 + 5}{-2}$

13a. _____

13b. _____

14. Remove grouping symbols and simplify:
$10x + 3\{2x - 4[3x - 3(2x - 1)]\}$

14. _____

15. Evaluate $2x^2 - 3xy + 4y^2$ when $x = -2$, $y = 3$.

15. _____

SCORE: _____

CHAPTER 8 PRETEST

NAME:
CLASS / SECTION: DATE:

ANSWERS

Answer as many of the following problems as you can before starting this chapter. When you finish the chapter, take the test at the end and compare your scores to see how much you have learned.

Solve for x.

1. _____   **1.** $4x = -28$ **2.** $3x - 4 = 10 - x$

2. _____   **3.** $3(2x - 5) = 2(3 - x) - 1$ **4.** $\dfrac{5}{8}x + \dfrac{3}{4} = \dfrac{1}{2}x - 2$

3. _____   **5.** $x - (3x + 4) = \dfrac{1}{2}(x + 1) + \dfrac{11}{2}$

4. _____   **6.** $5x - 8y = 2x + y$

    7. $2a(bx + c) + 3ac = 3abx$

5. _____   **8.** Graph $-2 \leq x < 5$ on the number line.

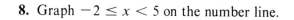

6. _____

Solve and graph the solution on the number line.

7. _____   **9.** $8x - 2 < 2(3x - 4)$

8. _____

    10. $\dfrac{3}{4}x - 5 \geq 3(x - 1)$

9. _____

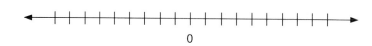

10. _____

SCORE: _____

358

CHAPTER 8

Equations and Inequalities Involving Signed Numbers

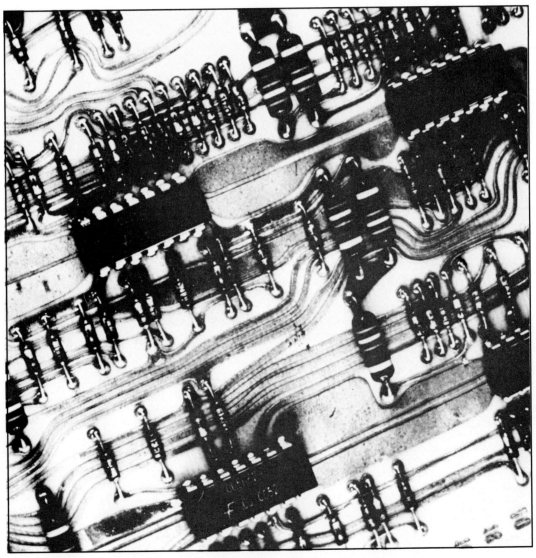

The formula $I = \dfrac{e + E}{nr + R}$ is used in electronics. Solve for E. Then find E when $I = .5$, $n = 3$, $r = 2$, $R = 4$, and $e = -2$.

 8-1 SOLVING EQUATIONS INVOLVING SIGNED NUMBERS

In chapter 5 we established rules for solving equations using the numbers of arithmetic. Now that we have learned the operations on signed numbers we will use those same rules to solve equations that involve negative numbers. We will also study techniques for solving and graphing inequalities having one unknown.

OBJECTIVES

In this section you will learn to solve equations involving signed numbers.

Example 1 Solve for x and check: $x + 5 = 3$

Using the same procedures learned in chapter 5, we subtract 5 from each side of the equation obtaining

$$x + 5 - 5 = 3 - 5$$
$$x = -2.$$

Check: $\quad x + 5 = 3$
$\quad\quad -2 + 5 = 3$
$\quad\quad\quad\quad 3 = 3$

Example 2 Solve for x and check: $-3x = 12$

Dividing each side by -3, we obtain

$$\frac{-3x}{-3} = \frac{12}{-3}$$
$$x = -4.$$

Check: $\quad -3x = 12$
$\quad\quad -3(-4) = 12$
$\quad\quad\quad\quad 12 = 12$

Example 3 Solve for x and check: $3x - 4 = 7x + 8$

$$3x - 4 = 7x + 8$$
$$3x - 4 + 4 = 7x + 8 + 4$$
$$3x = 7x + 12$$
$$3x - 7x = 7x - 7x + 12$$
$$-4x = 12$$
$$\frac{-4x}{-4} = \frac{12}{-4}$$
$$x = -3$$

Check: $\quad 3x - 4 = 7x + 8$
$\quad\quad 3(-3) - 4 = 7(-3) + 8$
$\quad\quad -9 - 4 = -21 + 8$
$\quad\quad\quad -13 = -13$

Example 4 Solve for x and check: $3(x + 2) = \frac{1}{4}(x - 3)$

$$3(x + 2) = \frac{1}{4}(x - 3)$$
$$3x + 6 = \frac{x}{4} - \frac{3}{4}$$
$$12x + 24 = x - 3$$
$$11x = -27$$
$$x = -\frac{27}{11}$$

Check:
$$3(x + 2) = \frac{1}{4}(x - 3)$$
$$3\left(-\frac{27}{11} + \frac{22}{11}\right) = \frac{1}{4}\left(-\frac{27}{11} - \frac{33}{11}\right)$$
$$3\left(-\frac{5}{11}\right) = \frac{1}{4}\left(-\frac{60}{11}\right)$$
$$-\frac{15}{11} = -\frac{15}{11}$$

▼ EXERCISE SET 8-1-1

Solve the following for x and check.

1. $x + 4 = 1$
2. $x + 8 = 3$

3. $x - 3 = -4$
4. $x - 5 = -6$

5. $5x = -20$
6. $2x = -8$

7. $-2x = 10$
8. $-3x = 15$

ANSWERS

1. _____

2. _____

3. _____

4. _____

5. _____

6. _____

7. _____

8. _____

ANSWERS

9. $8 - 3x = -4$

10. $11 - 4x = -9$

11. $3x + 5 = x + 7$

12. $4x - 3 = x - 9$

13. $3x - 1 = 2(x - 5)$

14. $3x + 2 = 5x - 8$

15. $x - 7 = 4x + 5$

16. $2x + 3 = 3x + 5$

17. $x - 6 = 5x - 14$

18. $3(x - 4) = 5 + 2(x + 1)$

19. $4x - 4 = 2x + 6$

20. $3x + 7 = 5x - 4$

21. $3x + 2(x - 5) = 7 - (x + 3)$

22. $5x - 3(x + 1) = 5$

9. _____

10. _____

11. _____

12. _____

13. _____

14. _____

15. _____

16. _____

17. _____

18. _____

19. _____

20. _____

21. _____

22. _____

23. $7x + 5 - 2(x - 1) = 21$

24. $5 + 6x - 3 = 2 + 4x$

25. $3(x + 1) + 3 = 7(x - 2)$

26. $5x - (2x - 3) = 4(x + 9)$

27. $\dfrac{x}{2} = \dfrac{1}{5} - x$

28. $4 - \dfrac{x + 2}{3} = 1$

29. $x - \dfrac{1}{2} = \dfrac{x}{3} + 7$

30. $\dfrac{x - 3}{4} - 9 = 11$

31. $5(x + 4) = \dfrac{1}{3}(x - 10)$

32. $\dfrac{2x + 5}{3} = 0$

33. $\dfrac{2}{3} - 3(x - 1) = \dfrac{3}{5}$

34. $2x - 3(x - 2) = \dfrac{1}{2}(x + 1)$

35. $\dfrac{2}{3} + 1 = x - \dfrac{5}{2}$

36. $4x - 2(x - 3) = \dfrac{1}{4}(x + 3)$

ANSWERS

23. _____

24. _____

25. _____

26. _____

27. _____

28. _____

29. _____

30. _____

31. _____

32. _____

33. _____

34. _____

35. _____

36. _____

37. $x - \dfrac{1}{5}x + 1 = \dfrac{1}{3}(x - 5)$ 38. $\dfrac{2}{3}x + \dfrac{x}{2} - \dfrac{1}{2} = \dfrac{1}{3}(x - 14)$

37. _____

38. _____   39. $\dfrac{2}{3}x - \dfrac{1}{2}x = x + \dfrac{1}{6}$ 40. $\dfrac{x}{5} - \dfrac{2}{3}x + \dfrac{1}{2} = \dfrac{1}{3}(x - 4)$

39. _____

40. _____

8-2 LITERAL EQUATIONS

In this section you will learn to:
1. Identify a literal equation.
2. Apply previously learned rules to solve literal equations.

An equation having more than one letter is sometimes called a **literal equation**. It is occasionally necessary to solve such an equation for one of the letters in terms of the others. The step-by-step procedure discussed and used in chapter 5 is still valid.

Example 1   Solve for c: $3(x + c) - 4y = 2x - 5c$

First remove parentheses.

$$3x + 3c - 4y = 2x - 5c$$

At this point we note that since we are solving for c we want to obtain c on one side and all other terms on the other side of the equation. Thus we obtain

$$3c + 5c = 2x - 3x + 4y$$
$$8c = 4y - x$$
$$c = \dfrac{4y - x}{8}.$$

Example 2   Solve for x: $3abx + cy = 2abx + 4cy$

First we subtract $2abx$ from both sides.

$$3abx + cy - 2abx = 2abx + 4cy - 2abx$$
$$abx + cy = 4cy$$

Subtract cy from both sides.

$$abx + cy - cy = 4cy - cy$$
$$abx = 3cy$$

Divide both sides by ab.

$$\frac{abx}{ab} = \frac{3cy}{ab}$$
$$x = \frac{3cy}{ab}$$

Example 3 Solve for x: $\frac{2}{3}(x + y) = 3(x + a)$

First remove parentheses.

$$\frac{2}{3}x + \frac{2}{3}y = 3x + 3a$$

Multiply each term by 3.

$$2x + 2y = 9x + 9a$$

Subtract $9x$ from both sides.

$$-7x + 2y = 9a$$

Subtract $2y$ from both sides.

$$-7x = 9a - 2y$$

Divide each term by -7.

$$x = \frac{9a - 2y}{-7}$$

Sometimes the form of an answer can be changed. In this example we could multiply both numerator and denominator of the answer by (-1) (this does not change the value of the answer) and obtain

$$x = \frac{9a - 2y}{-7} = \frac{(-1)(9a - 2y)}{(-1)(-7)} = \frac{-9a + 2y}{7} = \frac{2y - 9a}{7}.$$

The advantage of this last expression over the first is that there are not so many negative signs in the answer.

The most commonly used literal expressions are formulas from geometry, physics, business, electronics, and so forth.

Example 4 $A = \dfrac{1}{2}h(b + c)$ is the formula for the area of a trapezoid. Solve for c.

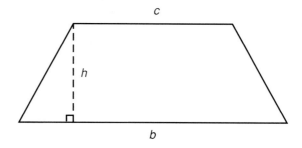

First remove parentheses.

$$A = \frac{1}{2}hb + \frac{1}{2}hc$$

Multiply each term by 2.

$$2A = hb + hc$$

Subtract hc from both sides.

$$2A - hc = hb$$

Subtract $2A$ from both sides.

$$-hc = hb - 2A$$

Divide each side by $-h$.

$$c = \frac{hb - 2A}{-h}$$

We can change the form of the answer.

$$c = \frac{hb - 2A}{-h} = \frac{(-1)(hb - 2A)}{(-1)(-h)} = \frac{-hb + 2A}{h} = \frac{2A - hb}{h}$$

Example 5 $I = \dfrac{prD}{365}$ is a formula giving interest (I) earned for a period of D days when the principal (p) and the yearly rate (r) are known. Find the yearly rate when the amount of interest, the principal, and the number of days are all known.

The problem requires solving $I = \dfrac{prD}{365}$ for r.

$$365I = prD$$
$$\frac{365I}{pD} = \frac{prD}{pD}$$
$$\frac{365I}{pD} = r$$

Notice in this example that r was left on the right side and thus the computation was simpler. We can rewrite the answer another way if we wish.

$$r = \frac{365I}{pD}$$

EXERCISE SET 8-2-1

Solve for x in each of the following.

1. $2x + y = x + 3y$

2. $3x - 2y = x + 4y$

3. $3x + 2y = 6y + x$

4. $2x + 8y = 5x - y$

5. $2(x + a) = x - 4a$

6. $3x + a = 7(x - a)$

7. $4(x - 2y) + y = 3x - 5y$

8. $3(2x + y) = 19y + 4x$

9. $4a - 2x = 3(4x - 8a)$

10. $11x + 15y = 5(3x - y)$

11. $3ax + 2bc = 4(ax - 5bc)$

12. $2ax + bc = 2(3ax + bc)$

13. $7abx - 3y = 4abx + 5a$

14. $3(2a - 5x) = 2(a + 3x)$

15. $4x + 3a = 5(2b - x)$

16. $5(x - a) = \frac{2}{3}(x + 1)$

ANSWERS

1. _____
2. _____
3. _____
4. _____
5. _____
6. _____
7. _____
8. _____
9. _____
10. _____
11. _____
12. _____
13. _____
14. _____
15. _____
16. _____

17. $3(2x + 5) = \dfrac{1}{5}(x + a)$

18. $\dfrac{2}{5}(x - a) = 4(x + a)$

19. $7x + a - 3b = 5x + 3a$

20. $2x + 3(x - a) = 7(x - a)$

21. $A = \dfrac{1}{2}bh$ is the formula for the area of a triangle. Solve for h.

22. The formula for simple interest is $I = prt$. Solve for r.

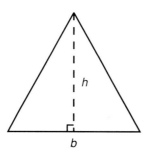

23. A distance formula from physics is $s = \dfrac{1}{2}gt^2$. Solve for g.

24. $F = \dfrac{9}{5}C + 32$ is a formula for changing temperature from Celsius to Fahrenheit degrees. Solve for C.

25. First solve the equation $I = \dfrac{prD}{365}$ for p (principal). Then find how much should be invested for 90 days at 12% annual rate of interest to yield $250 interest. Round answer to nearest cent.

26. The formula $I = \dfrac{e + E}{nr + R}$ is used in electronics. Solve for E. Then find E when $I = .5$, $n = 3$, $r = 2$, $R = 4$, and $e = -2$.

8-3 GRAPHING INEQUALITIES

We have already discussed the set of **rational numbers** as those that can be expressed as a ratio of two integers. There is also a set of numbers, called the **irrational numbers,** that cannot be expressed as the ratio of integers. This set includes such numbers as π, $\sqrt{5}$, $\sqrt[3]{7}$, and so on. The set composed of rational and irrational numbers is called the **real numbers.**

Given any two real numbers a and b, it is always possible to state that $a = b$ or $a \neq b$. Many times we are only interested in whether or not two numbers are equal, but there are situations where we also wish to represent the relative size of numbers that are not equal.

The symbols $<$ and $>$ are **inequality symbols** or **order relations** and are used to show the relative sizes of the values of two numbers. We usually read the symbol $<$ as "less than." For instance, $a < b$ is read as "a is less than b." We usually read the symbol $>$ as "greater than." For instance, $a > b$ is read as "a is greater than b." Notice that we have stated that we usually read $a < b$ as a is less than b. But this is only because we read from left to right. In other words, "a is less than b" is the same as saying "b is greater than a." Actually then, we have only one symbol that is written two ways only for convenience of reading. One way to remember the meaning of the symbol is that the pointed end is toward the *lesser* of the two numbers.

> **OBJECTIVES**
>
> In this section you will learn to:
> 1. Use the inequality symbol to represent the relative positions of two numbers on the number line.
> 2. Graph inequalities on the number line.

> **Definition** $a < b$, "a is less than b," if and only if there is a positive number c that can be added to a to give $a + c = b$.

In simpler words this definition states that a is less than b if we must add something to a to get b. Of course, the "something" must be positive.

If you think of the number line, you know that adding a positive number is equivalent to moving to the right on the number line. This gives rise to the following alternative definition, which may be easier to visualize.

> **Definition** $a < b$ means that a is to the left of b on the number line.

Example 1 $3 < 6$, because 3 is to the left of 6 on the number line.

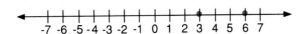

Example 2 $-4 < 0$, because -4 is to the left of 0 on the number line.

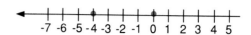

Example 3 $4 > -2$, because 4 is to the right of -2 on the number line.

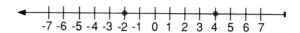

Example 4 $-6 < -2$, because -6 is to the left of -2 on the number line.

▼ EXERCISE SET 8-3-1

A N S W E R S

Locate the following numbers on the number line and replace the question mark with $<$ or $>$.

1. 6 ? 10

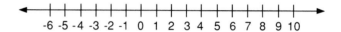

2. -6 ? -10

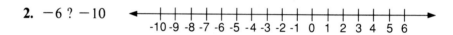

3. -3 ? 3

4. -4 ? -1

5. 4 ? 1

6. 1 ? 4

7. -2 ? -3

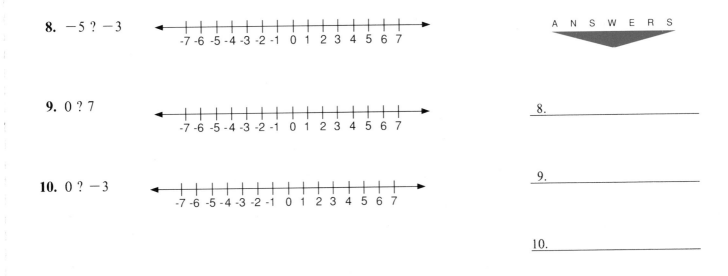

	ANSWERS
8.	
9.	
10.	

The mathematical statement $x < 3$, read as "x is less than 3," indicates that the variable x can be any number less than (or to the left of) 3. Remember, we are considering the real numbers and not just integers, so do not think of the values of x for $x < 3$ as only 2, 1, 0, -1, and so on.

As a matter of fact, to name the number x that is the largest number less than 3 is an impossible task. It can be indicated on the number line, however. To do this we need a symbol to represent the meaning of a statement such as $x < 3$.

Definition The symbols (and) used on the number line indicate that the endpoint is not included in the set.

Example 5 Graph $x < 3$ on the number line.

Note that the graph has an arrow indicating that the line continues without end to the left.

Example 6 Graph $x > 4$ on the number line.

Example 7 Graph $x > -5$ on the number line.

Example 8 Make a number line graph showing that $x > -1$ and $x < 5$. (The word "and" means that *both* conditions must apply.)

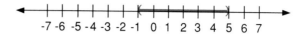

The statement $x > -1$ and $x < 5$ can be condensed to read $-1 < x < 5$.

Example 9 Graph $-3 < x < 3$.

If we wish to include the endpoint in the set, we use a different symbol, $\leq$ or $\geq$. We read these symbols as "equal to or less than" and "equal to or greater than."

Example 10 $x \geq 4$ indicates the number 4 *and* all real numbers to the right of 4 on the number line.

> **Definition** The symbols [and] used on the number line indicate that the endpoint is included in the set.

Example 11 Graph $x \geq 1$ on the number line.

This graph represents the number 1 and all real numbers greater than 1.

Example 12 Graph $x \leq -3$ on the number line.

Example 13 Write an algebraic statement represented by the following graph.

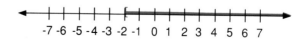

The statement is $x \geq -2$.

Example 14 Write an algebraic statement for the following graph.

$x \geq -4$ and $x \leq 5$ or $-4 \leq x \leq 5$.

Example 15 Write an algebraic statement for the following graph.

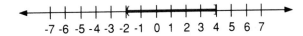

$x > -2$ and $x \leq 4$ or $-2 < x \leq 4$.

Example 16 Graph $x > 2\frac{1}{2}$ on the number line.

This example presents a small problem. How can we indicate $2\frac{1}{2}$ on the number line? If we estimate the point, then another person might misread the statement. Could you possibly tell if the point represents $2\frac{1}{2}$ or maybe $2\frac{7}{16}$? Since the purpose of a graph is to clarify, *always* label the endpoint.

▼ EXERCISE SET 8-3-2

Construct a graph on the number line for each of the following.

1. $x > 2$

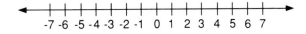

1. _____

2. $x < 5$

2. _____

3. $x < -1$

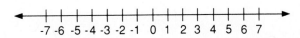

3. _____

4. $x > -3$

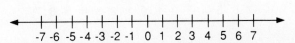

4. _____

ANSWERS

5. _____

6. _____

7. _____

8. _____

9. _____

10. _____

11. _____

12. _____

13. _____

14. _____

15. _____

5. $x \geq 1$

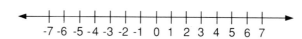

6. $x \leq 1$

7. $x \geq -4$

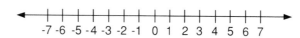

8. $x \leq -2$

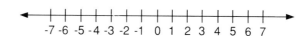

9. $3 < x < 7$

10. $-2 \leq x \leq 5$

11. $-5 \leq x \leq 0$

12. $x > 2\frac{3}{10}$

13. $x < 5.4$

14. $-2\frac{1}{2} \leq x < 3\frac{1}{2}$

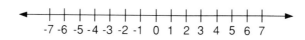

15. $6 < x < 7$

In each of the following write an algebraic statement for the graph.

16.

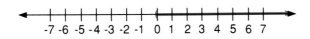

16. _____

17.

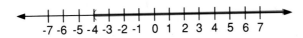

17. _____

18.

18. _____

19.

19. _____

20.

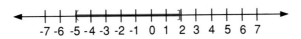

20. _____

8-4 SOLVING INEQUALITIES

The solutions for inequalities generally involve the same basic rules as equations. There is one exception, which we will soon discover. The first rule, however, is similar to that used in solving equations.

OBJECTIVES

In this section you will learn to solve inequalities involving one unknown.

> **Rule 1** If the same quantity is added to each side of an inequality, the results are unequal in the same order.

Example 1 If $5 < 8$, then $5 + 2 < 8 + 2$.

Example 2 If $7 < 10$, then $7 - 3 < 10 - 3$.

We can use this rule to solve certain inequalities.

Example 3 Solve for x: $x + 6 < 10$

If we add -6 to each side, we obtain

$$x + 6 - 6 < 10 - 6$$
$$x < 4.$$

Graphing this solution on the number line, we have

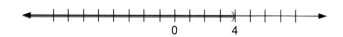

Example 4 Solve the inequality $2(x + 2) \leq x - 5$ and graph the solution on the number line.

$$2(x + 2) \leq x - 5$$

First remove parentheses.

$$2x + 4 \leq x - 5$$

Next add -4 to each side.

$$2x + 4 - 4 \leq x - 5 - 4$$
$$2x \leq x - 9$$

Finally add $-x$ to each side.

$$2x - x \leq x - x - 9$$
$$x \leq -9$$

Graphing this solution, we have

EXERCISE SET 8-4-1

Use the addition rule to solve the following and graph the solutions.

1. $x + 3 < 7$

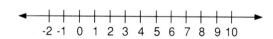

2. $x - 5 < 0$

3. $3x + 4 < 2x - 1$

4. $5x < 4x + 1$

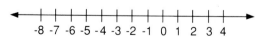

5. $2(x + 3) \leq x + 9$

```
<----+---+---+---+---+---+---+---+---+---+---+---+---+---->
    -6  -5  -4  -3  -2  -1   0   1   2   3   4   5   6
```

6. $-3(x - 1) \geq 2(1 - 2x)$

```
<----+---+---+---+---+---+---+---+---+---+---+---+---+---->
    -6  -5  -4  -3  -2  -1   0   1   2   3   4   5   6
```

7. $5(x + 2) > 2(2x + 3)$

```
<----+---+---+---+---+---+---+---+---+---+---+---+---+---->
    -8  -7  -6  -5  -4  -3  -2  -1   0   1   2   3   4
```

8. $2x + 3(x + 2) \leq 2(2x + 1) - 3$

```
<----+---+---+---+---+---+---+---+---+---+---+---+---+---->
   -10 -9  -8  -7  -6  -5  -4  -3  -2  -1   0   1   2
```

9. $3x + 5 \geq 2(x + 5) - 10$

```
<----+---+---+---+---+---+---+---+---+---+---+---+---+---->
   -10 -9  -8  -7  -6  -5  -4  -3  -2  -1   0   1   2
```

10. $3x + 5 - 2(x - 1) \leq 7$

```
<----+---+---+---+---+---+---+---+---+---+---+---+---+---->
    -6  -5  -4  -3  -2  -1   0   1   2   3   4   5   6
```

A N S W E R S

5. _____

6. _____

7. _____

8. _____

9. _____

10. _____

We will now use the addition rule to illustrate an important concept concerning multiplication or division of inequalities.

Suppose $x > a$.
Now add $-x$ to both sides by the addition rule.

$$x - x > a - x$$
$$0 > a - x$$

Now add $-a$ to both sides.

$$0 - a > a - a - x$$
$$-a > -x$$

The last statement, $-a > -x$, can be rewritten as $-x < -a$. Therefore we can say, "If $x > a$, then $-x < -a$." This translates into the following rule.

> **Rule 2** If an inequality is multiplied or divided by a **negative** number, the results will be unequal in the **opposite** order.

Example 5 Solve for x and graph the solution: $-2x > 6$

To obtain x on the left side we must divide each term by -2. Notice that since we are dividing by a negative number, we must change the direction of the inequality.

$$\frac{-2x}{-2} < \frac{6}{-2}$$
$$x < -3$$

<-|--|--|--|--|--|--|--|--|--|--|--|--|--|->
-7 -6 -5 -4 -3 -2 -1 0 1 2 3 4 5 6 7

Take special note of rule 2. Each time you divide or multiply by a negative number, you must change the direction of the inequality symbol. This is the *only* difference between solving equations and solving inequalities.

> **WARNING**
>
> Division or multiplication by a positive number will *not* change the sense of the inequality, *but* the same operations by a negative number *will* change it. This one point is a source of many errors—so be very careful!

Let us now review the step-by-step method from chapter 5 and note the difference when solving inequalities.

Step 1 Remove all parentheses. (No change)

Step 2 Eliminate fractions by multiplying all terms by the least common denominator of all fractions. (No change when we are multiplying by a positive number.)

Step 3 Simplify by combining like terms on each side of the inequality. (No change)

Step 4 Add or subtract quantities to obtain the unknown on one side and the numbers on the other. (No change)

Step 5 Divide each term of the inequality by the coefficient of the unknown. If the coefficient is positive, the inequality will remain the same. If the coefficient is negative, the inequality will be reversed. (This is the important difference between equations and inequalities.)

Example 6 Solve for x and graph the results: $2(x - 3) \geq 3x - 4$

$$2(x - 3) \geq 3x - 4$$

We first remove parentheses.

$$2x - 6 \geq 3x - 4$$

Then
$$2x - 6 + 6 \geq 3x - 4 + 6$$
$$2x \geq 3x + 2$$
$$2x - 3x \geq 3x + 2 - 3x$$
$$-x \geq 2.$$

We now divide each term by -1.
$$\frac{-x}{-1} \leq \frac{2}{-1}$$
$$x \leq -2$$

Example 7 Solve for x and graph the results.
$$\frac{1}{3}x + \frac{2}{3} < \frac{5}{6}x - 1$$

We first multiply each term by 6, obtaining
$$2x + 4 < 5x - 6$$
$$-3x < -10$$
$$\frac{-3x}{-3} > \frac{-10}{-3}$$
$$x > \frac{10}{3}.$$

▼ EXERCISE 8-4-2

Solve the following inequalities and graph the results on the number line.

1. $3x < 6$

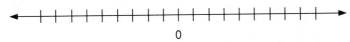

1. _____

2. $2x \leq -8$

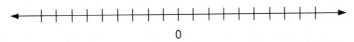

2. _____

3. $-5x < -10$

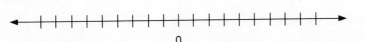

3. _____

ANSWERS

4. $\frac{1}{2}x < 2$

4. _____

5. $\frac{2}{3}x > 4$

5. _____

6. $-\frac{1}{3}x \geq -1$

6. _____

7. $5x - 1 > 3(x + 1)$

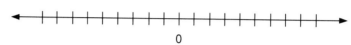

7. _____

8. $2(x + 3) \leq 7(x + 2) + 2$

8. _____

9. $\frac{2}{3}x + 1 \geq 2x - \left(\frac{x}{2} - 6\right)$

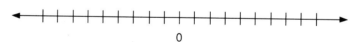

9. _____

10. $6x + 2\left(\frac{1}{3}x - 10\right) > 0$

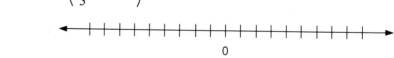

10. _____

11. $\frac{4}{5}(x - 2) \leq \frac{1}{2}(x - 1) + \frac{1}{4}$

11. _____

12. $3(x - 4) + 5 \leq 5x - 11$

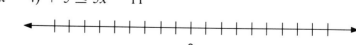

12. _____

13. $\frac{2}{3}\left(x + \frac{3}{4}\right) < \frac{5}{6}$

13. _____

14. $3x + 7 \leq 2x + \dfrac{1}{3}(2x + 1)$

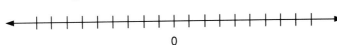

15. $2(x + 3) + 4x > 4\left(x - \dfrac{1}{2}\right)$

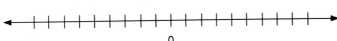

16. $3x + 2(x - 5) \leq 7 - (x + 3)$

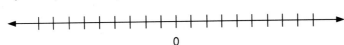

17. $\dfrac{5}{8}x - \dfrac{1}{2} < \dfrac{3}{5}x + 3$

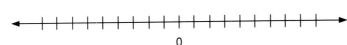

18. $x + \dfrac{3}{8} \geq 3x + \dfrac{3}{4}$

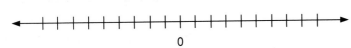

19. $7x + 5 > 2(x - 1) - 21$

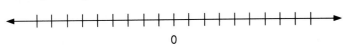

20. $x + \dfrac{6}{7} + \dfrac{2}{3}x < 3x - \dfrac{22}{7}$

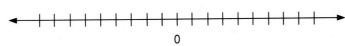

21. $\dfrac{2}{3}(2x - 7) + 2 \geq \dfrac{1}{2}(4x - 13)$

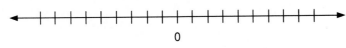

22. $3(x + 1) + 3 > 7(x - 2)$

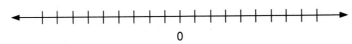

23. $\dfrac{4}{5}x - 10 > \dfrac{2}{3}x + 2 - \dfrac{1}{5}x$

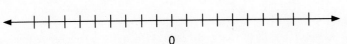

ANSWERS

14. _____

15. _____

16. _____

17. _____

18. _____

19. _____

20. _____

21. _____

22. _____

23. _____

ANSWERS

24. $\frac{5}{6}(x+6) < \frac{1}{2}(10-x)$

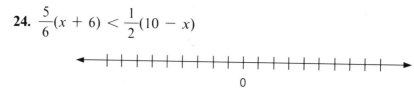

24. _____

25. $\frac{3}{5}(4-x) - 6 > x + 6$

25. _____

CHAPTER 8 SUMMARY

Equations

- The step-by-step method for solving equations, discussed in chapter 5, is used to solve equations involving signed numbers.

Literal Equations

- A **literal** equation is an equation involving more than one letter.
- To solve a literal equation for one letter in terms of the others follow the step-by-step method from chapter 5.

Inequalities

- The statement $a < b$ (read as "a is less than b") means that a is to the left of b on the number line.
- $a < b$ is equivalent to $b > a$ (read as "b is greater than a").
- To solve an inequality follow the same steps as when solving an equation. The *only* difference may arise when you divide by the coefficient of the unknown quantity.
- When an inequality is divided or multiplied by a **negative** number, the results will be unequal in the **opposite** order.

CHAPTER 8 REVIEW

Solve for x.

1. $x + 8 = 3$

2. $x - 1 = -4$

3. $3x = -12$

4. $-5x = 35$

5. $-2x = -16$

6. $4x + 1 = 5x + 7$

7. $x - 3 = 3x + 1$

8. $2x + 3 = 3(2x + 5)$

9. $2(x - 5) = 6 + 4(x + 3)$

10. $\dfrac{x}{2} - \dfrac{2x}{3} = 7$

11. $3x - 5(x - 2) = \dfrac{1}{3}(x + 2)$

12. $\dfrac{2}{3}x - \dfrac{x}{2} + \dfrac{1}{4} = \dfrac{2}{3}(x - 4)$

13. $5 - 3x = a + 7x$

14. $4y - x = 3x - y$

15. $x - y - 3(x - 2y) = 5$

16. $3(x - 2a) + 3a = 5(2x + a)$

17. $2(xy + 3) = 5xy + z$

18. $3a(x - b) = 2ax - bc$

19. $3(x + y) = 5x - 4y + 7$

20. $\dfrac{1}{2}x + 4y = 12$

21. If $P = 2\ell + 2w$, solve for ℓ.

22. If $a = \frac{1}{2}h(a + b)$, solve for b.

23. Graph the statement $x > -2$ on the number line.

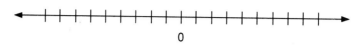

24. Graph the statement $x \leq 5$ on the number line.

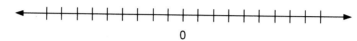

25. Graph the statement $-3 < x \leq 4$ on the number line.

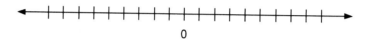

26. Write an algebraic statement represented by the following graph.

Solve for x and graph the results on the number line.

27. $3x - 2 < 2x + 3$

28. $5x - 6 \leq 7x + 4$

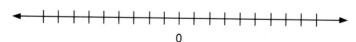

29. $x + 2 \geq 3x - 6$

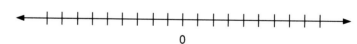

30. $3(x + 4) - 2x \geq 9$

31. $5(x + 1) - 7 \geq 3(x - 2)$

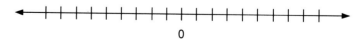

32. $\frac{2}{3}(x - 4) + \frac{1}{2} < \frac{9}{2}$

<------+--+--+--+--+--+--+--+--+--+--+--+--+--+--+--+--+--+------>
                        0

33. $\frac{x}{4} + 2x \geq 7x - 3$

<------+--+--+--+--+--+--+--+--+--+--+--+--+--+--+--+--+--+------>
                        0

34. $\frac{1}{2}(4x - 3) + \frac{2}{3} > x + 1$

<------+--+--+--+--+--+--+--+--+--+--+--+--+--+--+--+--+--+------>
                        0

ANSWERS

32. _____

33. _____

34. _____

SCORE: _____

CHAPTER 8 TEST

NAME: _____

CLASS / SECTION: _____ DATE: _____

ANSWERS

Solve for x.

1. $-7x = 28$
2. $3x + 4 = 10 - x$

1. _____

3. $2(x + 3) = 7(x + 2) + 2$
4. $\dfrac{3}{5}(2x - 1) + 4 = 5 - (x + 7)$

2. _____

5. $\dfrac{5}{8}x - \dfrac{1}{2} = \dfrac{3}{5}x + 3$
6. $3x + 2y = -2x - 13y$

3. _____

4. _____

7. If $A = \dfrac{1}{2}h(a + b)$, solve for a.

5. _____

6. _____

8. Graph $x \geq -3$ on the number line.

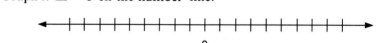

7. _____

Solve and graph the solution on the number line.

8. _____

9. $3(2x + 3) > 4x - 1$

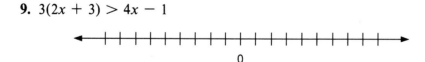

9. _____

10. $\dfrac{1}{2}x + \dfrac{1}{3} \leq 2x - \dfrac{2}{3}$

10. _____

SCORE: _____

386

CHAPTERS 1-8 CUMULATIVE TEST

1. Subtract: 10,813
              −   954

2. Multiply: 1,043
              × 145

3. Reduce: $\dfrac{105}{126}$

4. Divide: $\dfrac{3}{25} \div \dfrac{21}{45}$

5. Add: $5\dfrac{1}{3} + 3\dfrac{5}{6}$

6. Multiply: 9.17
              ×23.4

7. Divide: $16.3\overline{)91.932}$

8. Write .0000043 in scientific notation.

9. A formula from mathematics is $L = a + (n - 1)d$. Find L when $a = 3$, $n = 9$, $d = \dfrac{1}{2}$.

10. Combine like terms: $x^2y + 5xy + 4x^2y − 3x^2y − 3xy$

11. Solve for x: $\dfrac{2}{3}x = 18$

12. Solve for x: $\dfrac{4}{5}x − \dfrac{1}{2} = \dfrac{3}{4}x + 2$

13. An automobile uses 4 gallons of gasoline to travel 92 miles. How many miles could be traveled on 11 gallons?

14. Change $\dfrac{17}{5}$ to a percent.

ANSWERS

1. _____
2. _____
3. _____
4. _____
5. _____
6. _____
7. _____
8. _____
9. _____
10. _____
11. _____
12. _____
13. _____
14. _____

ANSWERS

15. The interest on a $31,000 loan for one year was $2,635. What was the annual rate of interest?

16. Simplify:
$7 - 15 + 4 - (-6) - 3$

15. _____

17. Evaluate: $\dfrac{(-3)(-4)}{-2}$

18. Remove grouping symbols and simplify:
$14x - 2\{5x - 3[12x - 6(3x - 5)]\}$

16. _____

17. _____

19. Solve for x:
$\dfrac{1}{3}(x - 5) + \dfrac{1}{2} = \dfrac{2}{3} - (x + 1)$

18. _____

19. _____

20. Solve and graph the solution on the number line:
$2x - 6 \geq 4(2x + 3)$

20. _____

SCORE: _____

APPENDIX A Useful Facts from Plane Geometry

The study of plane geometry is a model of a logical system. We make no attempt here at such a study, but rather list some definitions and formulas that are derived from such a study. These definitions and formulas relate to the world around us and are often useful in our everyday activities.

A-1 DEFINITIONS

Some terms that we use in geometry cannot be formally defined. Since we cannot define them, we speak of them descriptively to arrive at an idea. For instance, we describe a point as "position without measurement." This is not a true definition since measurement isn't defined without using the word "point." A plane can be described as a flat surface (such as a table top) that extends forever in all directions. Plane geometry is the study of figures on such a plane. Following are definitions of some plane figures.

A-1-1 The Circle

Definition A **circle** is the set of all points on a plane that are an equal distance from a fixed point of the plane called the center of the circle.

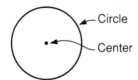

Definition The **diameter** of a circle is a straight line segment connecting two points of the circle and passing through the center.

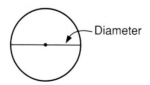

Definition The **radius** of a circle is a straight line segment from the center to a point on the circle.

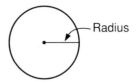

Definition The **circumference** of a circle is the distance around the circle.

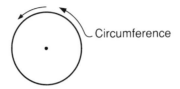

Definition The **area** of a circle is the portion of the plane enclosed by the circle.

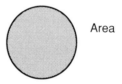

A-1-2 Angles and Their Measure

Definition If a straight line segment is rotated on a plane about a fixed point of the segment, the figure formed is called an **angle**.

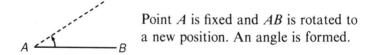

Point A is fixed and AB is rotated to a new position. An angle is formed.

Angles are usually named using a lowercase letter, a number, or three capital letters. If three capital letters are used, the middle letter is the fixed point.

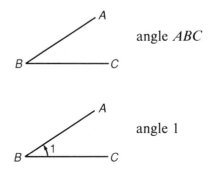

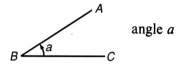

angle a

The symbol used for an angle is ∠. We could have written ∠ABC, 1, and a. Angles are measured by the amount of rotation involved. Degree measurement is most commonly used.

Definition One degree is $\frac{1}{360}$ of a complete revolution; or one complete revolution is 360 degrees.

The symbol for degree is a small circle (°) usually written after and slightly above a number. (Ten degrees is written as 10°.)

Definition A **right angle** is an angle that is $\frac{1}{4}$ of a revolution and measures 90 degrees.

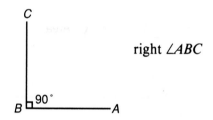

right ∠ABC

Definition An **acute angle** is an angle that measures between 0° and 90°.

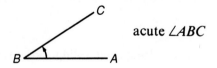

acute ∠ABC

Definition An **obtuse angle** is an angle that measures between 90° and 180°.

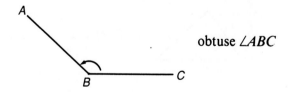

obtuse ∠ABC

Definition A **straight angle** is an angle of 180°.

straight ∠ABC

Definition   If two lines meet at an angle of 90°, they are said to be **perpendicular lines.**

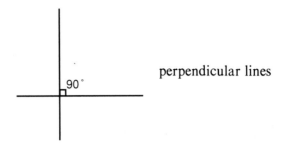

Definition   Two lines in the same plane that do not intersect are called **parallel lines.**

A-1-3 Triangles

Definition   A **triangle** is a closed plane figure having three sides and three angles.

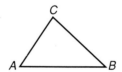

Definition   The point where two sides of a triangle intersect is called a **vertex** of the triangle.

Triangles are divided into three groups according to the size of their angles.

Definition   An **acute triangle** has all three angles measuring less than 90° each.

Definition   A **right triangle** has one angle whose measure is 90°.

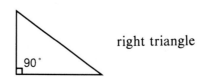

Definition In a right triangle the side opposite the 90° angle is called the **hypotenuse**.

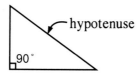

Definition An **obtuse triangle** has one angle that measures greater than 90°.

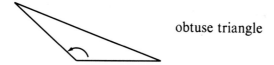

Definition If two sides of a triangle are equal in length, the triangle is **isosceles**.

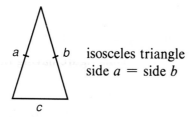

Definition If all three sides of a triangle are equal in length, the triangle is **equilateral**.

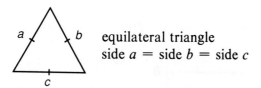

Definition The **altitude** of a triangle is the distance measured along a perpendicular from any vertex to the opposite side.

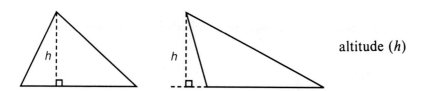

Definition The side to which an altitude is measured is called the **base** of the triangle.

Definition   The **area** of a triangle is the portion of the plane enclosed by the triangle.

Definition   The **perimeter** of a triangle is the distance around the triangle.

A–1–4 Quadrilaterals

Definition   A **quadrilateral** is a closed plane figure with four sides.

Some quadrilaterals, called regular quadrilaterals, have special names based on relationships of their sides.

Definition   A **parallelogram** is a quadrilateral in which opposite sides are parallel.

parallelogram
AB is parallel to *CD*
and *AC* is parallel to *BD*.

Definition   A **rectangle** is a parallelogram in which all four angles are right angles.

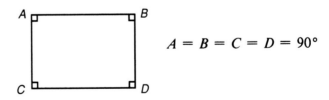

$A = B = C = D = 90°$

Definition   A **square** is a rectangle in which all four sides are equal.

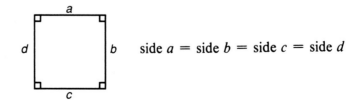

side a = side b = side c = side d

Definition The **altitude** of a parallelogram is the distance, measured along a perpendicular line, between two parallel sides.

Definition The side from which the altitude is measured is called the **base** (*b*) of the parallelogram.

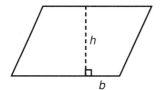 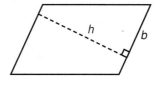

Definition A **trapezoid** is a quadrilateral having only two parallel sides.

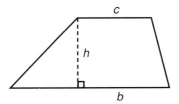

Definition The two parallel sides of a trapezoid are called the **bases** of the trapezoid.

Definition The distance between the parallel sides of a trapezoid is the **altitude** of the trapezoid.

A-2 FACTS AND FORMULAS

Many of the facts and formulas proven in plane geometry are useful in everyday life situations.

A-2-1 Facts from Plane Geometry

Fact The diameter (*d*) of a circle is twice the length of a radius (*r*).

$$d = 2r$$

Definition The ratio of the circumference (*C*) of a circle to the diameter (*d*) of the circle is the number designated by the symbol π (Greek lowercase letter pi).

$$\pi = \frac{C}{d}$$

π is a constant and is *not* a rational number. That is, it cannot be expressed as a ratio of two integers. It is approximately equal to 3.14.

Fact Opposite sides of a parallelogram are equal.

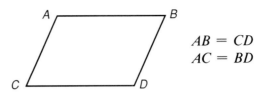

$AB = CD$
$AC = BD$

Fact In any right triangle the area of a square on the hypotenuse is equal to the sum of the areas of the squares on the other two sides.

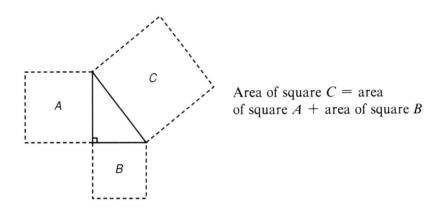

Area of square C = area of square A + area of square B

A-2-2 Formulas from Plane Geometry

1. The circumference (C) of a circle with radius (r) and diameter (d) is given by the formula $C = 2\pi r$ or $C = \pi d$.

2. The area (A) of a circle with radius (r) is given by the formula $A = \pi r^2$.

3. The area (A) of a triangle with base (b) and altitude or height (h) is given by the formula $A = \frac{1}{2}bh$.

4. The area (A) of a parallelogram with base (b) and altitude (h) is given by the formula $A = bh$.

5. Since a rectangle is also a parallelogram, the formula for the area of a rectangle is also $A = bh$. This is sometimes written as $A = \ell w$ where ℓ stands for the length and w for the width.

6. The area (A) of a square with side (s) is $A = s^2$.

7. The area (A) of a trapezoid with bases (b) and (c) and altitude (h) is given by the formula $A = \frac{1}{2}h(b + c)$.

8. The perimeter (P) of a parallelogram with sides ℓ and w is given by $P = 2\ell + 2w$.

9. Since a rectangle is a parallelogram, the formula for the perimeter of a rectangle is also $P = 2\ell + 2w$.

10. The perimeter (P) of a square with side (s) is given by the formula $P = 4s$.

A-3 EXAMPLES AND APPLICATIONS

The facts and formulas from plane geometry are used in many situations of our everyday life. Following are some worked-out examples and then some exercises for you to practice.

A-3-1 Examples

Example 1   Mr. Jones wishes to cover his dining room floor with a type of floor covering that is sold in squares one foot by one foot. He has measured the floor and found it to be 15 feet long and 12 feet wide. How many pieces of floor covering will he need?

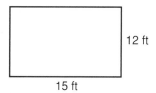

Here we wish to find the area of a rectangle so we use the formula $A = \ell w$.

$$A = \ell w$$
$$= (15)(12)$$
$$= 180 \text{ sq ft}$$

Mr. Jones will need 180 pieces of floor covering.

Example 2   Suppose Mr. Jones (example 1) has chosen a covering that sells for $1.63 per piece. What will be the cost of the material needed?

Since he needs 180 pieces, we find the cost (C) of 180 pieces at $1.63 per piece by multiplying the number of pieces and the cost of each.

$$C = (180)(\$1.63)$$
$$= \$293.40$$

Example 3   Mrs. Jones (still using examples 1 and 2) tells her husband she wants oak parquet covering that sells for $4.22 per piece. What will the material cost if she prevails? How much will Mr. Jones save if he can persuade his wife to accept his choice?

Cost of oak parquet:

$$C = (180)(\$4.22)$$
$$= \$759.60$$

Mrs. Jones' choice:	$759.60
Mr. Jones' choice:	$293.40
Savings:	$466.20

Example 4 Farmer Brown wishes to enclose a rectangular piece of land for a small pasture. He measured the land and found it to be 680 feet long and 425 feet wide. How many feet of fence will he need?

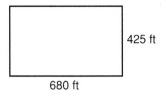

$$P = 2\ell + 2w$$
$$= 2(680) + 2(425)$$
$$= 1{,}360 + 850$$
$$= 2{,}210 \text{ ft}$$

Thus he will need 2,210 feet of fence.

Example 5 Farmer Brown (in example 4) wishes to fertilize his pasture. If each bag of fertilizer covers 1,700 square feet, how many bags will he need?

$$A = \ell w$$
$$= (680)(425)$$
$$= 289{,}000 \text{ sq ft}$$

Thus the area of the pasture is 289,000 sq ft. To find the number of bags of fertilizer needed we must divide the area by 1,700 (amount covered by each bag).

$$289{,}000 \div 1{,}700 = 170 \text{ bags}$$

Example 6 A rose garden is in the form of a circle with a radius of 7 feet with a walkway around it that is 2 feet wide. What is the area of the walkway?

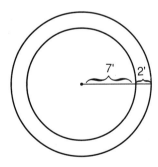

To find the area of the walkway we must subtract the area of the inner circle from the area of the outer circle.

Area of inner circle
$A = \pi r^2$
$= (3.14)(7)^2$
$= (3.14)(49)$
$= 153.86$ sq ft

Area of outer circle
(Here note that $r = 7 + 2 = 9$.)
$A = \pi r^2$
$= (3.14)(9)^2$
$= 254.34$ sq ft

Walkway area $= 254.34 - 153.86$
$= 100.48$ sq ft

A-3-2 Applications

1. A rectangular garden is 30 feet long and 20 feet wide. What is the area of the garden?

2. If the garden in question 1 is to be enclosed by a fence, how many feet of fencing are needed?

3. A rectangular room measures 18 feet by 14 feet. How many feet of baseboard molding are needed to go around the room?

4. What is the area of the floor of the room in question 3?

5. A fenced enclosure in the form of a triangle has sides of 120 feet, 135 feet, and 98 feet. What is the perimeter of the triangle?

6. If the cost of fencing is $4.50 per foot, what is the total cost of fencing the enclosure in question 5?

7. A circular fish pond has a diameter of 12 feet. What is the circumference of the pool? Use $\pi = 3.14$.

8. What is the area of the pond in question 7? Use $\pi = 3.14$.

9. A sail on a boat is in the form of a right triangle with a base of 10 feet and a height of 17 feet. What is the area of the sail?

10. A triangular platform has one side that measures 20 feet. The altitude to the opposite vertex is 15 feet. Find the area of the platform.

ANSWERS

1. _____
2. _____
3. _____
4. _____
5. _____
6. _____
7. _____
8. _____
9. _____
10. _____

ANSWERS

11. _____

12. _____

13. _____

14. _____

15. _____

16. _____

17. _____

18. _____

19. _____

20. _____

11. A square-shaped garden is to be fenced. The length of one side is 16 feet. How many feet of fencing are needed?

12. The perimeter of a triangular-shaped planter is 40 feet. If two sides measure 10 feet and 16 feet, respectively, what is the length of the third side?

13. A circular landing pad for a helicopter has a radius of 65 feet. What is the area of the pad? Use $\pi = 3.14$.

14. How much fencing is needed to enclose the pad in question 13? Use $\pi = 3.14$.

15. A side of a square-shaped flower garden measures 24 feet. The garden is surrounded by a 3-foot wide wooden walkway. Find the area of the walkway.

16. Inside the garden in question 15 is a circular fountain having a diameter of 10 feet. Find the area of the garden excluding the fountain. Use $\pi = 3.14$.

17. A rectangular-shaped swimming pool measuring 15 feet by 30 feet has a one-foot wide tile border around it. Find the area of the border.

18. What is the cost of the border in question 17 if the price of a one-foot square tile is $1.15?

19. A rectangular floor measures 18 feet by 15 feet. How much will it cost to carpet the floor if the price is $22.50 per square yard?

20. A lawn measures 200 feet by 90 feet. A bag of "weed and feed" costs $5.00 and covers 1,500 square feet of lawn. How much will it cost to "weed and feed" the entire lawn?

APPENDIX B Using a Calculator

There are many different makes of hand-held calculators on the market and some of them work differently than others. Most of the inexpensive calculators that perform the basic arithmetic operations, however, work essentially the same. Our discussion here will refer to these types of calculators. If your calculator is of a different type, you should refer to the instruction manual that accompanies it.

B-1 ARITHMETIC OPERATIONS

With most inexpensive calculators the numbers and operations are entered in the same order as you would write them on paper.

Example 1 Add: 12 + 38

To perform this addition on the calculator first enter the number 12 by pressing the $\boxed{1}$ button and the $\boxed{2}$ button in that order. Next press the $\boxed{+}$ button. Then enter the number 38 by pressing the $\boxed{3}$ button and the $\boxed{8}$ button in that order. Finally, press the $\boxed{=}$ button. To summarize these steps we have

$$12 \; \boxed{+} \; 38 \; \boxed{=} \; .$$

The number 50 (which represents the sum of 12 and 38) will appear in the display.

Example 2 Add: 53.6 + 9.51

To enter 53.6 first press the $\boxed{5}$ button, followed by the $\boxed{3}$ button, followed by the $\boxed{\cdot}$ button, followed by the $\boxed{6}$ button. Then press the $\boxed{+}$ button. Next enter 9.51 by pressing the $\boxed{9}$ button, followed by the $\boxed{\cdot}$ button, followed by the $\boxed{5}$ button, followed by the $\boxed{1}$ button. Finally, press the $\boxed{=}$ button. We thus have

$$53.6 \; \boxed{+} \; 9.51 \; \boxed{=} \; .$$

The sum 63.11 will appear in the display.

Multiple-step additions are performed in the same way.

Example 3 Add: 3.28 + 12.05 + .64 + 207.9

Again, we make the entries in the same order as they are written. Enter 3.28, followed by the $\boxed{+}$ button, followed by 12.05, followed by the $\boxed{+}$ button, followed by .64, followed by the $\boxed{+}$ button, followed by 207.9, followed by the $\boxed{=}$ button.

$$3.28 \boxed{+} 12.05 \boxed{+} .64 \boxed{+} 207.9 \boxed{=}$$

The sum 223.87 will appear in the display.

Example 4 Subtract: 2,351 − 874

Enter 2,351, followed by the $\boxed{-}$ button, followed by 874, followed by the $\boxed{=}$ button.

$$2,351 \boxed{-} 874 \boxed{=}$$

The difference 1,477 will appear in the display.

Example 5 Combine: 83.2 − 16.9 − 21.63 + 135.74 − 68.05

Performing these operations in order we have
83.2 $\boxed{-}$ 16.9 $\boxed{-}$ 21.63 $\boxed{+}$ 135.74 $\boxed{-}$ 68.05 $\boxed{=}$.
The result 112.36 will appear in the display.

Example 6 Multiply: 314 × 17.5

We have 314 $\boxed{\times}$ 17.5 $\boxed{=}$. The product 5,495 appears in the display.

Example 7 Divide: 2,401.64 ÷ 36.21. Round the answer to two decimal places.

We have 2,401.64 $\boxed{\div}$ 36.21 $\boxed{=}$. The quotient in the display rounded to two decimal places is 66.33.

Another function that your calculator most likely performs is that of finding the square root of a number. To do this we use the $\boxed{\sqrt{}}$ button.

For instance, to find the square root of 9 we enter $\boxed{9}$ and then press the $\boxed{\sqrt{}}$ button obtaining the result 3.

Example 8 Find $\sqrt{49}$

First enter 49. Then press the $\boxed{\sqrt{}}$ button. The result 7 will appear in the display.

We can also find the approximate square root of a number that is not a perfect square.

Example 9 Find $\sqrt{108}$ rounded to four decimal places.

We enter 108 followed by pressing the $\boxed{\sqrt{}}$ button. The result in the display rounded to four decimal places is 10.3923.

EXERCISE B-1-1

Perform the following operations using a calculator.

1. 28 + 49
2. 54 + 37
3. 81 + 79

4. 118 + 32
5. 138 − 59
6. 201 − 53

7. 136 + 407
8. 538 + 906
9. 2,105 − 921

10. 3,068 − 749
11. 29.7 + 31.8
12. 38.4 + 25.9

13. 42.6 − 18.4
14. 122.4 − 67.9
15. 159.63 + 216.08

16. 29.013 + 148.79
17. 91.3 − 46.72
18. 106.2 − 29.34

19. 31.16 + .25 + 9.65 + 123.8
20. 59.8 + 5.64 + 108.76 + .35

21. 68.3 + 71.8 − 86.4
22. 47.3 − 21.8 + 19.6

23. 71.3 − 16.51 + 13.64 − 8.99

24. 83.11 + 32.16 − 95.08 + 103.5

25. 64.3 − 15.5 − 21.32 + 63.55 − 81.09

26. 134.6 − 39.81 − 27.14 + 18.7 − 30.9

27. 48 × 56
28. 39 × 41

29. 214 × 6.3
30. 591 × 5.7

31. 26.12 × 18.34 (Round answer to two decimal places.)
32. 48.06 × 29.58 (Round answer to two decimal places.)

33. 1,216 ÷ 38
34. 656 ÷ 41

35. 181.9 ÷ 8.5
36. 98.58 ÷ 5.3

ANSWERS

1. _____
2. _____
3. _____
4. _____
5. _____
6. _____
7. _____
8. _____
9. _____
10. _____
11. _____
12. _____
13. _____
14. _____
15. _____
16. _____
17. _____
18. _____
19. _____
20. _____
21. _____
22. _____
23. _____
24. _____
25. _____
26. _____
27. _____
28. _____
29. _____
30. _____
31. _____
32. _____
33. _____
34. _____
35. _____
36. _____

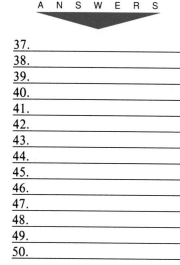

37. 521.03 ÷ 17.24 (Round answer to two decimal places.)

38. 1,214.17 ÷ 206.27 (Round answer to two decimal places.)

39. $\sqrt{169}$

40. $\sqrt{576}$

Round the answer to each of the following to four decimal places.

41. $\sqrt{2}$ **42.** $\sqrt{5}$ **43.** $\sqrt{13}$ **44.** $\sqrt{19}$ **45.** $\sqrt{28}$

46. $\sqrt{99}$ **47.** $\sqrt{109}$ **48.** $\sqrt{161}$ **49.** $\sqrt{17.1}$ **50.** $\sqrt{38.5}$

APPENDIX C Answers

Chapter 1 Pretest

The number in brackets after each answer refers to the section of the chapter that discusses that type of problem.
1. ten thousands [1–1]   2. thirty-eight thousand, twenty-six [1–1]   3. $7 \times 10^3 + 2 \times 10^2 + 5 \times 10^1 + 3 \times 10^0$ [1–2]
4. 620 [1–3]   5. 15,000 [1–3]   6. 1,002 [1–4]   7. 1,116 [1–4]   8. 6,907 [1–4]   9. 198 [1–5]   10. 108 [1–5]
11. 5,009 [1–5]   12. 21,812 [1–6]   13. 722,545 [1–6]   14. 58 [1–7]   15. 45 R18 [1–7]   16. 2,039 [1–7]
17. 10 [1–8]   18. $2 \times 3^2 \times 5 \times 11$ [1–9]   19. 21 [1–10]   20. 42 [1–10]

1-1-1

1. Ten thousands   2. Thousands   3. Hundreds   4. Hundred thousands   5. Millions   6. Ones   7. Ten millions   8. Tens
9. Three hundred twenty-one   10. Four hundred thirty-eight   11. Two thousand, one hundred twenty-nine
12. Five thousand, two hundred eighty   13. Ten thousand, one hundred twenty-four
14. Eleven thousand, four hundred thirty-two   15. One hundred thousand, two hundred forty-one
16. One hundred three thousand, three hundred seventy-five   17. Three hundred twenty thousand, four
18. Six hundred nineteen thousand, two hundred   19. Four hundred thousand, one
20. Five hundred sixty-two thousand, forty   21. Two million, five hundred three thousand, one hundred
22. Six million, twenty thousand, three hundred twenty-seven
23. Fifteen million, two hundred thousand, six hundred twenty-one
24. Three hundred fifty-two million, three hundred fifty-two thousand, three hundred fifty-two
25. Seven hundred sixty-three million, seven hundred sixty-three thousand, seven hundred sixty-three
26. Three hundred forty-nine million, two hundred sixteen thousand, one hundred twenty-three
27. Twenty-two billion, five hundred nineteen million, fifty-four thousand, one hundred eleven
28. Eighty-three billion, five hundred thousand, nine   29. 326   30. 408   31. 905   32. 1,040   33. 1,801   34. 13,419
35. 26,042   36. 125,004   37. 206,201   38. 500,003   39. 610,050   40. 3,040,619   41. 5,003,207   42. 31,411,916
43. 68,045,308   44. 122,610,005   45. 511,412,100   46. 58,020,014,312
47. Twenty-six thousand, seven hundred ninety-eight dollars   48. Twenty-nine thousand, twenty-eight feet
49. Thirty-one thousand, forty-five miles   50. Five thousand, eighty-four yards   51. Four thousand, five hundred fifty-two
52. Two hundred twenty-seven million, nine hundred sixty thousand, eight hundred four dollars   53. 2,596   54. 2,015
55. $52,904   56. 83,100   57. 250,000   58. 2,974,878

1-2-1

1. 9   2. 25   3. 16   4. 81   5. 1   6. 32   7. 8   8. 10   9. 100   10. 1,000   11. 10,000   12. 100,000

1-2-2

1. $2 \times 10^2 + 5 \times 10^1 + 1 \times 10^0$ 2. $3 \times 10^2 + 4 \times 10^1 + 1 \times 10^0$ 3. $6 \times 10^2 + 1 \times 10^1 + 9 \times 10^0$
4. $5 \times 10^2 + 1 \times 10^1 + 4 \times 10^0$ 5. $1 \times 10^2 + 8 \times 10^1 + 9 \times 10^0$ 6. $1 \times 10^2 + 3 \times 10^1 + 5 \times 10^0$
7. $2 \times 10^3 + 4 \times 10^2 + 5 \times 10^1 + 8 \times 10^0$ 8. $4 \times 10^3 + 3 \times 10^2 + 6 \times 10^1 + 2 \times 10^0$
9. $7 \times 10^3 + 6 \times 10^2 + 2 \times 10^1 + 5 \times 10^0$ 10. $2 \times 10^3 + 9 \times 10^2 + 5 \times 10^1 + 7 \times 10^0$
11. $3 \times 10^1 + 0 \times 10^0$ 12. $8 \times 10^1 + 0 \times 10^0$ 13. $2 \times 10^2 + 1 \times 10^1 + 0 \times 10^0$
14. $3 \times 10^2 + 5 \times 10^1 + 0 \times 10^0$ 15. $7 \times 10^2 + 0 \times 10^1 + 4 \times 10^0$ 16. $6 \times 10^2 + 0 \times 10^1 + 1 \times 10^0$
17. $5 \times 10^2 + 0 \times 10^1 + 0 \times 10^0$ 18. $9 \times 10^2 + 0 \times 10^1 + 0 \times 10^0$ 19. $3 \times 10^3 + 0 \times 10^2 + 4 \times 10^1 + 1 \times 10^0$
20. $4 \times 10^3 + 0 \times 10^2 + 2 \times 10^1 + 9 \times 10^0$ 21. $5 \times 10^3 + 3 \times 10^2 + 9 \times 10^1 + 8 \times 10^0$
22. $3 \times 10^3 + 7 \times 10^2 + 8 \times 10^1 + 2 \times 10^0$ 23. $6 \times 10^3 + 0 \times 10^2 + 0 \times 10^1 + 0 \times 10^0$
24. $9 \times 10^3 + 0 \times 10^2 + 0 \times 10^1 + 0 \times 10^0$ 25. $2 \times 10^4 + 3 \times 10^3 + 4 \times 10^2 + 2 \times 10^1 + 6 \times 10^0$
26. $3 \times 10^4 + 9 \times 10^3 + 5 \times 10^2 + 8 \times 10^1 + 2 \times 10^0$ 27. $4 \times 10^3 + 5 \times 10^2 + 0 \times 10^1 + 2 \times 10^0$

28. $7 \times 10^3 + 8 \times 10^2 + 0 \times 10^1 + 3 \times 10^0$ 29. $8 \times 10^4 + 1 \times 10^3 + 2 \times 10^2 + 9 \times 10^1 + 4 \times 10^0$
30. $6 \times 10^4 + 3 \times 10^3 + 5 \times 10^2 + 1 \times 10^1 + 4 \times 10^0$ 31. $1 \times 10^4 + 5 \times 10^3 + 8 \times 10^2 + 0 \times 10^1 + 4 \times 10^0$
32. $1 \times 10^4 + 9 \times 10^3 + 6 \times 10^2 + 0 \times 10^1 + 3 \times 10^0$ 33. $2 \times 10^4 + 3 \times 10^3 + 0 \times 10^2 + 3 \times 10^1 + 8 \times 10^0$
34. $4 \times 10^4 + 4 \times 10^3 + 0 \times 10^2 + 1 \times 10^1 + 6 \times 10^0$
35. $1 \times 10^5 + 7 \times 10^4 + 2 \times 10^3 + 2 \times 10^2 + 5 \times 10^1 + 4 \times 10^0$
36. $1 \times 10^5 + 3 \times 10^4 + 2 \times 10^3 + 6 \times 10^2 + 9 \times 10^1 + 3 \times 10^0$
37. $3 \times 10^5 + 5 \times 10^4 + 0 \times 10^3 + 4 \times 10^2 + 0 \times 10^1 + 0 \times 10^0$
38. $5 \times 10^5 + 2 \times 10^4 + 0 \times 10^3 + 5 \times 10^2 + 0 \times 10^1 + 0 \times 10^0$
39. $5 \times 10^6 + 4 \times 10^5 + 6 \times 10^4 + 1 \times 10^3 + 0 \times 10^2 + 0 \times 10^1 + 6 \times 10^0$
40. $8 \times 10^6 + 3 \times 10^5 + 1 \times 10^4 + 9 \times 10^3 + 0 \times 10^2 + 0 \times 10^1 + 1 \times 10^0$

1-3-1

1. 60   2. 90   3. 30   4. 40   5. 520   6. 390   7. 290   8. 790   9. 4,210   10. 1,390   11. 400   12. 600   13. 200
14. 200   15. 2,400   16. 5,200   17. 1,100   18. 3,000   19. 7,000   20. 5,000   21. 58,000   22. 23,000   23. 26,000
24. 89,000   25. 35,000   26. 36,000   27. 40,000   28. 71,000   29. 19,000   30. 60,000   31. 140,000   32. 320,000
33. 460,000   34. 630,000   35. 260,000   36. 580,000   37. 340,000   38. 140,000   39. 600,000   40. 290,000
41. 158,460   42. 158,000   43. 158,500   44. 160,000   45. 200,000   46. 296,000   47. 300,000   48. 296,000
49. 296,000   50. 300,000   51. 5,555,560   52. 5,600,000   53. 5,555,600   54. 5,556,000   55. 5,560,000   56. 6,000,000
57. 10,000   58. 10,000   59. 10,000   60. 10,000

1-4-1

1. 20   2. 20   3. 20   4. 20   5. 20   6. 25   7. 23   8. 29   9. 27   10. 28   11. 30   12. 24   13. 21   14. 28
15. 28   16. 29   17. 31   18. 25   19. 27   20. 29   21. 86   22. 99   23. 69   24. 59   25. 99   26. 60   27. 70
28. 80   29. 90   30. 80   31. 115   32. 117   33. 119   34. 129   35. 129   36. 104   37. 112   38. 100   39. 100
40. 107   41. 121   42. 124   43. 115   44. 123   45. 135   46. 109   47. 128   48. 182   49. 179   50. 157   51. 779
52. 939   53. 979   54. 999   55. 999   56. 761   57. 981   58. 863   59. 887   60. 985   61. 931   62. 912   63. 913
64. 906   65. 905   66. 1,064   67. 1,052   68. 1,444   69. 1,432   70. 1,167   71. 1,066   72. 1,219   73. 1,297
74. 1,288   75. 998   76. 1,359   77. 1,360   78. 1,300   79. 922   80. 1,317   81. 1,618   82. 1,539   83. 1,389
84. 1,370   85. 2,288   86. 1,551   87. 2,031   88. 1,408   89. 1,271   90. 1,795   91. 11,278   92. 8,455   93. 9,773
94. 14,470   95. 17,974   96. 10,576   97. 7,714   98. 127,720   99. 19,106   100. 55,700   101. 4,344   102. 9,681
103. 10,283   104. 7,574   105. 5,322   106. 6,530   107. 102,371   108. 10,828   109. 101,373   110. 1,012,486
111. $40   112. $531   113. 83   114. 3,563   115. $88   116. $3,015   117. 1,029   118. $231   119. 537
120. 4,942,291

1-5-1

1. 3   2. 5   3. 3   4. 2   5. 3   6. 1   7. 0   8. 0   9. 11   10. 13   11. 11   12. 12   13. 10   14. 20   15. 22
16. 32   17. 22   18. 24   19. 32   20. 12   21. 21   22. 13   23. 51   24. 32   25. 11   26. 11   27. 20   28. 20
29. 321   30. 212   31. 111   32. 414   33. 321   34. 171   35. 232   36. 314   37. 520   38. 220   39. 402   40. 402
41. 7   42. 12   43. 32   44. 121   45. 101   46. 401   47. 243   48. 221   49. 300   50. 300

1-5-2

1. 15   2. 7   3. 5   4. 27   5. 46   6. 56   7. 37   8. 46   9. 18   10. 17   11. 27   12. 17   13. 18   14. 18   15. 19
16. 19   17. 17   18. 23   19. 188   20. 286   21. 468   22. 287   23. 329   24. 567   25. 188   26. 179   27. 379
28. 278   29. 176   30. 287   31. 142   32. 166   33. 164   34. 144   35. 78   36. 68   37. 88   38. 79   39. 118
40. 129   41. 28   42. 35   43. 28   44. 17   45. 166   46. 353   47. 159   48. 168   49. 38   50. 16

1-5-3

1. 22   2. 33   3. 44   4. 16   5. 156   6. 465   7. 557   8. 656   9. 249   10. 277   11. 339   12. 129   13. 487
14. 649   15. 189   16. 188   17. 378   18. 269   19. 77   20. 16   21. 1,619   22. 1,778   23. 4,688   24. 2,787

25. 279 **26.** 3,649 **27.** 1,458 **28.** 566 **29.** 4,056 **30.** 3,019 **31.** 308 **32.** 209 **33.** 375 **34.** 462 **35.** 411
36. 208 **37.** 219 **38.** 539 **39.** 141 **40.** 263 **41.** 314 **42.** 213 **43.** 3,009 **44.** 5,089 **45.** 109 **46.** 308
47. 2,128 **48.** 1,128 **49.** 3,268 **50.** 4,201 **51.** 310 **52.** 201 **53.** 198 **54.** 198 **55.** 109 **56.** 109 **57.** 39,153
58. 30,016 **59.** 758 **60.** 279 **61.** 11,232 **62.** 11,145 **63.** 2,999 **64.** 1,999 **65.** 99 **66.** 97 **67.** 232 **68.** 508
69. 509 **70.** 308 **71.** 1,071 **72.** 1,093 **73.** 4,009 **74.** 2,008 **75.** 809 **76.** 837 **77.** 108 **78.** 109 **79.** 3,768
80. 2,658 **81.** 11,809 **82.** 51,086 **83.** 19 **84.** 79 **85.** 366 **86.** 259 **87.** 1,208 **88.** 1,406 **89.** 634
90. 919 **91.** 248 **92.** 658 **93.** 231 **94.** 211 **95.** 98 **96.** 1,298 **97.** 436 **98.** 595 **99.** 806 **100.** 97 **101.** 23
102. 62 **103.** 287 **104.** 78,936 **105.** 26 **106.** $88 **107.** 1,078 ft **108.** $1,099 **109.** 397 ft **110.** 14,534 ft
111. $8 **112.** 355 miles

1-6-1

1. 175 **2.** 78 **3.** 364 **4.** 312 **5.** 292 **6.** 180 **7.** 120 **8.** 162 **9.** 153 **10.** 200 **11.** 210 **12.** 236 **13.** 714
14. 512 **15.** 903 **16.** 1,395 **17.** 532 **18.** 666 **19.** 1,080 **20.** 910 **21.** 1,728 **22.** 1,512 **23.** 3,071 **24.** 4,897
25. 3,744 **26.** 12,504 **27.** 10,990 **28.** 5,450 **29.** 13,464 **30.** 9,879 **31.** 43,326 **32.** 47,922 **33.** 32,982
34. 52,212 **35.** 31,990 **36.** 61,200 **37.** 390 **38.** 410 **39.** 380 **40.** 230 **41.** 1,540 **42.** 5,230 **43.** 4,300
44. 6,100 **45.** 28,768 **46.** 51,182 **47.** 29,140 **48.** 50,676 **49.** 143,992 **50.** 212,670 **51.** 101,632 **52.** 817,355
53. 33,800 **54.** 129,970 **55.** 172,530 **56.** 167,478 **57.** 41,600 **58.** 32,800 **59.** 54,200 **60.** 26,400 **61.** 72,400
62. 81,300 **63.** 64,000 **64.** 32,000 **65.** 381,300 **66.** 334,884 **67.** 866,084 **68.** 993,896 **69.** 641,982 **70.** 1,599,576
71. 805,200 **72.** 401,440 **73.** 1,406,000 **74.** 1,226,433 **75.** 2,490,896 **76.** 2,509,884 **77.** 1,434,823 **78.** 1,055,868
79. 1,634,684 **80.** 3,324,195 **81.** 1,275,425 **82.** 393,568 **83.** 2,037,527 **84.** 2,116,224 **85.** 3,033,577 **86.** 8,154,664
87. 4,996,824 **88.** 6,528,976 **89.** 2,512,000 **90.** 3,416,000 **91.** 5,428,000 **92.** 7,038,000 **93.** 32,050 **94.** 52,140
95. 320,500 **96.** 521,400 **97.** 3,205,000 **98.** 5,214,000 **99.** 5,900,000 **100.** 6,300,000 **101.** 1,750 **102.** 11,590
103. 11,080 **104.** 73,570 **105.** 1,700 **106.** 19,500 **107.** 17,900 **108.** 28,400 **109.** 3,000 **110.** 21,000 **111.** 34,000
112. 27,000 **113.** 30,000 **114.** 40,000 **115.** 620,000 **116.** 1,420,000 **117.** 806,940 **118.** 176,033 **119.** 1,970
120. 99,435 **121.** $90 **122.** $7,150 **123.** 630 miles **124.** 480 miles **125.** $772 **126.** $9,024 **127.** 476 sq ft
128. 336 sq ft **129.** $11,808; $2,813 **130.** $512

1-6-2

1. 120 **2.** 810 **3.** 2,110 **4.** 6,010 **5.** 3,200 **6.** 31,600 **7.** 20,500 **8.** 121,600 **9.** 132,000 **10.** 204,000
11. 160,000 **12.** 3,450,000 **13.** 50,300,000 **14.** 6,200,000 **15.** 2,091,000 **16.** 5,117,000 **17.** 2,000 **18.** 80,000
19. 6,100,000 **20.** 1,000,000

1-7-1

1. 4 **2.** 2 **3.** 2 **4.** 6 **5.** 5 **6.** 3 **7.** 5 **8.** 8 **9.** 7 **10.** 7 **11.** 9 **12.** 9 **13.** 7 **14.** 8 **15.** 7 **16.** 8 **17.** 7
18. 7 **19.** 9 **20.** 9 **21.** 9 R3 **22.** 8 R2 **23.** 9 R3 **24.** 9 R5 **25.** 8 R1 **26.** 9 R1 **27.** 3 R2 **28.** 7 R3
29. 9 R1 **30.** 6 R1 **31.** 9 R2 **32.** 9 R6 **33.** 8 R1 **34.** 9 R5 **35.** 8 R5 **36.** 8 R5 **37.** 9 R2 **38.** 8 R1
39. 9 R5 **40.** 9 R2

1-7-2

1. 12 **2.** 14 **3.** 12 **4.** 11 **5.** 23 **6.** 23 **7.** 31 **8.** 21 **9.** 15 **10.** 24 **11.** 12 **12.** 12 **13.** 13 **14.** 14 **15.** 25
16. 29 **17.** 15 R4 **18.** 15 R1 **19.** 46 R1 **20.** 27 R1 **21.** 35 R5 **22.** 64 R1 **23.** 157 **24.** 133 R3 **25.** 254 R1
26. 233 R2 **27.** 160 R3 **28.** 226 R2 **29.** 64 R2 **30.** 140 **31.** 240 **32.** 152 **33.** 420 R2 **34.** 630 R1 **35.** 250 R4
36. 680 R2 **37.** 302 R4 **38.** 701 R1 **39.** 208 R6 **40.** 903 R6 **41.** 3,107 **42.** 4,205 **43.** 6,004 **44.** 3,007
45. 5,051 R2 **46.** 5,039 R3 **47.** 4,306 R5 **48.** 7,102 R1 **49.** $38 **50.** 6 cents **51.** 58 **52.** 32 **53.** $56 **54.** 23
55. 6 R2 **56.** 51 R2 **57.** 26 mpg **58.** $1,512 **59.** 4 R4 **60.** 24 R2

1-7-3

1. 36 **2.** 54 **3.** 23 **4.** 31 **5.** 63 **6.** 47 **7.** 32 **8.** 26 **9.** 42 R14 **10.** 23 R17 **11.** 28 R15 **12.** 22 R31
13. 304 R22 **14.** 601 R38 **15.** 803 R20 **16.** 405 R15 **17.** 340 **18.** 690 **19.** 580 **20.** 730 **21.** 34 **22.** 69

23. 144 **24.** 235 **25.** 563 **26.** 722 **27.** 341 **28.** 284 **29.** 115 R6 **30.** 221 R11 **31.** 213 R21 **32.** 158 R27 **33.** 403 R10 **34.** 205 R20 **35.** 500 R16 **36.** 200 R70 **37.** 2,203 **38.** 2,025 **39.** 3,014 **40.** 2,103 **41.** 53 **42.** 64 **43.** 68 **44.** 75 **45.** 83 R91 **46.** 76 R101 **47.** 47 R412 **48.** 52 R479 **49.** 285 **50.** 582 **51.** 350 **52.** 710 **53.** 3,004 **54.** 2,007 **55.** 4,008 **56.** 6,001 **57.** 2,005 R20 **58.** 2,004 R30 **59.** 3,007 R26 **60.** 2,009 R10 **61.** $3,012 **62.** $205 **63.** 56 cases, 6 cans **64.** 241 bottles, 8 ounces **65.** $173 **66.** 52 days **67.** 78 gallons **68.** $105 **69.** 21 gross **70.** 22 pkg **71.** 24 mpg **72.** 34 weeks

1-8-1

1. 1 **2.** 3 **3.** 7 **4.** 16 **5.** 10 **6.** 12 **7.** 2 **8.** 2 **9.** 21 **10.** 6 **11.** 3 **12.** 3 **13.** 38 **14.** 18 **15.** 24 **16.** 30 **17.** 15 **18.** 1 **19.** 42 **20.** 55 **21.** 24 **22.** 35 **23.** 98 **24.** 21 **25.** 1 **26.** 0 **27.** 3 **28.** 12 **29.** 44 **30.** 12

1-8-2

1. 15 **2.** 10 **3.** 0 **4.** 2 **5.** 2 **6.** 6 **7.** 16 **8.** 7 **9.** 40 **10.** 0 **11.** 18 **12.** 29 **13.** 24 **14.** 7 **15.** 20 **16.** 17 **17.** 14 **18.** 8 **19.** 26 **20.** 0 **21.** 17 **22.** 1 **23.** 23 **24.** 14 **25.** 47 **26.** 20 **27.** 2 **28.** 16 **29.** 22 **30.** 0

1-8-3

1. 6 **2.** 8 **3.** 8 **4.** 1 **5.** 4 **6.** 0 **7.** 64 **8.** 30 **9.** 9 **10.** 25 **11.** 9 **12.** 19 **13.** 11 **14.** 70 **15.** 31 **16.** 2 **17.** 5 **18.** 5 **19.** 40 **20.** 1 **21.** 28 **22.** 3 **23.** 28 **24.** 22 **25.** 8 **26.** 7 **27.** 46 **28.** 0 **29.** 5 **30.** 3

1-9-1

1. 1, 2, 4 **2.** 1, 2, 3, 6 **3.** 1, 3, 9 **4.** 1, 2, 4, 8 **5.** 1, 5 **6.** 1, 7 **7.** 1, 2, 5, 10 **8.** 1, 2, 7, 14 **9.** 1, 2, 11, 22 **10.** 1, 2, 3, 6, 9, 18 **11.** 1, 2, 3, 4, 6, 8, 12, 24 **12.** 1, 3, 11, 33 **13.** 1, 17 **14.** 1, 19 **15.** 1, 2, 3, 6, 7, 14, 21, 42 **16.** 1, 2, 23, 46 **17.** 1, 3, 9, 27 **18.** 1, 3, 17, 51 **19.** 1, 2, 29, 58 **20.** 1, 2, 3, 4, 5, 6, 10, 12, 15, 20, 30, 60 **21.** 1, 3, 23, 69 **22.** 1, 3, 9, 27, 81 **23.** 1, 2, 4, 8, 11, 22, 44, 88 **24.** 1, 2, 4, 7, 8, 14, 28, 56 **25.** 1, 2, 4, 19, 38, 76 **26.** 1, 2, 3, 6, 13, 26, 39, 78 **27.** 1, 3, 9, 11, 33, 99 **28.** 1, 2, 3, 4, 6, 9, 12, 18, 27, 36, 54, 108 **29.** 1, 2, 4, 31, 62, 124 **30.** 1, 3, 37, 111 **31.** 1, 2, 4, 5, 10, 11, 20, 22, 44, 55, 110, 220 **32.** 1, 2, 3, 6, 9, 13, 18, 26, 39, 78, 117, 234

1-9-2

2, 3, 5, 7, 11, 13, 17, 19, 23, 29, 31, 37, 41, 43, 47, 53, 59, 61, 67, 71, 73, 79, 83, 89, 97, 101, 103, 107, 109, 113, 127, 131, 137, 139, 149, 151, 157, 163, 167, 173, 179, 181, 191, 193, 197, 199

1-9-3

1. $2^2 \times 3$ **2.** 2×3^2 **3.** $2 \times 3 \times 5$ **4.** $2 \times 5 \times 7$ **5.** $2^2 \times 7$ **6.** $2 \times 3 \times 7$ **7.** $3^2 \times 5$ **8.** 3×5^2 **9.** $2^2 \times 3 \times 5$ **10.** $2^3 \times 7$ **11.** $3^2 \times 7$ **12.** $2^2 \times 7$ **13.** 3×17 **14.** 3×19 **15.** 2×47 **16.** 5×13 **17.** $2 \times 3 \times 37$ **18.** $3 \times 7 \times 11$ **19.** $2^3 \times 3^2$ **20.** $2^2 \times 3^3$ **21.** $2^2 \times 3^2 \times 5$ **22.** $2 \times 3^2 \times 5^2$ **23.** $2 \times 3^3 \times 7$ **24.** $2 \times 3 \times 5 \times 7 \times 11$ **25.** $2^2 \times 3 \times 31$ **26.** $2^3 \times 41$ **27.** $2^4 \times 3 \times 7$ **28.** $2^4 \times 3 \times 5$ **29.** $2^2 \times 3^2 \times 5^2 \times 7$ **30.** $2^2 \times 3^3 \times 7 \times 11$

1-9-4

1. 3×7 **2.** 3×5 **3.** $2^2 \times 3$ **4.** 2×3^2 **5.** 2^4 **6.** 3^4 **7.** $2^2 \times 3 \times 5$ **8.** $2 \times 3^2 \times 5$ **9.** $2^3 \times 7$ **10.** $2^3 \times 5$ **11.** $3^2 \times 5$ **12.** 3×5^2 **13.** $2 \times 5 \times 7$ **14.** $2 \times 5 \times 11$ **15.** 7×11 **16.** 7×13 **17.** $2^2 \times 7^2$ **18.** $2 \times 3^2 \times 7$ **19.** $2 \times 3 \times 5 \times 13$ **20.** $2 \times 3 \times 5 \times 11$ **21.** $3 \times 5 \times 7 \times 11$ **22.** $5 \times 7^2 \times 11$ **23.** 11×17 **24.** 11×19 **25.** $2 \times 11 \times 17$ **26.** $3 \times 11 \times 19$ **27.** $5 \times 17 \times 19$ **28.** $7 \times 11 \times 17$ **29.** $2^3 \times 11 \times 31$ **30.** $2 \times 3^2 \times 5 \times 37$ **31.** $2 \times 3^2 \times 7 \times 41$ **32.** $3 \times 5^2 \times 7 \times 19$ **33.** $2^2 \times 3^2 \times 7 \times 11$ **34.** $2^2 \times 3^2 \times 5 \times 11$ **35.** 2^7 **36.** $2^4 \times 3^3$ **37.** $3^2 \times 5 \times 11 \times 13$ **38.** $2 \times 7 \times 11 \times 43$ **39.** $2 \times 3^2 \times 17 \times 43$ **40.** $3^2 \times 5 \times 23 \times 31$

1-10-1

1. 6 **2.** 8 **3.** 7 **4.** 9 **5.** 18 **6.** 8 **7.** 3 **8.** 4 **9.** 8 **10.** 9

1-10-2

1. 4 **2.** 3 **3.** 2 **4.** 6 **5.** 1 **6.** 8 **7.** 42 **8.** 18 **9.** 18 **10.** 15 **11.** 1 **12.** 22 **13.** 21 **14.** 1 **15.** 18
16. 84 **17.** 9 **18.** 15 **19.** 4 **20.** 1 **21.** 12 **22.** 1 **23.** 42 **24.** 9 **25.** 1 **26.** 12 **27.** 36 **28.** 30 **29.** 63
30. 90

1-10-3

1. 6 **2.** 15 **3.** 12 **4.** 18 **5.** 12 **6.** 30 **7.** 90 **8.** 72 **9.** 72 **10.** 48 **11.** 36 **12.** 70 **13.** 72 **14.** 200
15. 120 **16.** 210 **17.** 12 **18.** 24 **19.** 24 **20.** 180 **21.** 120 **22.** 360 **23.** 252 **24.** 120 **25.** 270 **26.** 360
27. 594 **28.** 252 **29.** 7,140 **30.** 320 **31.** 144 **32.** 306 **33.** 90 **34.** 120 **35.** 72 **36.** 48 **37.** 360 **38.** 180
39. 108 **40.** 360

Chapter 1 Review

The number in brackets after each answer refers to the section of the chapter that discusses that type of problem.
1. Tens [1-1] **2.** Thousands [1-1] **3.** Ten thousands [1-1] **4.** Hundred millions [1-1] **5.** Ten thousand, three hundred twenty-five [1-1] **6.** Twenty-one thousand, three hundred sixty-four [1-1]
7. Seven hundred twenty-four million, eight hundred three thousand, two hundred forty-one [1-1]
8. One hundred five million, sixty-three thousand, four [1-1]
9. 523,201 [1-1] **10.** 4,086,021 [1-1] **11.** $8 \times 10^2 + 2 \times 10^1 + 7 \times 10^0$ [1-2] **12.** $4 \times 10^2 + 6 \times 10^1 + 8 \times 10^0$ [1-2]
13. $4 \times 10^4 + 0 \times 10^3 + 3 \times 10^2 + 9 \times 10^1 + 6 \times 10^0$ [1-2]
14. $2 \times 10^4 + 1 \times 10^3 + 5 \times 10^2 + 0 \times 10^1 + 8 \times 10^0$ [1-2]
15. 350 [1-3] **16.** 520 [1-3] **17.** 1,740 [1-3] **18.** 3,690 [1-3] **19.** 21,300 [1-3] **20.** 55,700 [1-3] **21.** 46,000 [1-3]
22. 39,000 [1-3] **23.** 460,000 [1-3] **24.** 260,000 [1-3] **25.** 28 [1-4] **26.** 24 [1-4] **27.** 97 [1-4] **28.** 79 [1-4]
29. 123 [1-4] **30.** 100 [1-4] **31.** 110 [1-4] **32.** 129 [1-4] **33.** 605 [1-4] **34.** 921 [1-4] **35.** 1,000 [1-4]
36. 1,000 [1-4] **37.** 1,282 [1-4] **38.** 1,316 [1-4] **39.** 8,417 [1-4] **40.** 9,029 [1-4] **41.** 10,719 [1-4]
42. 11,831 [1-4] **43.** 233 [1-5] **44.** 133 [1-5] **45.** 259 [1-5] **46.** 179 [1-5] **47.** 197 [1-5] **48.** 299 [1-5]
49. 3,178 [1-5] **50.** 3,188 [1-5] **51.** 22,179 [1-5] **52.** 8,085 [1-5] **53.** Commutative property of multiplication [1-6]
54. Distributive law of multiplication over addition [1-6] **55.** 498 [1-6] **56.** 600 [1-6] **57.** 782 [1-6] **58.** 918 [1-6]
59. 11,050 [1-6] **60.** 14,013 [1-6] **61.** 2,280 [1-6] **62.** 3,490 [1-6] **63.** 65,178 [1-6] **64.** 79,570 [1-6]
65. 54,800 [1-6] **66.** 63,400 [1-6] **67.** 360,460 [1-6] **68.** 1,760,850 [1-6] **69.** 620,464 [1-6] **70.** 1,129,920 [1-6]
71. 27 [1-7] **72.** 46 [1-7] **73.** 391 [1-7] **74.** 473 [1-7] **75.** 42 R3 [1-7] **76.** 35 R4 [1-7] **77.** 56 [1-7]
78. 63 [1-7] **79.** 25 R6 [1-7] **80.** 32 R17 [1-7] **81.** 3,005 [1-7] **82.** 4,006 [1-7] **83.** 4 [1-8] **84.** 25 [1-8]
85. 15 [1-8] **86.** 88 [1-8] **87.** 26 [1-8] **88.** 0 [1-8] **89.** 20 [1-8] **90.** 10 [1-8] **91.** 1, 2, 19, 38 [1-9]
92. 1, 3, 13, 39 [1-9] **93.** 1, 2, 3, 4, 6, 7, 12, 14, 21, 28, 42, 84 [1-9] **94.** 1, 2, 3, 4, 6, 8, 12, 16, 24, 32, 48, 96 [1-9]
95. $2^2 \times 3 \times 11$ [1-9] **96.** $2^2 \times 3 \times 7$ [1-9] **97.** $2^2 \times 3^2 \times 5 \times 7$ [1-9] **98.** $2 \times 3^3 \times 7 \times 11$ [1-9] **99.** 2 [1-10]
100. 4 [1-10] **101.** 4 [1-10] **102.** 42 [1-10] **103.** 6 [1-10] **104.** 12 [1-10] **105.** 20 [1-10] **106.** 24 [1-10]
107. 24 [1-10] **108.** 60 [1-10] **109.** 180 [1-10] **110.** 216 [1-10] **111.** 72 [1-10] **112.** 48 [1-10]
113. Three thousand, nine hundred fifteen [1-1] **114.** Ten thousand, five hundred sixty-one [1-1] **115.** 1,815 [1-1]
116. 521,000 [1-1] **117.** $373 [1-4] **118.** $128 [1-5] **119.** $14,940 [1-6] **120.** $2,016 [1-6] **121.** 13 [1-7]
122. $256 [1-7]

Chapter 1 Test

The number in brackets after each answer refers to the section of the chapter that discusses that type of problem.
1. Thousands [1-1] **2.** Fifty-two thousand, six hundred nineteen [1-1] **3.** $3 \times 10^3 + 4 \times 10^2 + 2 \times 10^1 + 8 \times 10^0$ [1-2]
4. 380 [1-3] **5.** 12,000 [1-3] **6.** 1,003 [1-4] **7.** 1,347 [1-4] **8.** 5,364 [1-4] **9.** 295 [1-5] **10.** 409 [1-5]
11. 4,007 [1-5] **12.** 10,192 [1-6] **13.** 501,410 [1-6] **14.** 64 [1-7] **15.** 47 R16 [1-7] **16.** 3,017 [1-7] **17.** 174 [1-8]
18. $2^3 \times 3 \times 17$ [1-9] **19.** 14 [1-10] **20.** 60 [1-10]

Chapter 2 Pretest

The number in brackets following each answer refers to the section of the chapter that discusses that type of problem.

1. $\frac{2}{7}$ [2–1] 2. $\frac{119}{12}$ [2–5] 3. 12 [2–4] 4. 45 [2–4] 5. $\frac{2}{15}$ [2–2] 6. $\frac{18}{35}$ [2–3] 7. $\frac{9}{11}$ [2–4] 8. $\frac{1}{4}$ [2–4] 9. 9 [2–5]
10. $\frac{53}{56}$ [2–4] 11. $\frac{3}{22}$ [2–3] 12. $\frac{7}{20}$ [2–4] 13. $1\frac{8}{9}$ [2–4] 14. $\frac{3}{22}$ [2–2] 15. $3\frac{5}{12}$ [2–5] 16. $\frac{10}{21}$ [2–3] 17. $\frac{4}{9}$ [2–2]
18. $12\frac{5}{36}$ [2–5] 19. $\frac{9}{16}$ [2–5] 20. $5\frac{13}{24}$ [2–5]

2-1-1

1. $\frac{2}{3}$ 2. $\frac{2}{3}$ 3. $\frac{3}{4}$ 4. $\frac{3}{4}$ 5. $\frac{2}{5}$ 6. $\frac{2}{5}$ 7. $\frac{3}{5}$ 8. $\frac{4}{5}$ 9. $\frac{5}{6}$ 10. $\frac{5}{6}$ 11. $\frac{2}{3}$ 12. $\frac{3}{7}$ 13. $\frac{3}{7}$ 14. $\frac{4}{5}$
15. $\frac{2}{3}$ 16. $\frac{2}{3}$ 17. $\frac{1}{2}$ 18. $\frac{1}{3}$ 19. $\frac{1}{4}$ 20. $\frac{1}{2}$ 21. $\frac{1}{3}$ 22. $\frac{1}{3}$ 23. $\frac{1}{9}$ 24. $\frac{1}{2}$ 25. $\frac{1}{4}$ 26. $\frac{1}{4}$ 27. $\frac{1}{2}$
28. $\frac{1}{3}$ 29. $\frac{3}{4}$ 30. $\frac{1}{3}$ 31. $\frac{2}{3}$ 32. $\frac{6}{7}$ 33. $\frac{6}{7}$ 34. $\frac{2}{9}$ 35. $\frac{3}{8}$ 36. $\frac{15}{32}$ 37. $\frac{10}{27}$ 38. $\frac{5}{6}$ 39. $\frac{4}{11}$ 40. $\frac{2}{13}$
41. $\frac{28}{39}$ 42. $\frac{2}{5}$ 43. $\frac{4}{13}$ 44. $\frac{3}{8}$ 45. $\frac{7}{12}$ 46. $\frac{8}{11}$ 47. $\frac{4}{9}$ 48. $\frac{3}{7}$ 49. $\frac{2}{3}$ 50. $\frac{36}{91}$ 51. $\frac{3}{14}$ 52. $\frac{4}{5}$ 53. $\frac{63}{68}$
54. $\frac{3}{8}$ 55. $\frac{3}{4}$ 56. $\frac{4}{9}$ 57. $\frac{6}{11}$ 58. $\frac{8}{13}$ 59. $\frac{7}{12}$ 60. $\frac{11}{18}$

2-2-1

1. $\frac{3}{10}$ 2. $\frac{2}{15}$ 3. $\frac{1}{8}$ 4. $\frac{1}{18}$ 5. $\frac{10}{21}$ 6. $\frac{6}{55}$ 7. $\frac{9}{20}$ 8. $\frac{4}{15}$ 9. $\frac{2}{9}$ 10. $\frac{8}{25}$ 11. $\frac{1}{4}$ 12. $\frac{1}{3}$ 13. $\frac{2}{7}$ 14. $\frac{1}{8}$
15. $\frac{5}{11}$ 16. $\frac{3}{7}$ 17. $\frac{15}{64}$ 18. $\frac{6}{25}$ 19. $\frac{3}{5}$ 20. $\frac{3}{8}$ 21. $\frac{2}{5}$ 22. $\frac{4}{7}$ 23. $\frac{10}{3}$ 24. $\frac{3}{13}$ 25. $\frac{4}{11}$ 26. $\frac{6}{13}$ 27. $\frac{1}{5}$
28. $\frac{5}{16}$ 29. $\frac{3}{40}$ 30. $\frac{2}{105}$ 31. $\frac{3}{10}$ 32. $\frac{1}{75}$ 33. $\frac{1}{2}$ 34. $\frac{1}{4}$ 35. $\frac{4}{25}$ 36. $\frac{3}{5}$ 37. $\frac{1}{18}$ 38. $\frac{1}{8}$ 39. $\frac{6}{11}$ 40. $\frac{3}{4}$
41. $\frac{1}{8}$ 42. $\frac{1}{15}$ 43. $\frac{27}{160}$ 44. $\frac{1}{24}$ 45. $\frac{4}{15}$ 46. $\frac{32}{57}$ 47. $\frac{3}{10}$ 48. $\frac{5}{6}$ 49. 1 50. 1 51. 4 52. 2 53. 3 54. 6
55. 9 56. 12 57. 5 58. $\frac{9}{4}$ 59. 1 60. 1 61. 60 62. 14 63. 9 64. 12 65. $\frac{3}{7}$ 66. $\frac{3}{7}$ 67. 1 68. $\frac{22}{61}$
69. 13 70. $\frac{1}{3}$ 71. 18 ounces 72. 32 pounds 73. $\frac{1}{2}$ liter 74. $\frac{1}{8}$ of the business
75. $\frac{1}{3}$ cup sugar, $\frac{1}{8}$ teaspoon vanilla, $\frac{1}{4}$ cup flour 76. $\frac{1}{4}$ square miles 77. $75 78. 150 miles 79. $90 80. $1,200

2-3-1

1. $\frac{8}{9}$ 2. $\frac{14}{15}$ 3. $\frac{3}{4}$ 4. $\frac{4}{9}$ 5. 20 6. $\frac{9}{2}$ 7. $\frac{1}{20}$ 8. $\frac{2}{9}$ 9. $\frac{2}{3}$ 10. $\frac{5}{3}$ 11. 1 12. $\frac{4}{5}$ 13. $\frac{2}{3}$ 14. $\frac{2}{5}$ 15. $\frac{3}{2}$
16. $\frac{5}{2}$ 17. $\frac{3}{2}$ 18. $\frac{4}{3}$ 19. $\frac{2}{3}$ 20. $\frac{3}{4}$ 21. $\frac{25}{18}$ 22. $\frac{5}{9}$ 23. 1 24. $\frac{8}{3}$ 25. $\frac{35}{36}$ 26. $\frac{15}{64}$ 27. $\frac{3}{2}$ 28. $\frac{2}{3}$
29. $\frac{5}{7}$ 30. $\frac{4}{3}$ 31. 10 32. 9 33. $\frac{6}{7}$ 34. $\frac{15}{8}$ 35. $\frac{48}{49}$ 36. $\frac{20}{81}$ 37. $\frac{25}{2}$ 38. $\frac{27}{2}$ 39. $\frac{2}{25}$ 40. $\frac{2}{27}$ 41. 2
42. $\frac{1}{3}$ 43. $\frac{1}{112}$ 44. $\frac{1}{80}$ 45. 18 46. 25 47. $\frac{9}{20}$ 48. $\frac{7}{2}$ 49. $\frac{1}{4}$ 50. $\frac{4}{5}$ 51. $\frac{1}{10}$ 52. $\frac{3}{13}$ 53. $\frac{2}{3}$ 54. $\frac{28}{9}$

55. $\frac{3}{10}$ 56. $\frac{3}{2}$ 57. $\frac{5}{24}$ 58. $\frac{5}{4}$ 59. $\frac{8}{15}$ 60. $\frac{1}{4}$ 61. 9 62. 1 63. 6 64. $\frac{9}{11}$ 65. $\frac{10}{3}$ 66. $\frac{16}{27}$ 67. $\frac{7}{15}$ 68. $\frac{10}{3}$
69. $\frac{7}{9}$ 70. $\frac{24}{25}$ 71. 100 students 72. 48 hamburgers 73. 15 bows 74. 16 bottles 75. 6 furlongs 76. 64 tablets
77. 32 pails 78. $\frac{1}{8}$ acre 79. 16 loads 80. 20 minutes

2-4-1

1. $\frac{3}{5}$ 2. $\frac{2}{5}$ 3. $\frac{3}{4}$ 4. 1 5. $\frac{2}{3}$ 6. $\frac{1}{2}$ 7. 2 8. 1 9. $\frac{11}{3}$ 10. $\frac{5}{6}$ 11. $\frac{3}{2}$ 12. $\frac{1}{2}$ 13. $\frac{8}{9}$ 14. 3 15. $\frac{13}{20}$
16. 1 17. $\frac{8}{3}$ 18. 2 19. $\frac{16}{19}$ 20. $\frac{3}{8}$

2-4-2

1. 6 2. 15 3. 6 4. 6 5. 30 6. 36 7. 18 8. 24 9. 48 10. 180 11. 84 12. 42 13. 72 14. 60 15. 18
16. 24 17. 72 18. 30 19. 144 20. 180

2-4-3

1. 3 2. 4 3. 4 4. 2 5. 3 6. 5 7. 16 8. 20 9. 12 10. 28 11. 16 12. 9 13. 18 14. 56 15. 12
16. 18 17. 65 18. 52 19. 48 20. 133

2-4-4

1. $\frac{8}{15}$ 2. $\frac{9}{14}$ 3. $\frac{17}{12}$ 4. $\frac{19}{15}$ 5. $\frac{7}{24}$ 6. $\frac{6}{35}$ 7. $\frac{9}{10}$ 8. $\frac{3}{2}$ 9. $\frac{1}{2}$ 10. $\frac{3}{8}$ 11. $\frac{29}{60}$ 12. $\frac{13}{30}$ 13. $\frac{1}{36}$ 14. $\frac{23}{36}$
15. $\frac{29}{40}$ 16. $\frac{29}{30}$ 17. $\frac{5}{12}$ 18. $\frac{3}{14}$ 19. $\frac{9}{10}$ 20. $\frac{19}{18}$ 21. $\frac{1}{36}$ 22. $\frac{1}{18}$ 23. $\frac{1}{6}$ 24. $\frac{7}{60}$ 25. $\frac{7}{8}$ 26. 1 27. $\frac{29}{20}$
28. $\frac{47}{24}$ 29. $\frac{71}{72}$ 30. $\frac{19}{15}$ 31. $\frac{1}{3}$ 32. $\frac{1}{5}$ 33. $\frac{79}{96}$ 34. $\frac{73}{72}$ 35. $\frac{31}{90}$ 36. $\frac{10}{21}$ 37. $\frac{101}{210}$ 38. $\frac{259}{288}$ 39. $\frac{5}{24}$ 40. $\frac{41}{36}$
41. $\frac{1}{10}$ 42. $\frac{9}{50}$ 43. $\frac{11}{12}$ 44. 1 45. $\frac{1}{10}$ 46. $\frac{1}{4}$ 47. $\frac{1}{8}$ 48. $\frac{34}{15}$ 49. $\frac{17}{24}$ cup 50. $\frac{7}{8}$ ounce 51. $\frac{1}{16}$ pound
52. $\frac{5}{8}$ ounce 53. $\frac{3}{10}$ mile 54. $\frac{7}{12}$ acre 55. $\frac{37}{24}$ acre 56. $\frac{5}{12}$ day 57. $\frac{15}{8}$ tons 58. $\frac{31}{60}$ of her salary 59. $\frac{1}{8}$ cup
60. $\frac{1}{3}$ gallon

2-5-1

1. $\frac{9}{2}$ 2. $\frac{17}{8}$ 3. $\frac{11}{3}$ 4. $\frac{27}{4}$ 5. $\frac{53}{8}$ 6. $\frac{29}{3}$ 7. $\frac{47}{6}$ 8. $\frac{76}{9}$ 9. $\frac{49}{4}$ 10. $\frac{125}{9}$ 11. $2\frac{1}{2}$ 12. $2\frac{2}{3}$ 13. $4\frac{1}{2}$ 14. $2\frac{2}{5}$
15. $2\frac{4}{7}$ 16. $4\frac{1}{6}$ 17. $6\frac{4}{5}$ 18. $7\frac{1}{3}$ 19. $13\frac{1}{4}$ 20. $9\frac{5}{7}$

2-5-2

1. $10\frac{5}{6}$ 2. $18\frac{2}{3}$ 3. $7\frac{7}{8}$ 4. $1\frac{7}{20}$ 5. $1\frac{19}{22}$ 6. $2\frac{2}{29}$ 7. 45 8. $2\frac{4}{5}$ 9. $2\frac{2}{3}$ 10. 15 11. 30 12. $1\frac{1}{2}$ 13. $13\frac{5}{7}$

14. $7\frac{11}{15}$ 15. $5\frac{45}{56}$ 16. 4 17. $3\frac{5}{9}$ 18. 9 19. $16\frac{4}{5}$ 20. $\frac{7}{8}$ 21. $1\frac{2}{5}$ 22. $1\frac{103}{112}$ 23. $5\frac{1}{15}$ 24. $5\frac{7}{9}$

25. $32\frac{1}{2}$ hours 26. 213 miles 27. $34\frac{2}{15}$ mpg 28. 209 sq ft 29. $1\frac{1}{3}$ pounds 30. $4\frac{3}{4}$ pounds

2-5-3

1. $7\frac{2}{3}$ 2. $12\frac{3}{4}$ 3. $5\frac{3}{4}$ 4. $7\frac{13}{15}$ 5. $6\frac{11}{15}$ 6. $13\frac{5}{12}$ 7. $12\frac{1}{8}$ 8. $15\frac{3}{40}$ 9. $17\frac{7}{12}$ 10. $26\frac{3}{10}$ 11. $8\frac{23}{24}$ 12. $6\frac{13}{24}$
13. $10\frac{2}{63}$ 14. $13\frac{5}{18}$ 15. $17\frac{29}{60}$ 16. $23\frac{23}{24}$ 17. $15\frac{5}{12}$ cups 18. $24\frac{1}{4}$ square yards 19. $20\frac{7}{12}$ hours 20. $39\frac{1}{2}$ hours

2-5-4

1. $2\frac{1}{2}$ 2. $1\frac{1}{4}$ 3. $2\frac{2}{3}$ 4. $2\frac{2}{5}$ 5. $4\frac{1}{2}$ 6. $4\frac{3}{4}$ 7. $3\frac{1}{8}$ 8. $2\frac{1}{4}$ 9. $1\frac{1}{12}$ 10. $\frac{33}{40}$ 11. $10\frac{22}{35}$ 12. $7\frac{61}{72}$ 13. $1\frac{1}{3}$
14. $2\frac{1}{4}$ 15. $8\frac{7}{12}$ 16. $6\frac{19}{24}$ 17. $22\frac{7}{12}$ ounces 18. $1\frac{7}{12}$ hours 19. $7\frac{3}{4}$ yards 20. $57\frac{5}{6}$ acres 21. $14\frac{5}{12}$ hours
22. $64\frac{5}{6}$ yards

Chapter 2 Review

The number in brackets following each answer refers to the section of the chapter that discusses that type of problem.

1. $\frac{3}{7}$ [2–1] 2. $\frac{2}{5}$ [2–1] 3. $\frac{3}{8}$ [2–1] 4. $\frac{3}{7}$ [2–1] 5. $\frac{6}{7}$ [2–1] 6. $\frac{3}{35}$ [2–1] 7. $\frac{2}{7}$ [2–1] 8. $\frac{21}{22}$ [2–1] 9. 15 [2–4]
10. 48 [2–4] 11. 66 [2–4] 12. 49 [2–4] 13. $\frac{69}{7}$ [2–5] 14. $\frac{91}{8}$ [2–5] 15. $7\frac{5}{8}$ [2–5] 16. $11\frac{1}{4}$ [2–5] 17. $\frac{5}{18}$ [2–2]
18. $\frac{3}{4}$ [2–2] 19. $\frac{2}{3}$ [2–3] 20. $\frac{3}{4}$ [2–3] 21. $\frac{11}{12}$ [2–4] 22. $\frac{17}{18}$ [2–4] 23. $\frac{1}{4}$ [2–4] 24. $\frac{2}{5}$ [2–4] 25. $\frac{12}{13}$ [2–4]
26. $\frac{10}{11}$ [2–4] 27. $\frac{2}{7}$ [2–2] 28. $\frac{3}{5}$ [2–2] 29. $\frac{1}{24}$ [2–3] 30. $\frac{2}{35}$ [2–3] 31. $\frac{13}{20}$ [2–2] 32. $\frac{15}{23}$ [2–2] 33. $1\frac{3}{5}$ [2–4]
34. $1\frac{1}{2}$ [2–4] 35. $\frac{1}{21}$ [2–4] 36. $\frac{7}{30}$ [2–4] 37. $1\frac{1}{2}$ [2–4] 38. $1\frac{5}{14}$ [2–4] 39. 20 [2–2] 40. 12 [2–2] 41. $4\frac{2}{3}$ [2–3]
42. $\frac{7}{10}$ [2–3] 43. $2\frac{11}{48}$ [2–5] 44. $1\frac{7}{10}$ [2–5] 45. $\frac{29}{36}$ [2–4] 46. $1\frac{7}{36}$ [2–4] 47. $6\frac{1}{2}$ [2–2] 48. $4\frac{1}{2}$ [2–2] 49. $1\frac{1}{2}$ [2–5]
50. $1\frac{1}{2}$ [2–5] 51. $6\frac{1}{2}$ [2–5] 52. $9\frac{1}{12}$ [2–5] 53. $\frac{13}{36}$ [2–4] 54. $\frac{1}{6}$ [2–4] 55. 10 [2–5] 56. $15\frac{5}{8}$ [2–5] 57. 9 [2–3]
58. 16 [2–3] 59. $1\frac{1}{80}$ [2–4] 60. $1\frac{29}{36}$ [2–4] 61. $6\frac{13}{24}$ [2–5] 62. $2\frac{23}{24}$ [2–5] 63. $4\frac{4}{5}$ [2–5] 64. $16\frac{1}{5}$ [2–5]
65. $\frac{25}{27}$ [2–3] 66. $1\frac{13}{32}$ [2–3] 67. $1\frac{11}{20}$ [2–4] 68. $1\frac{13}{48}$ [2–4] 69. $\frac{19}{24}$ [2–5] 70. $1\frac{5}{36}$ [2–5] 71. $5\frac{5}{12}$ [2–5]
72. $4\frac{1}{2}$ [2–5] 73. $\frac{2}{27}$ [2–2] 74. $\frac{1}{10}$ [2–2] 75. $6\frac{3}{8}$ [2–4] 76. $8\frac{2}{5}$ [2–4] 77. $9\frac{29}{72}$ [2–5] 78. $7\frac{25}{72}$ [2–5]
79. 2 [2–5] 80. $1\frac{1}{3}$ [2–5] 81. $6\frac{41}{42}$ [2–5] 82. $3\frac{1}{144}$ [2–5] 83. $12\frac{1}{2}$ [2–5] 84. $45\frac{1}{2}$ [2–5] 85. $3\frac{3}{5}$ [2–5]

86. $\frac{5}{9}$ [2–5] **87.** $150 [2–2] **88.** 16 gallons [2–2] **89.** 128 seconds [2–3] **90.** 160 cartons [2–3] **91.** $43\frac{1}{3}$ hours [2–5]
92. $20\frac{5}{12}$ hours [2–5] **93.** $15\frac{5}{12}$ yards [2–5] **94.** $3\frac{3}{4}$ liters [2–5] **95.** $\frac{3}{5}$ second [2–5] **96.** $2\frac{7}{8}$ [2–5]

Chapter 2 Test

The number in brackets following each answer refers to the section of the chapter that discusses that type of problem.

1. $\frac{3}{5}$ [2–1] **2.** $\frac{58}{5}$ [2–5] **3.** 15 [2–4] **4.** 24 [2–4] **5.** $\frac{2}{21}$ [2–2] **6.** $\frac{33}{50}$ [2–3] **7.** $\frac{11}{13}$ [2–4] **8.** $\frac{1}{3}$ [2–4] **9.** 12 [2–5]
10. $\frac{38}{45}$ [2–4] **11.** $\frac{3}{32}$ [2–3] **12.** $\frac{11}{42}$ [2–4] **13.** $1\frac{35}{36}$ [2–4] **14.** $\frac{2}{15}$ [2–2] **15.** $2\frac{13}{24}$ [2–5] **16.** $\frac{4}{15}$ [2–3] **17.** $\frac{3}{10}$ [2–2]
18. $13\frac{5}{8}$ [2–5] **19.** $\frac{5}{6}$ [2–5] **20.** $16\frac{11}{18}$ [2–5]

Chapters 1–2 Cumulative Test

The number in brackets after each answer refers to the chapter and section that discusses that type of problem.
1. Five thousand, three hundred nine [1–1] **2.** $2 \times 10^3 + 5 \times 10^2 + 4 \times 10^1 + 3 \times 10^0$ [1–2] **3.** 16,800 [1–3]
4. 1,201 [1–4] **5.** 676 [1–5] **6.** 9,288 [1–6] **7.** 129 R28 [1–7] **8.** 4 [1–8] **9.** $2^2 \times 3^2 \times 7$ [1–9] **10.** 90 [1–10]
11. $\frac{5}{13}$ [2–1] **12.** 48 [2–4] **13.** $\frac{1}{12}$ [2–2] **14.** $\frac{8}{9}$ [2–3] **15.** $\frac{1}{6}$ [2–4] **16.** $\frac{7}{10}$ [2–4] **17.** $8\frac{2}{5}$ [2–5] **18.** $\frac{11}{24}$ [2–4]
19. $2\frac{7}{8}$ [2–5] **20.** $3\frac{3}{4}$ [2–5]

Chapter 3 Pretest

The number in brackets after each answer refers to the section of the chapter that discusses that type of problem.
1. $3 \times 10 + 8 \times 1 + 7 \times \frac{1}{10} + 5 \times \frac{1}{100} + 2 \times \frac{1}{1,000}$ [3–1] **2.** Four hundred six and ninety-eight hundredths [3–1]
3. 5.69 [3–2] **4.** $\frac{3}{40}$ [3–7] **5.** 6.375 [3–7] **6.** 23.75 [3–3] **7.** 14.787 [3–3] **8.** 3.6 [3–5] **9.** 11.34 [3–6]
10. 1.58 [3–4] **11.** 29.31 [3–3] **12.** 14.285 [3–3] **13.** 12.227 [3–5] **14.** 11.46 [3–6] **15.** 145.306 [3–4]
16. 502.07 [3–3] **17.** 23.549 [3–3] **18.** 5.06 [3–5] **19.** 2.28068 [3–4] **20.** .007 [3–6] **21.** .0000514 [3–8]
22. 4.5×10^{10} [3–8] **23.** $3\sqrt{2}$ [3–9] **24.** $4\sqrt{11}$ [3–9]

3-1-1

1. $3 \times \frac{1}{10}$ **2.** $2 \times \frac{1}{10} + 4 \times \frac{1}{100}$ **3.** $0 \times \frac{1}{10} + 4 \times \frac{1}{100}$ **4.** $1 \times \frac{1}{10} + 3 \times \frac{1}{100} + 5 \times \frac{1}{1,000}$
5. $3 \times \frac{1}{10} + 0 \times \frac{1}{100} + 9 \times \frac{1}{1,000}$ **6.** $0 \times \frac{1}{10} + 0 \times \frac{1}{100} + 4 \times \frac{1}{1,000}$
7. $1 \times \frac{1}{10} + 0 \times \frac{1}{100} + 5 \times \frac{1}{1,000} + 3 \times \frac{1}{10,000}$ **8.** $6 \times 1 + 3 \times \frac{1}{10} + 9 \times \frac{1}{100}$
9. $3 \times 10 + 8 \times 1 + 1 \times \frac{1}{10} + 2 \times \frac{1}{100} + 4 \times \frac{1}{1,000}$
10. $1 \times 100 + 0 \times 10 + 4 \times 1 + 1 \times \frac{1}{10} + 0 \times \frac{1}{100} + 0 \times \frac{1}{1,000} + 5 \times \frac{1}{10,000}$
11. Six tenths **12.** Thirty-two hundredths **13.** One hundred twenty-nine thousandths **14.** Four hundredths **15.** Eight hundred two thousandths **16.** Three and five tenths **17.** Twenty seven and three hundredths **18.** Forty-one and two hundred seventy-six thousandths **19.** Sixteen and four ten thousandths **20.** One hundred and one hundred five ten thousandths **21.** .4

22. .105 **23.** .361 **24.** 21.07 **25.** 125.204 **26.** 503.75 **27.** 100.007 **28.** 300.0005 **29.** One hundred five and seventy-three hundredths dollars **30.** One thousand four and six hundredths dollars

3-2-1

1. .5 **2.** .48 **3.** 0 **4.** .481 **5.** 1 **6.** .837 **7.** .8 **8.** .84 **9.** .516 **10.** .5 **11.** .52 **12.** 1 **13.** 9 **14.** 9.054 **15.** 9.1 **16.** 9.05 **17.** 24.6 **18.** 24.55 **19.** 25 **20.** 24.553 **21.** 62 **22.** 61.961 **23.** 62.0 **24.** 61.96

3-3-1

1. 6.895 **2.** 11.05 **3.** 19.355 **4.** 13.00 **5.** 7.888 **6.** 7.3203 **7.** 21.183 **8.** 48.318 **9.** 70.9114 **10.** 42.7278 **11.** 28.03 **12.** 48.244 **13.** 44.5615 **14.** 54.026 **15.** 103.3195 **16.** 64.8224 **17.** 12.13 **18.** 2.87 **19.** 14.83 **20.** 6.06 **21.** 5.079 **22.** 3.832 **23.** 6.838 **24.** 2.7429 **25.** 3.738 **26.** 6.335 **27.** 25.48 **28.** 5.464 **29.** 5.76 **30.** 21.5539 **31.** 2.296 **32.** 1.368 **33.** $72.00 **34.** $42.70 **35.** $15.05 **36.** $67.62 **37.** $13.22 **38.** $61.31 **39.** $103.20 **40.** $3.45 **41.** $262.66 **42.** $43.36

3-4-1

1. 1.28 **2.** 1.70 **3.** .32 **4.** .0072 **5.** .924 **6.** 11.284 **7.** 1.692 **8.** .8834 **9.** 13.0192 **10.** 1.36396 **11.** 47.43 **12.** .00364 **13.** .04464 **14.** .00036 **15.** 31.46 **16.** 713.9 **17.** 13.217 **18.** 1.462 **19.** .6813 **20.** .04562 **21.** 8.917 **22.** 8.554 **23.** 7.0392 **24.** 4.5115 **25.** 19.2168 **26.** 13.5786 **27.** .0732 **28.** .3168 **29.** 222.794 **30.** 380.775 **31.** 24.7303 **32.** 84.4803 **33.** $3.30 **34.** $3.15 **35.** $17.90 **36.** $15.48 **37.** $10.41 **38.** $9.81 **39.** 158.88 sq ft **40.** 7,620.5 sq ft **41.** $10.41 **42.** $7.80

3-4-2

1. 35 **2.** 164 **3.** 161.3 **4.** 125.31 **5.** 241.7 **6.** 826.1 **7.** 30,120 **8.** 7,250 **9.** 4,210 **10.** 9,160 **11.** 82.59 **12.** 51.17 **13.** 31.9 **14.** 17.3 **15.** 4.1 **16.** 6.1 **17.** 213,510 **18.** 64,920 **19.** .14 **20.** .03

3-5-1

1. .7 **2.** 1.7 **3.** 3.5 **4.** 1.34 **5.** 7.04 **6.** .14 **7.** .017 **8.** .0018 **9.** 16.04 **10.** 10.09 **11.** 4.013 **12.** 5.027 **13.** 1.483 **14.** 2.864 **15.** .31265 **16.** .73468 **17.** 4.83 **18.** .47 **19.** .66 **20.** 2.19 **21.** 4.96 **22.** 3.27 **23.** 3.42 **24.** 2.10 **25.** .01 **26.** .01 **27.** 3.474 **28.** 2.519 **29.** 3.077 **30.** 2.941 **31.** 2.854 **32.** 4.917 **33.** .030 **34.** .024

3-5-2

1. 1.83 **2.** 14.32 **3.** 4.126 **4.** .1731 **5.** .136 **6.** .105 **7.** .1294 **8.** .1452 **9.** .0831 **10.** .0014 **11.** .0293 **12.** .5137 **13.** .0001 **14.** .00035 **15.** .01003 **16.** .0139 **17.** .03496 **18.** .008163 **19.** .00204 **20.** .000194

3-5-3

1. 83.8 **2.** 4.06 inches **3.** 21.3 points per game **4.** 1.14 points per game **5.** $6.21 **6.** $108.96 **7.** $18,871.67 **8.** $177.65 **9.** 27.0 years old **10.** 147.88 pounds **11.** 295.8 miles **12.** 7.3 hours

3-6-1

1. 3.5 **2.** 3.1 **3.** 25.1 **4.** 10.6 **5.** 2.14 **6.** 3.7 **7.** 8.22 **8.** 4.07 **9.** .014 **10.** .005 **11.** 31.5 **12.** 1,620 **13.** 33.4 **14.** 287.1 **15.** 10.8 **16.** 23.2 **17.** 20.9 **18.** 2.7 **19.** 11.43 **20.** 7.05 **21.** 5.34 **22.** 21.06 **23.** 4.27 **24.** 5.79 **25.** $.69 per pound **26.** $3.98 per pound **27.** 23.8 miles per gallon **28.** 32.4 miles per gallon **29.** 1.95 meters per second **30.** 11.5 meters per second

3-7-1

1. $\frac{7}{10}$ 2. $\frac{11}{100}$ 3. $\frac{2}{5}$ 4. $\frac{4}{25}$ 5. $\frac{17}{50}$ 6. $\frac{11}{20}$ 7. $2\frac{9}{10}$ 8. $4\frac{4}{5}$ 9. $\frac{31}{250}$ 10. $\frac{1}{40}$ 11. $\frac{119}{500}$ 12. $\frac{18}{125}$ 13. $\frac{52}{125}$
14. $\frac{183}{250}$ 15. $7\frac{13}{125}$ 16. $3\frac{62}{125}$ 17. $11\frac{31}{40}$ 18. $19\frac{2}{125}$ 19. $\frac{72}{625}$ 20. $18\frac{69}{625}$

3-7-2

1. .500 2. .250 3. .800 4. .200 5. .125 6. .375 7. .167 8. .833 9. .556 10. .444 11. .429 12. .143
13. 3.333 14. 4.400 15. 2.750 16. 1.875 17. .214 18. .438 19. 5.810 20. 7.720 21. 10.909 22. 13.056
23. 3.379 24. 6.974

3-7-3

1. 1.76 2. 11.75 3. 4.15 4. 5.05 5. 1.39 6. 6.68 7. 4.0306 8. 1.8 9. 19.1 10. 5.5

3-8-1

1. 1 2. 1 3. $\frac{1}{4}$ 4. $\frac{1}{8}$ 5. $\frac{1}{125}$ 6. $\frac{1}{36}$ 7. $\frac{1}{100,000}$ 8. $\frac{1}{1,000,000}$ 9. 8×10^{-1}
10. $2 \times 10^{-1} + 7 \times 10^{-2}$ 11. $9 \times 10^{-1} + 1 \times 10^{-2} + 5 \times 10^{-3}$ 12. $6 \times 10^{-1} + 0 \times 10^{-2} + 5 \times 10^{-3}$
13. $7 \times 10^{0} + 1 \times 10^{-1} + 4 \times 10^{-2}$ 14. $6 \times 10^{1} + 0 \times 10^{0} + 3 \times 10^{-1}$
15. $1 \times 10^{2} + 2 \times 10^{1} + 8 \times 10^{0} + 7 \times 10^{-1} + 2 \times 10^{-2}$
16. $5 \times 10^{3} + 1 \times 10^{2} + 2 \times 10^{1} + 4 \times 10^{0} + 1 \times 10^{-1} + 7 \times 10^{-2}$
17. 4 18. 7.3 19. 460 20. 85 21. 3,160 22. 72,500 23. 6.1 24. .34 25. .0605 26. .103 27. 794,000
28. 52,100,000 29. .00000917 30. .00000000432

3-8-2

1. yes 2. no 3. yes 4. yes 5. no 6. no 7. no 8. yes 9. 5×10^{3} 10. 5.28×10^{3} 11. 7.28×10^{5}
12. 3.46×10^{8} 13. 2.35×10^{-7} 14. 5.2×10^{-9} 15. 2.33×10^{14} 16. 7.39×10^{-11} 17. 320,100 18. .00000728
19. .00000000107 20. 623,000,000,000,000,000,000,000 21. 50,200,000,000 22. .358 23. .0000407 24. 537.62
25. 3.6 26. 99 27. 4.9×10^{7} miles 28. 8×10^{6} miles 29. 6,000,000,000,000 miles
30. 13,000,000,000,000,000,000,000 pounds 31. .00000001 cm 32. .0000000000035 cm 33. 1×10^{-3} cm
34. 4.3×10^{19} combinations

3-9-1

1. 0, 1, 4, 9, 16, 25, 36, 49, 64, 81, 100, 121, 144, 169, 196, 225, 256, 289, 324, 361, 400, 441, 484, 529, 576 2. 2 3. 3
4. 0 5. 9 6. 10 7. 11 8. 20 9. 15 10. 6 11. 12 12. 1 13. 8 14. 13 15. 16 16. 17 17. 21
18. 22 19. 23 20. 30 21. 100

3-9-3

1. $2\sqrt{3}$ 2. $2\sqrt{5}$ 3. $3\sqrt{2}$ 4. Simplified 5. $5\sqrt{3}$ 6. $4\sqrt{3}$ 7. $6\sqrt{5}$ 8. $4\sqrt{5}$ 9. $4\sqrt{6}$ 10. $6\sqrt{2}$ 11. $6\sqrt{3}$
12. Simplified 13. $8\sqrt{3}$ 14. $4\sqrt{11}$ 15. $5\sqrt{7}$ 16. $6\sqrt{11}$ 17. $9\sqrt{3}$ 18. $7\sqrt{6}$ 19. $9\sqrt{6}$ 20. $10\sqrt{10}$

Chapter 3 Review

The number in brackets after each answer refers to the section of the chapter that discusses that type of problem.

1. $2 \times \frac{1}{10} + 4 \times \frac{1}{100}$ [3–1]   2. $0 \times \frac{1}{10} + 7 \times \frac{1}{100}$ [3–1]

3. $7 \times 1 + 3 \times \frac{1}{10} + 0 \times \frac{1}{100} + 7 \times \frac{1}{1,000} + 4 \times \frac{1}{10,000}$ [3–1]

4. $1 \times 10 + 2 \times 1 + 1 \times \frac{1}{10} + 3 \times \frac{1}{100} + 0 \times \frac{1}{1,000} + 8 \times \frac{1}{10,000}$ [3–1]

5. Forty-nine hundredths [3–1]   6. Five hundred thirteen thousandths [3–1]   7. Fifteen and seventy-five thousandths [3–1]
8. Eight and seventy-nine thousandths [3–1]   9. .403 [3–1]   10. .0208 [3–1]   11. .81 [3–2]   12. 4 [3–2]   13. 6.397 [3–2]
14. 4.1 [3–2]   15. 1.00 [3–2]   16. .45 [3–2]   17. .5 [3–2]   18. .150 [3–2]   19. 10 [3–2]   20. 1.556 [3–2]
21. $\frac{83}{100}$ [3–7]   22. $\frac{7}{100}$ [3–7]   23. $\frac{36}{125}$ [3–7]   24. $\frac{38}{125}$ [3–7]   25. $17\frac{12}{25}$ [3–7]   26. $24\frac{1}{8}$ [3–7]   27. .958 [3–7]
28. .324 [3–7]   29. 5.778 [3–7]   30. 11.375 [3–7]   31. .0628 [3–4]   32. 102.51 [3–3]   33. .180 [3–5]   34. 97.000 [3–6]
35. 59.3 [3–3]   36. .2092 [3–4]   37. 83.714 [3–6]   38. .240 [3–5]   39. 27.66 [3–3]   40. 12.26 [3–3]   41. 15.84 [3–3]
42. 14.761 [3–3]   43. 3.150 [3–5]   44. .125 [3–6]   45. 2.223 [3–4]   46. 16.01 [3–3]   47. .0004 [3–4]   48. 4.620 [3–5]
49. 3.400 [3–6]   50. 2.0006 [3–4]   51. 23.189 [3–3]   52. .00098 [3–4]   53. 6.44 [3–3]   54. 231.42 [3–3]
55. 3,247 [3–4]   56. .003 [3–6]   57. 460.4 [3–3]   58. 10.92 [3–3]   59. 37.77 [3–3]   60. 6,417 [3–4]   61. .008 [3–6]
62. 128.001 [3–3]   63. 2.329 [3–5]   64. 180.89 [3–3]   65. 79.39188 [3–4]   66. 213.724 [3–3]   67. 61.828 [3–3]
68. 3.409 [3–5]   69. 21.004 [3–3]   70. 35.58413 [3–4]   71. 29.159 [3–6]   72. 3.614 [3–6]   73. 513.541 [3–3]
74. 19.009 [3–3]   75. 19.51248 [3–4]   76. 110.78322 [3–4]   77. 26.475 [3–5]   78. 526.174 [3–3]   79. 144.99 [3–3]
80. 119.39 [3–3]   81. 247.637 [3–3]   82. 22.346 [3–5]   83. 17.798 [3–3]   84. 16.242 [3–5]   85. 11.04356 [3–4]
86. 11.036 [3–6]   87. 3.081 [3–5]   88. 18.296 [3–3]   89. 3.074 [3–6]   90. 32.6304 [3–4]   91. $12.52 [3–3]
92. $155.85 [3–3]   93. $247.59 [3–3]   94. $7.09 [3–3]   95. $109.59 [3–3]   96. $17.03 [3–4]   97. $16.71 [3–4]
98. $279.92 [3–4]   99. $38.39 [3–4]   100. 80.27 [3–5]   101. 1.35 inches [3–5]   102. 99.2 yards [3–5]   103. $2.09 [3–6]
104. 8.12 meters per second [3–5]   105. 42,600 [3–8]   106. 1,840,000 [3–8]   107. .0139 [3–8]   108. .00004271 [3–8]
109. 5.18×10^3 [3–8]   110. 6.02×10^4 [3–8]   111. 8.145×10^{-1} [3–8]   112. 9.99×10^{-2} [3–8]   113. 6 and 7 [3–9]
114. 5 and 6 [3–9]   115. $2\sqrt{7}$ [3–9]   116. $2\sqrt{11}$ [3–9]   117. $2\sqrt{14}$ [3–9]   118. $3\sqrt{7}$ [3–9]   119. $2\sqrt{34}$ [3–9]
120. $4\sqrt{19}$ [3–9]

Chapter 3 Test

The number in brackets following each answer refers to the section of the chapter that discusses that type of problem.

1. $5 \times 10 + 2 \times 1 + 6 \times \frac{1}{10} + 4 \times \frac{1}{100} + 9 \times \frac{1}{1,000}$ [3–1]   2. Two hundred ninety-eight and thirty-seven hundredths [3–1]

3. 8.7 [3–2]   4. $\frac{14}{125}$ [3–7]   5. 13.714 [3–7]   6. 42.42 [3–3]   7. 58.768 [3–3]   8. 7.400 [3–5]   9. 15.230 [3–6]
10. 5.016 [3–4]   11. 35.661 [3–3]   12. 17.392 [3–3]   13. 21.919 [3–5]   14. 8.340 [3–6]   15. 1,344.84 [3–4]
16. 386.211 [3–3]   17. 23.869 [3–3]   18. 7.08 [3–5]   19. 1.37347 [3–4]   20. .009 [3–6]   21. .0003164 [3–8]
22. 2.14×10^{-6} [3–8]   23. $2\sqrt{3}$ [3–9]   24. $4\sqrt{7}$ [3–9]

Chapter 4 Pretest

The number in brackets after each answer refers to the section of the chapter that discusses that type of problem.

1. $(a + b)(x - y)$ [4–1]   2. The quotient of a, and the difference of a and b [4–1]   3. $\frac{22 - 7}{(3)(5)}$ [4–1]   4. 3 [4–1]
5. 37 [4–1]   6. 36 [4–2]   7. 227 [4–2]   8. 62.8 [4–2]   9. $2x + 7y$ [4–3]   10. $8a^2b + 7ab$ [4–3]

4–1–1

1. $x + y$ 2. $c - d$ 3. ap or $(a)(p)$ or $a(p)$ 4. xy or $(x)(y)$ or $x(y)$ 5. $a(x + y)$ 6. $\frac{a}{x + y}$ 7. $(x + y)(x - y)$
8. $\frac{ab}{x + y}$ 9. $3y - 5x$ 10. $\frac{7a}{11c}$ 11. $\frac{a - b}{a + b}$ 12. $\frac{a}{b} - (a + b)$ 13. The sum of a and b 14. The product of x and y
15. The difference of x and y 16. a divided by the sum of b and c 17. The quotient of the difference of a and b, and c
18. The difference of a, and the quotient of b and c 19. The quotient of the sum of a and b, and the difference of a and b.
20. The product of x, and the sum of x and y 21. The product of the sum of a and b, and the difference of a and b

22. The quotient of the difference of x and y, and the sum of x and y **23.** $\frac{3}{4} + \frac{7}{8} = \frac{13}{8}$ **24.** $17 - 12$ **25.** $\frac{19 + 23}{(7)(3)} = 2$

26. $\frac{2}{3} \div \frac{1}{2} = \frac{2}{3} \times \frac{2}{1} = \frac{4}{3}$ **27.** $\frac{2}{3} \times \frac{3}{7} \div \frac{1}{2} = \frac{2}{3} \times \frac{3}{7} \times \frac{2}{1} = \frac{4}{7}$ **28.** $\frac{1}{2}(9 + 19) = \frac{1}{2}(28) = 14$

29. $\left(\frac{2}{3}\right)\left(\frac{5}{8}\right) + \left(\frac{2}{3} \div \frac{5}{8}\right) = \frac{5}{12} + \frac{16}{15} = \frac{89}{60}$ **30.** $10 - \left(\frac{1}{2} + \frac{2}{3}\right) = \frac{60}{6} - \frac{7}{6} = \frac{53}{6}$

4–1–2

1. 2   **2.** 2   **3.** 1   **4.** 1   **5.** 2   **6.** 2   **7.** 1   **8.** 2   **9.** 3   **10.** 3   **11.** 3   **12.** 1   **13.** 4   **14.** 3   **15.** 4   **16.** 2
17. 1   **18.** 1   **19.** 2   **20.** 1   **21.** 9   **22.** 25   **23.** 8   **24.** 343   **25.** 625   **26.** 16   **27.** 243   **28.** 1,024   **29.** $\frac{1}{8}$
30. $\frac{1}{81}$ **31.** 50   **32.** 100   **33.** 7,776   **34.** 486   **35.** 405   **36.** 50,625   **37.** 35   **38.** 49   **39.** 77   **40.** 4

4–2–1

1. 18   **2.** 4   **3.** 20   **4.** 16   **5.** 0   **6.** 5   **7.** 9   **8.** 11   **9.** 5   **10.** 3   **11.** 16   **12.** 243   **13.** 75   **14.** 512   **15.** 56
16. 10   **17.** 414   **18.** 981   **19.** 261   **20.** 148   **21.** 8   **22.** 8   **23.** 42   **24.** 90   **25.** 288   **26.** 100   **27.** 44
28. 61   **29.** 73   **30.** 1,008   **31.** 648   **32.** 32   **33.** 35   **34.** 1,847   **35.** 1,402   **36.** 0   **37.** 2,500   **38.** 3,909   **39.** 66
40. 196   **41.** 38   **42.** 27   **43.** 32   **44.** 12.25   **45.** 44   **46.** 154   **47.** 20   **48.** 86   **49.** 1,390   **50.** 16

4–3–1

1. $11x$ **2.** $13a$ **3.** $4x$ **4.** $5a$ **5.** $4a$ **6.** $6x$ **7.** $8x^2$ **8.** $4x^2$ **9.** $6a^3$ **10.** $5a^3$ **11.** $16xy$ **12.** $7ab$ **13.** xy
14. $11a^2b$ **15.** $4a + 6b$ **16.** $11x - 4y$ **17.** $9a^2b + 11ab^2$ **18.** $8xy^2 - 7x^2y$ **19.** $22abc + ab$ **20.** $9xy + 6x^2y$
21. $ab + ac$ **22.** $10x^2y + 4xy^2$ **23.** $3a^2 + 8a$ **24.** xz **25.** $8x^3y^2$ **26.** $a^2b + 6ab + 2ab^2$ **27.** $7x^2y - 5xy^2 + 9x^2y^2$
28. $9ab^2 + 13a^2b - 5ab$ **29.** $3m^2n + 6mn^2$ **30.** 0   **31.** $M + 8m$ **32.** $T + t$ **33.** $2\ell + 2w$ **34.** $9x$

Chapter 4 Review

The number in brackets after each answer refers to the section of the chapter that discusses that type of problem.

1. $x + y$ [4–1]   **2.** df [4–1]   **3.** $\frac{3a}{4b}$ [4–1]   **4.** $(a + b)(x - 3)$ [4–1]   **5.** $\frac{x + y}{ac}$ [4–1]   **6.** $\frac{8x}{9 - a}$ [4–1]   **7.** $6\frac{11}{12}$ [4–1]
8. $36\frac{1}{4}$ [4–1]   **9.** 1 [4–1]   **10.** $4\frac{1}{7}$ [4–1]   **11.** 3 [4–1]   **12.** 1 [4–1]   **13.** 216 [4–1]   **14.** 80 [4–1]   **15.** 109 [4–1]
16. 72 [4–2]   **17.** 116 [4–2]   **18.** 92 [4–2]   **19.** 60 [4–2]   **20.** 1,210 [4–2]   **21.** 21 [4–2]   **22.** 3 [4–2]   **23.** 9 [4–2]
24. 105 [4–2]   **25.** $9a - 6b$ [4–3]   **26.** $5x^2 + 9x - 4y$ [4–3]   **27.** $7xy + 2x^2y$ [4–3]   **28.** $a^2b + 4a^2$ [4–3]   **29.** xy [4–3]
30. $13x^2y^3 + 4x^2y - x^2y^2$ [4–3]

Chapter 4 Test

The number in brackets after each answer refers to the section of the chapter that discusses that type of problem.

1. $\frac{xy}{a - b}$ [4–1]   **2.** The product of a, and the sum of a and b [4–1]   **3.** $\frac{24 - 9}{2 + 3} = 3$ [4–1]   **4.** 1 [4–1]   **5.** 128 [4–1]
6. 48 [4–2]   **7.** 84 [4–2]   **8.** 27 [4–2]   **9.** $3a + 4b$ [4–3]   **10.** $2x^2y$ [4–3]

Chapters 1–4 Cumulative Test

The number in brackets after each answer refers to the chapter and section that discusses that type of problem.

1. 8,310 [1–3]   **2.** 5,202 [1–4]   **3.** 548,488 [1–6]   **4.** 14 [1–10]   **5.** 48 [1–10]   **6.** $\frac{14}{15}$ [2–1]   **7.** 24 [2–4]   **8.** $\frac{5}{12}$ [2–3]

9. $\frac{1}{3}$ [2-4]   10. 15 [2-5]   11. $\frac{33}{125}$ [3-7]   12. 10.626 [3-4]   13. 4.37 [3-6]   14. 6.31 × 10⁻⁵ [3-8]   15. $3\sqrt{5}$ [3-9]
16. $\frac{a+b}{cd}$ [4-1]   17. 36 [4-1]   18. 115 [4-2]   19. 51 [4-2]   20. $9x^2y + 4y^2$ [4-3]

Chapter 5 Pretest

The number in brackets after each answer refers to the section of the chapter that discusses that type of problem.

1. Conditional [5-1]   2. Yes [5-1]   3. 30 [5-3]   4. $\frac{2}{3}$ [5-5]   5. $8\frac{1}{3}$ [5-4]   6. 16 [5-2]   7. 4 [5-4]   8. $4\frac{2}{7}$ [5-6]
9. $1\frac{6}{19}$ [5-6]   10. Base = 12 in.; height = 8 in. [5-2]

5-1-1

1. Identity   2. Conditional   3. Conditional   4. Identity   5. Identity   6. Identity   7. Identity   8. Conditional
9. Conditional   10. Conditional   11. 3   12. 3   13. 11   14. 9   15. 4   16. $\frac{1}{2}$ 17. 10   18. 4   19. 6   20. 2
21. $\frac{3}{5}$ 22. 5   23. $\frac{1}{3}$ 24. 13   25. 0   26. 9   27. 7   28. 3   29. 6   30. 5   31. Equivalent
32. Not equivalent   33. Not equivalent   34. Equivalent   35. Not equivalent   36. Equivalent   37. Not equivalent
38. Not equivalent   39. Equivalent   40. Equivalent

5-2-1

1. 12   2. 9   3. 4   4. 5   5. 3   6. 6   7. $\frac{1}{3}$ 8. $\frac{1}{2}$ 9. $\frac{1}{2}$ 10. $\frac{1}{3}$ 11. 3   12. 8   13. $\frac{1}{3}$ 14. $\frac{1}{3}$ 15. 1
16. 1   17. 3   18. 5   19. $\frac{1}{2}$ 20. $\frac{1}{5}$ 21. 15   22. 22   23. $\frac{2}{5}$ 24. $\frac{5}{6}$ 25. $\frac{1}{4}$ 26. $\frac{1}{5}$ 27. 25   28. 30
29. 3.2   30. 6.5   31. 15   32. 30   33. 8   34. 27   35. $\frac{3}{4}$ 36. $\frac{4}{5}$ 37. 17 meters   38. 13 inches   39. 31 inches
40. $69\frac{2}{5}$ km/hr

5-3-1

1. 2   2. 7   3. 8   4. 8   5. 4.7   6. .7   7. $3\frac{13}{20}$ 8. $1\frac{5}{6}$ 9. 37   10. 9   11. 0   12. 0   13. 1   14. 7   15. 1   16. 8
17. 8   18. 9   19. 6   20. 4   21. 4   22. 6   23. 2   24. 3   25. 1   26. 1   27. 3   28. 2   29. 5   30. 1   31. 5
32. 7   33. 0   34. 4   35. 2   36. 15.4 m   37. 12.4 cm   38. $9.45   39. $38.75   40. 30 degrees C

5-4-1

1. 10   2. 9   3. 6   4. 16   5. $4\frac{1}{2}$ 6. $1\frac{4}{15}$ 7. 6.3   8. 16.1   9. 4   10. 5   11. 8   12. 7   13. $\frac{2}{3}$ 14. $\frac{3}{5}$ 15. 1
16. $\frac{1}{2}$ 17. 2   18. 3   19. 1   20. 3   21. $\frac{5}{6}$ 22. $\frac{1}{4}$ 23. 3   24. 3   25. 2   26. $\frac{3}{4}$ 27. $44.40   28. $13.89
29. 89.6° F   30. 44.1 m

5-5-1

1. 12 **2.** 36 **3.** 15 **4.** 15 **5.** 5 **6.** 16 **7.** $10\frac{2}{3}$ **8.** $7\frac{1}{2}$ **9.** 48 **10.** 24 **11.** $32\frac{2}{3}$ **12.** $1\frac{1}{2}$ **13.** $2\frac{4}{7}$ **14.** $\frac{15}{16}$
15. $1\frac{7}{18}$ **16.** $\frac{2}{5}$ **17.** $\frac{4}{5}$ **18.** $\frac{2}{3}$ **19.** $1\frac{1}{4}$ **20.** $6\frac{7}{8}$ **21.** $4\frac{2}{3}$ **22.** $1\frac{3}{7}$ **23.** $\frac{5}{8}$ **24.** $5\frac{1}{17}$ **25.** $22\frac{1}{2}$ **26.** $\frac{1}{4}$
27. 8 in. **28.** $4\frac{9}{10}$ **29.** 100° C **30.** 12 feet

5-6-1

1. 2 **2.** 3 **3.** $\frac{17}{2}$ **4.** $\frac{3}{10}$ **5.** $\frac{1}{2}$ **6.** $\frac{9}{20}$ **7.** 1 **8.** 4 **9.** $\frac{40}{21}$ **10.** $\frac{27}{26}$ **11.** $\frac{2}{3}$ **12.** $\frac{44}{21}$ **13.** $\frac{27}{2}$ **14.** $\frac{51}{64}$ **15.** $\frac{13}{34}$
16. 140 **17.** $\frac{25}{3}$ **18.** $\frac{165}{56}$ **19.** $\frac{312}{5}$ **20.** 36 **21.** $\frac{112}{113}$ **22.** $\frac{294}{13}$ **23.** $\frac{16}{33}$ **24.** $\frac{240}{7}$ **25.** $\frac{35}{2}$ **26.** $\frac{47}{21}$ **27.** 75 inches
28. $84.00 **29.** $8,526.00 **30.** 18 in., 9 in., 12 in.

Chapter 5 Review

The number in brackets after each answer refers to the section of the chapter that discusses that type of problem.
1. Conditional [5–1] **2.** Identity [5–1] **3.** Identity [5–1] **4.** Conditional [5–1] **5.** Conditional [5–1] **6.** Identity [5–1]
7. Equivalent [5–1] **8.** Equivalent [5–1] **9.** Not equivalent [5–1] **10.** Not equivalent [5–1] **11.** Equivalent [5–1]
12. Equivalent [5–1] **13.** 8 [5–2] **14.** 19 [5–3] **15.** 48 [5–4] **16.** 52 [5–2] **17.** 39 [5–3] **18.** 75 [5–4]
19. $8\frac{1}{3}$ [5–2] **20.** 4 [5–6] **21.** 6 [5–5] **22.** $\frac{41}{5}$ [5–2] **23.** $4\frac{1}{2}$ [5–3] **24.** 100 [5–5] **25.** $2\frac{4}{9}$ [5–6] **26.** 13 [5–2]
27. 6 [5–3] **28.** $1\frac{3}{7}$ [5–5] **29.** $\frac{19}{35}$ [5–4] **30.** $\frac{9}{14}$ [5–6] **31.** $4\frac{1}{5}$ [5–6] **32.** 12 [5–2] **33.** 138 [5–6] **34.** $4\frac{1}{4}$ [5–6]
35. 5 [5–4] **36.** $2\frac{2}{9}$ [5–5] **37.** 70 [5–6] **38.** $\frac{45}{62}$ [5–6] **39.** $15\frac{1}{11}$ [5–2] **40.** 6 [5–6] **41.** $\frac{5}{12}$ [5–6] **42.** 6 [5–6]
43. 0 [5–6] **44.** 5 [5–6] **45.** $1\frac{5}{21}$ [5–5] **46.** 5 [5–6] **47.** $1\frac{31}{35}$ [5–6] **48.** $1\frac{1}{6}$ [5–6] **49.** $7\frac{1}{2}$ [5–6] **50.** $4\frac{1}{16}$ [5–6]
51. 18 ft [5–2] **52.** 18 m [5–5] **53.** $5.80 per hour [5–2] **54.** 74.3° F [5–6] **55.** 84.2 cm [5–6] **56.** $4\frac{1}{6}$ m [5–6]

Chapter 5 Test

The number in brackets after each answer refers to the section of the chapter that discusses that type of problem.
1. Identity [5–1] **2.** No [5–1] **3.** 9 [5–3] **4.** 9 [5–4] **5.** 13 [5–2] **6.** $\frac{1}{6}$ [5–5] **7.** $1\frac{1}{2}$ [5–6] **8.** 30 [5–6]
9. $1\frac{1}{5}$ [5–6] **10.** $\ell = 12\frac{1}{2}$ in., $w = 5$ in. [5–6]

Chapter 6 Pretest

The number in brackets after each answer refers to the section of the chapter that discusses that type of problem.
1. $\frac{2}{7}$ [6–1] **2.** No [6–1] **3.** True [6–2] **4.** True [6–2] **5.** 4 [6–3] **6.** 9 [6–3] **7.** 10.5 [6–3] **8.** 8 students [6–3]
9. 336 miles [6–3] **10.** 12 teaspoons [6–3] **11.** 104.6 km/hr [6–4] **12.** 1.85 gal [6–4] **13. a.** 37.5% [6–5] **b.** .2% [6–5]
14. 42.5% [6–6] **15.** $89.25 [6–6]

6-1-1

1. $\dfrac{5}{7}$ 2. $\dfrac{2}{5}$ 3. $\dfrac{8}{9}$ 4. $\dfrac{1}{4}$ 5. $\dfrac{2}{5}$ 6. $\dfrac{2}{11}$ 7. $\dfrac{3}{17}$ 8. $\dfrac{2}{5}$ 9. $\dfrac{1}{3}$ 10. $\dfrac{4}{5}$ 11. $\dfrac{6}{5}$ 12. $\dfrac{5}{9}$ 13. $\dfrac{1}{9}$ 14. $\dfrac{1}{30}$
15. $\dfrac{4}{5}$ 16. $\dfrac{2}{13}$ 17. $\dfrac{7}{3}$ 18. Yes 19. Yes 20. Yes 21. Yes 22. No 23. Yes 24. Yes 25. No 26. No
27. Yes 28. $\dfrac{2}{3}$ inches per hour 29. 40 mph 30. 9.63 cents per ounce 31. .305 32. 8.7 meters per second

6-2-1

1. True 2. True 3. True 4. True 5. True 6. True 7. False 8. False 9. True 10. False 11. False 12. True
13. True 14. True 15. False 16. True 17. False 18. True 19. True 20. False 21. No 22. Yes 23. Yes
24. No 25. Yes 26. Yes 27. Yes 28. Yes 29. Yes 30. No

6-3-1

1. 4 2. 3 3. 20 4. 8 5. 20 6. 16 7. 30 8. 36 9. 4 10. 30 11. 9 12. 2 13. 20 14. 8 15. 15
16. 1.2 17. 10.8 18. 12 19. 10 20. 14 21. $56 22. $225 23. 18 cans 24. 24 feet 25. 250 miles
26. 12.5 bags 27. $6.45 28. 25 teeth 29. 440 miles 30. $7.98 31. $605 32. $192\dfrac{1}{2}$ miles 33. 60 bulbs
34. $5,400 35. 7.9 gallons 36. $25,200 in bonds 37. 219 games 38. 48 39. 3,125 black bass 40. 6.5 inches

6-4-1

1. 11.8 in. 2. 91.4 cm 3. 36.6 m 4. 10.9 yd 5. 137.8 in. 6. .9 kg 7. 1.9 liters 8. 196.9 in. 9. 4.6 m
10. 82.0 yd 11. 48.3 km 12. 29.5 in. 13. 10.9 oz 14. 113.3 g 15. 1,553.5 mi 16. 2.3 kg 17. 109.4 yd
18. 91.4 m 19. 1.9 qt 20. 275.6 lb 21. 590.4 mi 22. 453.3 g 23. 14.3 qt 24. 62.2 km/hr 25. Answers will vary
26. Answers will vary 27. 11.9 gallons 28. 58.7 liters

6-5-1

1. $\dfrac{1}{2}$ 2. $\dfrac{1}{5}$ 3. $\dfrac{13}{20}$ 4. $\dfrac{3}{2}$ 5. $\dfrac{49}{200}$ 6. $\dfrac{39}{125}$ 7. $\dfrac{5}{6}$ 8. $\dfrac{7}{400}$ 9. $\dfrac{7}{8}$ 10. $\dfrac{141}{125}$ 11. .15 12. .3 13. .03 14. .018
15. .156 16. .0019 17. 2.38 18. 1 19. .0009 20. .1864 21. 92% 22. 50% 23. 320% 24. 6% 25. 40.3%
26. 800% 27. 7.2% 28. 203% 29. .4% 30. 101.8% 31. 50% 32. 20% 33. 75% 34. 90% 35. 7% 36. 16.7%
37. 37.5% 38. 57.1% 39. 175% 40. 333.3%

6-6-1

1. 16 2. 24.5 3. 9.6 4. 42 5. 151.8 6. 348.6 7. 10 8. 40 9. 124 10. 450 11. 225 12. 1,800 13. 75%
14. 46.7% 15. 20.8% 16. 32% 17. 89.3% 18. 76.4% 19. $700 20. $64 21. 75 members 22. $150,000
23. $24,000 24. $360 25. 3,915 miles 26. 88.6 inches 27. 15% 28. 17.6% 29. 43.6% 30. 63.6% 31. 5.6%
32. 20%

Chapter 6 Review

The number in brackets after each answer refers to the section of the chapter that discusses that type of problem.

1. $\dfrac{3}{8}$ [6–1] 2. $\dfrac{2}{3}$ [6–1] 3. $\dfrac{4}{9}$ [6–1] 4. $\dfrac{3}{7}$ [6–1] 5. Yes [6–1] 6. Yes [6–1] 7. No [6–1] 8. Yes [6–1]
9. True [6–2] 10. False [6–2] 11. True [6–2] 12. True [6–2] 13. False [6–2] 14. True [6–2] 15. True [6–2]
16. True [6–2] 17. 20 [6–3] 18. 20 [6–3] 19. 14 [6–3] 20. 12 [6–3] 21. 3.2 [6–3] 22. 22.5 [6–3] 23. 2 [6–3]
24. 40.5 [6–3] 25. 6.9 m [6–4] 26. 17.0 liters [6–4] 27. 84.7 in. [6–4] 28. 1.6 oz [6–4] 29. 12.4 mi [6–4] 30. 61.2

kg [6-4]   **31.** 226.6 g [6-4]   **32.** 9.5 liters [6-4]   **33.** .37 [6-5]   **34.** .09 [6-5]   **35.** .138 [6-5]   **36.** 2.06 [6-5]
37. .007 [6-5]   **38.** .1204 [6-5]   **39.** 55% [6-5]   **40.** 30% [6-5]   **41.** 6% [6-5]   **42.** 520% [6-5]   **43.** 1.6% [6-5]
44. 419.8% [6-5]   **45.** 40% [6-5]   **46.** 18.8% [6-5]   **47.** 70% [6-5]   **48.** 52.5% [6-5]   **49.** 250% [6-5]   **50.** 240% [6-5]
51. 36 [6-6]   **52.** 21 [6-6]   **53.** 134.4 [6-6]   **54.** 48 [6-6]   **55.** 20.0% [6-6]   **56.** 18.2% [6-6]   **57.** 17.5% [6-6]
58. 16.7% [6-6]   **59.** 5 [6-6]   **60.** 16.87 [6-6]   **61.** 1,163.64 [6-6]   **62.** 31.64 [6-6]   **63.** 15 [6-3]   **64.** 38 mph [6-1]
65. Yes [6-2]   **66.** 225 miles [6-3]   **67.** 14 teeth [6-3]   **68.** 9.5 gallons [6-3]   **69.** 1.9 qt [6-4]   **70.** 1,640.4 yd [6-4]
71. 97 people [6-6]   **72.** $3,780 [6-6]   **73.** 21.5% [6-6]   **74.** 81.7% [6-6]   **75.** 11,420 [6-6]   **76.** $13,500 [6-6]

Chapter 6 Test

The number in brackets after each answer refers to the section of the chapter that discusses that type of problem.

1. $\frac{5}{6}$ [6-1]   **2.** Yes [6-1]   **3.** True [6-2]   **4.** False [6-2]   **5.** 10 [6-3]   **6.** 35 [6-3]   **7.** 4.8 [6-3]   **8.** 18 boys [6-3]
9. 7 gallons [6-3]   **10.** 45 people [6-3]   **11.** 88.5 kph [6-4]   **12.** 53 liters [6-4]   **13. a.** 62.5% [6-5]   **b.** 130% [6-5]
14. 41.25% [6-6]   **15.** 9.5% [6-6]

Chapters 1-6 Cumulative Test

The number in brackets after each answer refers to the chapter and section that discusses that type of problem.

1. 2,487 [1-5]   **2.** 253 [1-7]   **3.** 14 [1-8]   **4.** $\frac{5}{7}$ [2-1]   **5.** $\frac{4}{9}$ [2-2]   **6.** $1\frac{7}{24}$ [2-5]   **7.** 25.14 [3-2]   **8.** .448 [3-4]
9. 44.966 [3-3]   **10.** .000705 [3-8]   **11.** $\frac{a-b}{x+y}$ [4-1]   **12.** 16 [4-2]   **13.** $9a^3b + 2ab^2$ [4-3]   **14.** 14 [5-2]   **15.** $\frac{9}{10}$ [5-5]
16. $\frac{4}{7}$ [5-6]   **17.** 32° F [5-6]   **18.** 20 feet [6-3]   **19.** 37.5% [6-5]   **20.** $1,092 [6-6]

Chapter 7 Pretest

The number in brackets after each answer refers to the section of the chapter that discusses that type of problem.

1. a. 5.9 [7-1]   **b.** $-\frac{3}{7}$ [7-1]   **2.** 13 [7-1]   **3.** +6 [7-2]   **4. a.** -13 [7-3]   **b.** -22 [7-3]   **5. a.** -29 [7-3]
b. 14 [7-3]   **6.** 9 [7-3]   **7.** 29 [7-3]   **8.** $7x^2y$ [7-4]   **9.** 8 [7-5]   **10. a.** -70 [7-6]   **b.** 48 [7-6]   **11.** -36 [7-6]
12. a. 4 [7-6]   **b.** $-\frac{3}{2}$ [7-6]   **13. a.** -4 [7-6]   **b.** 3 [7-6]   **14.** $100 - 43x$ [7-7]   **15.** -54 [7-7]

7-1-1

1. -3   **2.** +4   **3.** $-\frac{1}{2}$ **4.** +5.3   **5.** $-y$ **6.** $+b$ **7.** -7   **8.** $+4\frac{1}{3}$ **9.** 0   **10.** -100   **11.** $-5\frac{3}{4}$ **12.** +7.38
13. -4   **14.** -2.05   **15.** $-\frac{1}{2}$ **16.** $+\frac{3}{4}$ **17.** $A = -7, B = -3, C = +7$

18. number line with point at -3

19. number line with point at 1

20. number line with point at -4

21. number line with point at -2

22. -20   **23.** +10   **24.** -5   **25.** -13   **26.** -6

7-1-2

1. $-x - y$ 2. $-a - 4$ 3. $x - 5$ 4. $a - b$ 5. $-a + b + c$ 6. $-x - y + 3$ 7. $a - b + 8$ 8. $a - c - 4$
9. $-a - b + c$ 10. $-x + y + 15$ 11. $-x - 4 + z$ 12. $-a + b - c + 5$ 13. $a - b$ 14. $-b + a$ or $a - b$
15. $a - 2b + c$ 16. $-a + b - 4c$ 17. $x - y$ 18. $2x + y$ 19. $a + 6b - 4$ 20. $3a - b + 16$
21. $-5x + y - 4z + 3$ 22. $x - 8y - 2z - 10$ 23. $-a - 2b + 5c - 6$ 24. $2a - 3b + 4c + 9$

7-2-1

1. $+7$ 2. -3 3. -2 4. -9 5. $+2$ 6. $+22$ 7. $+7$ 8. -8 9. -24 10. -6 11. -4 12. -7 13. $+4$
14. -3 15. -8 16. $+6$ 17. $+53°$ 18. $+56°$ 19. $+1$ 20. $0°$

7-2-2

1. $+8$ 2. -7 3. $+6$ 4. $+3$ 5. $+5$ 6. -6 7. 0 8. -7 9. -6 10. $+7$ 11. -9 12. $+9$ 13. 0 14. 0
15. 0 16. -4 17. -14 18. -14 19. 0 20. 0 21. $+4$ 22. $-3°$ 23. $+7$ 24. -8 25. $+23°$

7-3-1

1. $+12$ 2. $+41$ 3. $+17$ 4. -8 5. -23 6. -21 7. $+61$ 8. -32 9. -31 10. $+35$ 11. $+25$ 12. $+23$
13. -28 14. -30 15. $+17$ 16. -24 17. -27 18. $+74$ 19. $+29$ 20. -38

7-3-2

1. $+5$ 2. $+4$ 3. $+9$ 4. -6 5. -2 6. -14 7. 0 8. $+1$ 9. 0 10. $+3$ 11. -8 12. -2 13. -1
14. -9 15. $+14$ 16. -33 17. 0 18. $+25$ 19. -1 20. $+26$

7-3-3

1. 6 2. 0 3. 1 4. -10 5. 3 6. -6 7. 1 8. 1 9. 9 10. -1 11. -3 12. -2 13. 14 14. 25 15. -9
16. 4 17. 17 18. 8 19. 40 20. 9

7-3-4

1. 3 2. -1 3. 6 4. -5 5. 11 6. 8 7. 12 8. 12 9. 2 10. 2 11. 3 12. 8 13. 12 14. 13 15. 27
16. 20 17. -1 18. -4 19. 0 20. 7 21. 7 22. 11 23. -6 24. -5 25. 5 26. 1 27. -6 28. -5
29. 4 30. 10

7-4-1

1. $13x$ 2. $2a$ 3. $-3x$ 4. $18x^2$ 5. $-4xy$ 6. Cannot be combined 7. Cannot be combined 8. $6abc$
9. Cannot be combined 10. x^2y^2 11. 0 12. $-8y$ 13. $-2ab$ 14. x 15. $-15a^2b + 5ab^2$ 16. Cannot be combined
17. $ab + 3ac$ 18. $8a - 3xy$ 19. 0 20. $4xyz + 15xy + 7xy^2$ 21. $3x - 1$ 22. $2x + 1$ 23. 1 24. $5x + 3$
25. $2x - 3$ 26. $2a - 4$ 27. $10x - 9$ 28. $7a + 1$ 29. $5x + 4$ 30. $2a + 9$ 31. $10a - b$ 32. $9x - 6y$
33. $a + 3b$ 34. $6x + 2$ 35. $x^2 - 7x$

7-5-1

1. $6 + (-4) = 2$ 2. $9 + (-12) = -3$ 3. $13 + 2 = 15$ 4. $-16 + 9 = -7$ 5. $22 + 6 = 28$ 6. $3 + (-5) = -2$
7. $-11 + 1 = -10$ 8. $24 + (-24) = 0$ 9. $-25 + 3 = -22$ 10. $18 + (-18) = 0$ 11. $8 + (-3) = 5$
12. $8 + (-21) = -13$ 13. -1 14. 21 15. $a + 4$ 16. $-x^2 + 2x - 3$ 17. -4 18. 13 19. -41 20. 6
21. 22 22. -10 23. 6 24. -1 25. $4x + 2y - 5$ 26. $-3x^2y - 4xy + 6y^2$

7-6-1

1. -40 2. -40 3. -21 4. -24 5. -182 6. -4 7. $-\dfrac{3}{5}$ 8. -7.5 9. -11.73 10. $-\dfrac{1}{4}$ 11. -2.1
12. -40 13. -48 14. -60 15. -35 16. 0 17. -6 18. $-9\dfrac{1}{2}$ 19. -165.6 20. 0

7-6-2

1. $5x + 20$ 2. $2x - 18$ 3. $6x + 8$ 4. $-6x - 15$ 5. $42x - 28$ 6. $-16x - 40$ 7. $6x^2 + 12x + 6$
8. $10x^2 - 5x + 15$ 9. $-2x^2 - 6x - 2$ 10. $-12a - 8b - 4c$ 11. $5x + 15$ 12. $5x - 6$ 13. $12x + 2$
14. $-8x + 3$ 15. -3 16. $14a$ 17. $7x - 5y$ 18. $-3x + 2y$ 19. $13x - 18$ 20. $-2x^2 - x - 14$

7-6-3

1. 6 2. 8 3. 21 4. 45 5. 15 6. $\dfrac{1}{5}$ 7. -40 8. -21 9. 6.0 10. 60 11. -12 12. 60 13. -1 14. 0
15. 150 16. 30 17. 168 18. -300 19. -80 20. -37

7-6-4

1. 3 2. -3 3. 5 4. -5 5. -4 6. 7 7. 4 8. -27 9. -7 10. 4 11. $-\dfrac{14}{15}$ 12. -2 13. $\dfrac{7}{2}$ 14. $-\dfrac{1}{21}$
15. $-\dfrac{5}{2}$ 16. -2 17. 12 18. -2 19. 2 20. -3

7-7-1

1. $x + 6y$ 2. $11a - 16b$ 3. $-38x + 65y$ 4. $11a$ 5. $4x + 9$ 6. $21a - 42$ 7. $8x + 105$ 8. $-2a - 22$
9. $-x + 10y$ 10. $3a + 22b$ 11. $-19x - 20y + 122$ 12. $12a - 13x$ 13. $-3x - 8$ 14. $-85x + 40$
15. $-32x + 22y$ 16. -16 17. 22 18. 25 19. -25 20. 25 21. -24 22. -216 23. 28 24. 14 25. 8
26. -16 27. 17 28. 19 29. -15 30. 14

Chapter 7 Review

The number in brackets after each answer refers to the section of the chapter that discusses that type of problem.

1. -10 [7–1]   2. $-\dfrac{2}{3}$ [7–1]   3. $+\dfrac{3}{8}$ [7–1]   4. $+5.1$ [7–1]   5. -2 [7–3]   6. -22 [7–3]   7. $+\dfrac{1}{3}$ [7–3]
8. $+\dfrac{1}{6}$ [7–3]   9. -7 [7–3]   10. -5 [7–3]   11. -1 [7–3]   12. 9 [7–3]   13. 2 [7–3]   14. 6 [7–3]   15. -2 [7–3]
16. -3 [7–3]   17. 22 [7–3]   18. 14 [7–3]   19. -48 [7–6]   20. -2 [7–6]   21. $\dfrac{3}{14}$ [7–6]   22. 8.05 [7–6]
23. 280 [7–6]   24. -120 [7–6]   25. 12 [7–6]   26. -4 [7–6]   27. $-\dfrac{2}{3}$ [7–6]   28. 25 [7–6]   29. 3 [7–6]   30. 4 [7–6]
31. -60 [7–6]   32. 180 [7–6]   33. -6 [7–6]   34. 3 [7–6]   35. $x - 12$ [7–4]   36. $14a + 7$ [7–4]   37. $9x + 3$ [7–4]
38. $10 - 4x$ [7–4]   39. $17 - 6x$ [7–7]   40. $3a - 14$ [7–7]   41. $14 - 7a$ [7–7]   42. $15x - 42$ [7–7]
43. $23x + 20$ [7–7]   44. $11a - 1$ [7–7]   45. $4a - 12$ [7–7]   46. $45 - 11x$ [7–7]   47. $-3a - 20$ [7–7]
48. $184 - 21x$ [7–7]   49. -375 [7–7]   50. 8 [7–7]   51. 45 [7–7]   52. -7 [7–7]   53. -51 [7–7]   54. -16 [7–7]

Chapter 7 Test

The number in brackets after each answer refers to the section of the chapter that discusses that type of problem.

1. **a.** $\dfrac{5}{8}$ [7–1]   **b.** -6.9 [7–1]   2. $+28$ [7–1]   3. $+40$ [7–2]   4. **a.** -14 [7–3]   **b.** -39 [7–3]   5. **a.** -20 [7–3]

b. 4 [7–3]   **6.** −7 [7–3]   **7.** 36 [7–3]   **8.** 4ab + 3a² [7–4]   **9.** 21 [7–5]   **10. a.** 55 [7–6]   **b.** −96 [7–6]   **11.** −192 [7–6]   **12. a.** 13 [7–6]   **b.** $-\frac{1}{10}$ [7–6]   **13. a.** −12 [7–6]   **b.** 8 [7–6]   **14.** 52x − 36 [7–7]   **15.** 62 [7–7]

Chapter 8 Pretest

The number in brackets after each answer refers to the section of the chapter that discusses that type of problem.

1. −7 [8–1]   **2.** $\frac{7}{2}$ [8–1]   **3.** $\frac{5}{2}$ [8–1]   **4.** −22 [8–1]   **5.** −4 [8–1]   **6.** 3y [8–2]   **7.** $\frac{5c}{b}$ [8–2]

8. 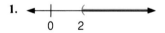 [8–3]   **9.** ⟵─────┼────→ x < −3 [8–4]   **10.** ⟵──┤├──→ x ≤ $-\frac{8}{9}$ [8–4]

8–1–1

1. −3   **2.** −5   **3.** −1   **4.** −1   **5.** −4   **6.** −4   **7.** −5   **8.** −5   **9.** 4   **10.** 5   **11.** 1   **12.** −2   **13.** −9   **14.** 5
15. −4   **16.** −2   **17.** 2   **18.** 19   **19.** 5   **20.** $\frac{11}{2}$ **21.** $\frac{7}{3}$ **22.** 4   **23.** $\frac{14}{5}$ **24.** 0   **25.** 5   **26.** −33   **27.** $\frac{2}{15}$
28. 7   **29.** $\frac{45}{4}$ **30.** 83   **31.** −5   **32.** $-\frac{5}{2}$ **33.** $\frac{46}{45}$ **34.** $\frac{11}{3}$ **35.** $\frac{25}{6}$ **36.** −3   **37.** $-\frac{40}{7}$ **38.** −5   **39.** $-\frac{1}{5}$
40. $\frac{55}{24}$

8–2–1

1. 2y   **2.** 3y   **3.** 2y   **4.** 3y   **5.** −6a   **6.** 2a   **7.** 2y   **8.** 8y   **9.** 2a   **10.** 5y   **11.** $\frac{22bc}{a}$ **12.** $-\frac{bc}{4a}$
13. $\frac{5a + 3y}{3ab}$ **14.** $\frac{4a}{21}$ **15.** $\frac{10b - 3a}{9}$ **16.** $\frac{15a + 2}{13}$ **17.** $\frac{a - 75}{29}$ **18.** $-\frac{11a}{9}$ **19.** $\frac{2a + 3b}{2}$ **20.** 2a   **21.** $h = \frac{2A}{b}$
22. $r = \frac{I}{pt}$ **23.** $g = \frac{2s}{t^2}$ **24.** $C = \frac{5F - 160}{9}$ **25.** $p = \frac{365I}{rD}$, $8,449.07   **26.** $E = I(nr + R) - e$, $E = 7$

8–3–1

1. 6 < 10   **2.** −6 > −10   **3.** −3 < 3   **4.** −4 < −1   **5.** 4 > 1   **6.** 1 < 4   **7.** −2 > −3   **8.** −5 < −3   **9.** 0 < 7
10. 0 > −3

8–3–2

1. ⟵──┼──(────→
       0   2

2. ⟵────────┼────────→
              0   5

3. ⟵────)──┼────→
         −1 0

4. ⟵──(────┼────→
       −3   0

5. ⟵──┼[───→
       0 1

6.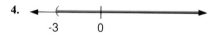

7. ⟵──[────┼────→
       −4   0

8.

9.

10.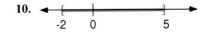

11. ◄——[————]►
     -5   0

12. ◄————|——(————►
     0   2 3/10

13. ◄————|————)►
         0   5.4

14. ◄——[——|——]——►
        -2½ 0 3½

15. ◄—|————(—)—►
       0   6 7

16. $x > 0$

17. $x > -4$ 18. $x \leq -4$ 19. $-4 \leq x \leq 4$ 20. $-5 < x \leq 2$

8-4-1

1. $x < 4$ ◄——|——|)——►
              0 4

2. $x < 5$ ◄——|——|)——►
              0   5

3. $x < -5$ ◄(|——|——►
             -5 0

4. $x < 1$ ◄——|)——►
              0 1

5. $x \leq 3$ ◄——|——|]——►
                0 3

6. $x \geq -1$ ◄——[|——|——►
                 -1 0

7. $x > -4$ ◄——(|——|——►
                -4 0

8. $x \leq -7$ ◄——]|——|——►
                 -7   0

9. $x \geq -5$ ◄——[|——|——►
                 -5 0

10. $x \leq 0$ ◄——|]——►
                  0

8-4-2

1. $x < 2$ ◄——|——|)——►
              0   2

2. $x \leq -4$ ◄——]|——|——►
                 -4 0

3. $x > 2$ ◄——|——(|——►
              0   2

4. $x < 4$ ◄——|——|)——►
              0   4

5. $x > 6$ ◄——|——(|——►
              0   6

6. $x \leq 3$ ◄——|——|]——►
                0   3

7. $x > 2$ ◄——|——(|——►
              0   2

8. $x \geq -2$ ◄——[|——|——►
                 -2 0

9. $x \leq -6$
10. $x > 3$
11. $x \leq 4\frac{1}{2}$
12. $x \geq 2$
13. $x < \frac{1}{2}$
14. $x \leq -20$
15. $x > -4$
16. $x \leq \frac{7}{3}$
17. $x < 140$
18. $x \leq -\frac{3}{16}$
19. $x > -5\frac{3}{5}$
20. $x > 3$
21. $x \leq 5\frac{3}{4}$
22. $x < 5$
23. $x > 36$
24. $x < 0$
25. $x < -6$

Chapter 8 Review

The number in brackets after each answer refers to the section of the chapter that discusses that type of problem.

1. -5 [8–1] 2. -3 [8–1] 3. -4 [8–1] 4. -7 [8–1] 5. 8 [8–1] 6. -6 [8–1] 7. -2 [8–1] 8. -3 [8–1]
9. -14 [8–1] 10. -42 [8–1] 11. 4 [8–1] 12. $\frac{35}{6}$ [8–1] 13. $\frac{5-a}{10}$ [8–2] 14. $\frac{5y}{4}$ [8–2] 15. $\frac{5y-5}{2}$ [8–2]
16. $-\frac{8a}{7}$ [8–2] 17. $\frac{6-z}{3y}$ [8–2] 18. $\frac{3ab-bc}{a}$ [8–2] 19. $\frac{7y-7}{2}$ [8–2] 20. $24-8y$ [8–2] 21. $\ell = \frac{P-2w}{2}$ [8–2]
22. $b = \frac{2a-ha}{h}$ [8–2] 23. [8–3] 24. [8–3]
25. [8–3] 26. $-5 \leq x < 6$ [8–3] 27. $x < 5$ [8–4]
28. $x \geq -5$ [8–4] 29. $x \leq 4$ [8–4]
30. $x \geq -3$ [8–4] 31. $x \geq -2$ [8–4]
32. $x < 10$ [8–4] 33. $x \leq \frac{12}{19}$ [8–4] 34. $x > \frac{11}{6}$ [8–4]

Chapter 8 Test

The number in brackets after each answer refers to the section of the chapter that discusses that type of problem.

1. -4 [8-1]   2. $1\frac{1}{2}$ [8-1]   3. -2 [8-1]   4. $-2\frac{5}{11}$ [8-1]   5. 140 [8-1]   6. $-3y$ [8-2]   7. $a = \frac{2A - hb}{h}$ [8-2]

8. ⟵———|———⟶ [8-3]   9. $x > -5$ ⟵——|———⟶ [8-4]
    -3   0   -5   0

10. $x \geq \frac{2}{3}$ ⟵——|——⟶ [8-4]
            0 $\frac{2}{3}$

Chapters 1-8 Cumulative Test

The number in brackets after each answer refers to the chapter section that discusses that type of problem.

1. 9,859 [1-5]   2. 151,235 [1-6]   3. $\frac{5}{6}$ [2-1]   4. $\frac{9}{35}$ [2-3]   5. $9\frac{1}{6}$ [2-5]   6. 214.578 [3-4]   7. 5.64 [3-6]
8. 4.3×10^{-6} [3-8]   9. 7 [4-2]   10. $2x^2y + 2xy$ [4-3]   11. 27 [5-5]   12. 50 [5-6]   13. 253 miles [6-3]
14. 340% [6-5]   15. 8.5% [6-6]   16. -1 [7-3]   17. -6 [7-6]   18. $180 - 32x$ [7-7]   19. $\frac{5}{8}$ [8-1]
20. $x \leq -3$ ⟵——|———⟶ [8-4]
              -3   0

Appendixes
A-3-2

1. 600 sq ft   2. 100 ft   3. 64 ft   4. 252 sq ft   5. 353 ft   6. $1,588.50   7. 37.68 ft   8. 113.04 sq ft   9. 85 sq ft
10. 150 sq ft   11. 64 ft   12. 14 ft   13. 13,266.5 sq ft   14. 408.2 ft   15. 324 sq ft   16. 497.5 sq ft   17. 94 sq ft
18. $108.10   19. $675.00   20. $60.00

B-1-1

1. 77   2. 91   3. 160   4. 150   5. 79   6. 148   7. 543   8. 1,444   9. 1,184   10. 2,319   11. 61.5   12. 64.3
13. 24.2   14. 54.5   15. 375.71   16. 177.803   17. 44.58   18. 76.86   19. 164.86   20. 174.55   21. 53.7   22. 45.1
23. 59.44   24. 123.69   25. 9.94   26. 55.45   27. 2,688   28. 1,599   29. 1,348.2   30. 3,368.7   31. 479.04   32. 1,421.61
33. 32   34. 16   35. 21.4   36. 18.6   37. 30.22   38. 5.89   39. 13   40. 24   41. 1.4142   42. 2.2361   43. 3.6056
44. 4.3589   45. 5.2915   46. 9.9499   47. 10.4403   48. 12.6886   49. 4.1352   50. 6.2048

INDEX

A

Acute
    angle, 391
    triangle, 392
Addition
    associative property of, 19
    commutative property of, 18
    of decimal numbers, 173
    of fractions, 122
    of mixed numbers, 143
    rule for equations, 265
    of whole numbers, 17
Addition algorithm, 17
Algebraic expressions, 226
Algorithm, 17
    addition, 17
    long division, 48
    long multiplication, 36
Altitude
    of a parallelogram, 395
    of a trapezoid, 395
    of a triangle, 393
Angle, 390
    acute, 391
    obtuse, 391
    right, 391
    straight, 391
Area
    of a circle, 236, 390, 396
    of a parallelogram, 396
    of a rectangle, 236, 259, 396
    of a square, 236, 396
    of a trapezoid, 366, 396
    of a triangle, 259, 368, 394, 396
Associative property of addition, 19
Average, 190

B

Base, 230
    of a parallelogram, 395
    of a trapezoid, 395
    of a triangle, 393
Base ten, 10
Binary operation, 18
Borrowing, 27
Braces, 60
Brackets, 60

C

Calculator, 401
Carrying, 37
Celsius temperature
    changing to Fahrenheit, 237, 368
    relationship to Fahrenheit, 269, 273, 281
Circle, 389
    area, 236, 390, 396
    circumference, 236, 390, 396
    diameter, 389
    radius, 390
Circumference of a circle, 236, 390, 396
Coefficient, 230
Common fraction, 94
Commutative property
    of addition, 18
    of multiplication, 36
Complex fractions, 110
Conditional equation, 250

D

Decimal, 168
    changing to fraction, 197
    expanded form of, 168
    repeating, 199
    rounding, 171
Decimal fraction, 168
Decimal numbers, 168
    addition of, 173
    changing to a percent, 301
    division of, 183, 192
    multiplication of, 177
    subtraction of, 173
Decimal point, 168
Degree, 391
Denominator, 94
Diameter of a circle, 389
Directed numbers, 318
Distributive property of multiplication over addition, 36, 343
Dividend, 46
Division
    algorithm, 48
    of decimal numbers, 183, 192
    of fractions, 109
    of mixed numbers, 139
    rule for equations, 254
    of whole numbers, 45
Divisor, 46

E

Equations, 250
    addition rule, 265
    conditional, 250
    division rule, 254
    equivalent, 252
    literal, 364
    multiplication rule, 269
    roots of, 250
    solutions of, 250
    subtraction rule, 260
Equilateral triangle, 393
Equivalent equations, 252
Euclid, 69
Expanded form, 10
Exponent, 10, 202, 230

F

Factor, 67
Factoring, 71
Factors of an algebraic expression, 229
Fahrenheit temperature
    changing to Celsius, 237
    relationship to Celsius, 269, 273, 281

Fractions, 94
    addition of, 122
    changing to decimals, 198
    changing to a percent, 301
    common, 94
    complex, 110
    division of, 109
    fundamental principle of, 95
    improper, 94
    like, 122
    multiplication of, 99
    proper, 94
    reduced form, 95
    simplified form, 95
    simplifying, 94
    subtraction of, 122
Fundamental principle of
    fractions, 95
Fundamental Theorem of
    Arithmetic, 72

G

Graphing inequalities, 369
Greatest common factor, 77
Grouping principle, 19
Grouping symbols, 60
    removing, 322, 334, 349

H

Hand-held calculator, 401
Hypotenuse, 393

I

Identity, 250
Improper fraction, 94
Inequalities, 369
    graphing, 369
    solving, 375
Inequality symbols, 369
Integers, 318
Interest, 366
    simple, 368
Invert, 109
Irrational numbers, 369
Isosceles triangle, 393

J

Juxtaposition, 226

L

Least common denominator, 124
Least common multiple, 79, 271
Like fractions, 122
Like terms, 237
    combining, 336
Lines
    parallel, 392
    perpendicular, 392
Literal equations, 364
Literal expressions, 227
    evaluating, 232
Literal numbers, 227
Long division algorithm, 48
Long multiplication algorithm, 36

M

Metric measurement, 296
Minuend, 27
Mixed numbers, 136
    addition of, 143
    division of, 139
    multiplication of, 139
    subtraction of, 147
Multiples, 79
Multiplication
    algorithm, 36
    commutative property of, 36
    of decimal numbers, 177
    of fractions, 99
    long, 36
    of mixed numbers, 139
    rule for equations, 269
    of whole numbers, 36
Multiplicative inverse, 347

N

Negative numbers, 318
Negative of a number, 319
Number line, 318
Numbers, 4
    decimal, 168
    directed, 318
    negative, 318
    positive, 318
    rational, 318
    signed, 318
    whole, 4
Numbers of arithmetic, 94
Numeral, 4
Numerator, 94

O

Obtuse
    angle, 391
    triangle, 393
Order relations, 369

P

Parallel lines, 392
Parallelogram, 394
    altitude, 395
    area, 396
    base, 395
    perimeter, 396
Parentheses, 60
Percent, 299
    change to a decimal, 300
Perfect squares, 208
Perimeter
    of a parallelogram, 396
    of a rectangle, 234, 262, 396
    of a square, 236, 259, 396
    of a triangle, 394
Perpendicular lines, 392
Pi, 395
Place value, 4
Positive numbers, 318
Power, 10
Prime factorization, 71
Prime number, 69
Principle of substitution, 232
Proper fraction, 94
Proportions, 289
    problems involving, 292

Q

Quadrilateral, 394
Quotient, 46
    trial, 48

R

Radius of a circle, 390
Rate, 287
Rational number line, 318
Rational numbers, 318, 369
Ratios, 286
    equal, 286
Real numbers, 369
Reciprocal, 109, 202, 347
Rectangle, 394
    area, 236, 259, 396
    perimeter, 234, 262, 396

Rectangular solid, volume of, 273
Reduced form of a fraction, 95
Relatively prime numbers, 77
Remainder, 46
Repeating decimal, 199
Right
    angle, 391
    triangle, 392
Rounding
    decimals, 171
    whole numbers, 14

S

Scientific notation, 202
Sieve of Eratosthenes, 69
Signed numbers, 318
    combining, 323
    division of, 347
    multiplication of, 341
Simple interest, 368
Simplified form
    of a fraction, 95
    of a radical, 210
Square, 394
    area, 236, 396
    perimeter, 236, 259, 396

Square roots, 208
    finding, using a calculator, 402
    simplified form, 210
Straight angle, 391
Subtraction, 338
    of decimal numbers, 173
    of fractions, 122
    of mixed numbers, 147
    rule for equations, 260
    of whole numbers, 24
Subtrahend, 27

T

Terms, 229
Trapezoid, 395
    altitude, 395
    area, 396
    bases, 395
Tree method of prime factorization, 72
Trial quotient, 48
Triangle, 392
    acute, 392
    altitude, 393
    area, 259, 394, 396
    base, 393
    equilateral, 393
    hypotenuse, 393
    isosceles, 393
    obtuse, 393
    perimeter, 394
    vertex, 392

V

Variables, 250
Vertex of a triangle, 392
Volume of a rectangular solid, 273

W

Whole numbers, 4
    addition of, 17
    division of, 45
    expanded form of, 10
    multiplication of, 36
    reading, 4
    rounding, 14
    subtraction of, 24
    writing, 4

Z

Zero
    exponent, 202
    power, 11